The Art of Measuring in the Thermal Sciences

Heat Transfer

Series Editor:
Afshin J. Ghajar
Regents Professor, School of Mechanical and Aerospace Engineering,
Oklahoma State University

A SERIES OF REFERENCE BOOKS AND TEXTBOOKS

The Art of Measuring in the Thermal Sciences

Edited by
Josua P. Meyer and Michel De Paepe

CRC Press
Taylor & Francis Group
Boca Raton London New York

CRC Press is an imprint of the
Taylor & Francis Group, an **informa** business

MATLAB® is a trademark of The MathWorks, Inc. and is used with permission. The MathWorks does not warrant the accuracy of the text or exercises in this book. This book's use or discussion of MATLAB® software or related products does not constitute endorsement or sponsorship by The MathWorks of a particular pedagogical approach or particular use of the MATLAB® software.

First edition published 2021
by CRC Press
6000 Broken Sound Parkway NW, Suite 300, Boca Raton, FL 33487-2742

and by CRC Press
2 Park Square, Milton Park, Abingdon, Oxon, OX14 4RN

Visit the Taylor & Francis Web site at
http://www.taylorandfrancis.com

and the CRC Press Web site at
http://www.crcpress.com

ISBN: 978-0-367-19290-7 (hbk)
ISBN: 978-0-429-20162-2 (ebk)

Typeset in Palatino
by codeMantra

Contents

Preface

We are proud to present our first book resulting from years of cooperation. The idea for this book emerged from the sabbatical stay of Michel De Paepe at the University of Pretoria with Josua P. Meyer in 2018.

This book is aimed at providing scholars, students, and people from industry a scholarly, state-of-the-art overview of good practices for conducting correct and sound measurements in thermal sciences. All the material in the book is original, was written by academic experts in the field, and has not been published before. Every chapter was peer-reviewed for originality by independent reviewers, who are distinguished subject specialists in the field of the relevant chapter.

It is a guide for people who want to conduct advanced thermal experiments. The types of experiments discussed in the book are common in modern research and the development of thermal systems in the industries of energy generation, transport, manufacturing, mining, processes, HVAC, biomedical engineering, etc. Most of what we currently know about thermal sciences has been learnt by observation and proper data reduction (deduction).

These days, many thermal science problems are being solved numerically on large grids using the tools of computational fluid dynamics. Even for these computerized simulations of physical systems, the numerical simulations in most cases depend on experimental data for validation and comparison purposes. Therefore, there is a big need for good and accurate data with known uncertainties. Since the introduction of digital measurement equipment, a lot has changed in the common practice of conducting measurements. Computers have changed the way and the amount of data we process. The possibilities and accuracies have grown much due to the advancement of electronics, but so has the complexity of our work.

Little information is found in the recent literature on how to conduct scientifically sound and correct measurements. Institutional memory and practices are being transferred from one person to another in laboratories. At conferences and in archival journals, little attention is given to this kind of craftsmanship and at most, the uncertainties and calibration techniques are reported. Fortunately, most scholarly journals require the reporting of equipment errors and uncertainties. However, this information is normally provided so succinctly that it is challenging to learn from it.

This book brings together the expertise of several research teams of the scientific thermal science community. It gives a unique overview and insight into advanced measurement techniques and systems, which are normally never published. The book goes into the fundamentals where possible and focuses on advanced measurement techniques currently used in thermal systems. It also highlights the challenges and future research opportunities for measuring practices.

We are grateful to our colleagues in the thermal engineering research field for sharing their knowledge and expertise. We consider ourselves blessed to have been able to work at and with institutes of this high level. But mostly we need to thank our students for teaching us, by hard work in our laboratories, how to make measurements come to a good end.

MATLAB® is a registered trademark of The MathWorks, Inc. For product information,
    please contact:
    The MathWorks, Inc.
    3 Apple Hill Drive
    Natick, MA 01760-2098 USA
    Tel: 508-647-7000
    Fax: 508-647-7001
    E-mail: info@mathworks.com
    Web: www.mathworks.com

Editors

Josua P. Meyer is a Professor, Head of the Department of Mechanical and Aeronautical Engineering, and Chair of the School of Engineering in the Faculty of Engineering, Built Environment and Information Technology. His area of research is convective heat transfer, which relies on the engineering sciences of heat transfer, fluid mechanics, and thermodynamics. He and his students and colleagues have made it possible to predict the heat transfer characteristics in the previously unknown transitional flow regime. He has published more than 800 articles, conference papers, and book chapters in the field of thermal sciences and has successfully supervised more than 100 PhD and MSc students. He established the Clean Energy Research Group at the University of Pretoria, which now has 40 full-time postgraduate students and 13 staff members. The group members have developed, designed, and built more than ten unique, state-of-the-art experimental setups, which are being used for leading-edge heat transfer research. No other similar experimental setup exists in the world. The group conducts joint research and publishes with scholars at EPFL, MIT, Ghent, Duke, and INSA Toulouse.

Michel De Paepe is a Professor of Thermodynamics and Heat Transfer at the Department of Flow, Heat and Combustion Mechanics of the Faculty of Engineering and Architecture of the Ghent University. He is the Program Director of the Master Electromechanical Engineering at the Ghent University. He was supervisor/promotor of 19 PhDs and 150 MScs defended at the Ghent University and is the coauthor of more than 400 articles and conference papers in the field of thermal sciences. In 2002, Professor De Paepe founded the research group Applied Thermodynamics and Heat Transfer. Research by this team, with about 15 PhD students and 3 staff members, focuses on thermodynamics of new energy systems, performance of HVAC systems, and complex heat transfer phenomena in industrial applications, as in compact heat exchangers, combustion engines, refrigerant two-phase flow, and electronics cooling. The team is internationally recognized as an authority in thermal measurement techniques. Several sensors developed by this team are being used by high-level laboratories around the world: the University of Pretoria, INSA Lyon, TU, EPFL, Imperial College London, and the University of Oxford.

Contributors

Marco Azzolin
Department of Industrial
 Engineering
University of Padova
Padova, Italy

Christian K. Bach
Department of Mechanical and
 Aerospace Engineering
Building, Environmental and
 Thermal Systems Research Group
Oklahoma State University
Stillwater, Oklahoma

Hans-Jörg Bauer
Institute of Thermal
 Turbomachinery (ITS)
Karlsruhe Institute of Technology
Karlsruhe, Germany

Tadeusz Bohdal
Faculty of Mechanical Engineering
Department of Energy Engineering
Koszalin University of Technology
Koszalin, Poland

Jocelyn Bonjour
Univ Lyon, CNRS, INSA-Lyon,
 Université Claude Bernard Lyon 1
CETHIL UMR5008
Villeurbanne, France

Stefano Bortolin
Department of Industrial
 Engineering
University of Padova
Padova, Italy

Craig R. Bradshaw
Department of Mechanical and
 Aerospace Engineering
Building, Environmental and
 Thermal Systems Research
 Group
Oklahoma State University
Stillwater, Oklahoma

J. C. Chai
Department of Engineering and
 Technology
School of Computing and
 Engineering
University of Huddersfield
Huddersfield, United Kingdom

Henryk Charun
Faculty of Mechanical Engineering
Department of Energy Engineering
Koszalin University of Technology
Koszalin, Poland

Serge Cioulachtjian
Univ Lyon, CNRS, INSA-Lyon,
 Université Claude Bernard
 Lyon 1
CETHIL UMR5008
Villeurbanne, France

X. Cui
Institute of Building Environment
 and Sustainable Technology
School of Human Settlements and
 Civil Engineering
Xi'an Jiaotong University
Xi'an, China

Davide Del Col
Department of Industrial
    Engineering
University of Padova
Padova, Italy

T. E. Diller
Mechanical Engineering
    Department
Virginia Tech
Blacksburg, Virginia

Maximilian Elfner
Institute of Thermal
    Turbomachinery (ITS)
Karlsruhe Institute of
    Technology
Karlsruhe, Germany

Marilize Everts
Department of Mechanical and
    Aeronautical Engineering
University of Pretoria
Pretoria, South Africa

Solomon Giwa
Faculty of Engineering
Department of Mechanical and
    Aeronautical Engineering
University of Pretoria
Hatfield, South Africa

Iztok Golobič
Faculty of Mechanical
    Engineering
University of Ljubljana
Ljubljana, Slovenia

Eckhard A. Groll
Ray W. Herrick Laboratories
School of Mechanical Engineering
Purdue University
West Lafayette, Indiana

Fabian Hufgard
Institute of Space Systems
University of Stuttgart

L. W. Jin
Institute of Building Environment
    and Sustainable Technology
School of Human Settlements and
    Civil Engineering
Xi'an Jiaotong University
Xi'an, China

Emmanuel Kakaras
Laboratory of Steam Boilers and
    Thermal Plants
National Technical University of
    Athens
Athens, Greece

Sotirios Karellas
Laboratory of Steam Boilers and
    Thermal Plants
National Technical University of
    Athens
Athens, Greece

Roger Kempers
Department of Mechanical
    Engineering
York University
Toronto, Canada

Jungho Kim
Department of Mechanical
    Engineering
University of Maryland
College Park, Maryland

Thorsten Klahm
GESMEX Exchangers GmbH
Schwerin, Germany

Orkan Kurtulus
Ray W. Herrick Laboratories
School of Mechanical Engineering
Purdue University
West Lafayette, Indiana

Frédéric Lefèvre
Univ Lyon, CNRS, INSA-Lyon,
    Université Claude Bernard Lyon 1
CETHIL UMR5008
Villeurbanne, France

Stéphane Lips
Univ Lyon, CNRS, INSA-Lyon,
    Université Claude Bernard
    Lyon 1
CETHIL UMR5008
Villeurbanne, France

Stefan Loehle
Institute of Space Systems
University of Stuttgart
Stuttgart, Germany

C. F. Ma
Institute of Building Environment
    and Sustainable Technology
School of Human Settlements and
    Civil Engineering
Xi'an Jiaotong University
Xi'an, China

Denis Maillet
Laboratoire Énergies & Mécanique
    Théorique et Appliquée (LEMTA)
University of Lorraine, CNRS
Nancy, France

Romit Maulik
Department of Mechanical and
    Aerospace Engineering
Building, Environmental and
    Thermal Systems Research Group
Oklahoma State University
Stillwater, Oklahoma

X. Z. Meng
Institute of Building Environment
    and Sustainable Technology
School of Human Settlements and
    Civil Engineering
Xi'an Jiaotong University
Xi'an, China

Josua P. Meyer
Department of Mechanical and
    Aeronautical Engineering
University of Pretoria
Pretoria, South Africa

Ana S. Moita
IN+ Center for Innovation,
    Technology and Policy Research
Mechanical Engineering
    Department
Instituto Superior Técnico,
    Universidade de Lisboa
Lisboa, Portugal

António L. N. Moreira
IN+ Center for Innovation,
    Technology and Policy Research
Mechanical Engineering
    Department
Instituto Superior Técnico,
    Universidade de Lisboa
Lisboa, Portugal

Platon Pallis
Laboratory of Steam Boilers and
  Thermal Plants
National Technical University of
  Athens
Athens, Greece

Miguel R. O. Panão
ADAI-LAETA, Mechanical
  Engineering Department
University of Coimbra
Coimbra, Portugal

Pedro Pontes
IN+ Center for Innovation,
  Technology and Policy Research
Mechanical Engineering
  Department
Instituto Superior Técnico,
  Universidade de Lisboa
Lisboa, Portugal

Rémi Revellin
Laboratoire CETHIL, INSA-Lyon
  University

Anthony Robinson
Trinity College Dublin
Dublin, Ireland

Romuald Rullière
Univ Lyon, CNRS, INSA-Lyon,
  Université Claude Bernard Lyon 1
CETHIL UMR5008
Villeurbanne, France

Omer San
Department of Mechanical and
  Aerospace Engineering
Building, Environmental and
  Thermal Systems Research Group
Oklahoma State University
Stillwater, Oklahoma

Achmed Schulz
Retired

Mohsen Sharifpur
Faculty of Engineering
Department of Mechanical and
  Aeronautical Engineering
University of Pretoria
Hatfield, South Africa

Jaime Sieres
Área de Máquinas y Motores
  Térmicos
University of Vigo
Vigo, Spain

Małgorzata Sikora
Faculty of Mechanical Engineering
Department of Energy Engineering
Koszalin University of Technology
Koszalin, Poland

Ioannis-Alexandros Sofras
Laboratory of Steam Boilers and
  Thermal Plants
National Technical University of
  Athens
Athens, Greece

S. M. Sohel Murshed
Department of Mechanical
  Engineering
Center for Innovation, Technology
  and Policy Research
Instituto Superior Técnico
  Universidade de Lisboa
Lisboa, Portugal

Janez Štrancar
Condensed Matter Physics
  Department
Jožef Stefan Institute
Ljubljana, Slovenia

Christophe T'Joen
Department of Electromechanical,
    Systems and Metal Engineering
Sustainable Thermo-Fluid Energy
    Systems
Ghent University
Ghent, Belgium

Emanuele Teodori
ASML Holding N.V
Veldhoven, The Netherlands

Panagiotis Vourliotis
Laboratory of Steam Boilers and
    Thermal Plants
National Technical University of
    Athens
Athens, Greece

L. C. Wei
Institute of Building Environment
    and Sustainable Technology
School of Human Settlements and
    Civil Engineering
Xi'an Jiaotong University
Xi'an, China
and
Shenzhen Envicool Technology Co.,
    Ltd.
Shenzhen, China

Davide Ziviani
Ray W. Herrick Laboratories
School of Mechanical Engineering
Purdue University
West Lafayette, Indiana

Part A

Measuring in Thermal Systems

Reducing Errors and Error Analysis

1

Measuring the Right Data—Verifying Experimental Boundary Conditions

Christophe T'Joen

Ghent University

CONTENTS

1.1 Introduction

As the cost (in terms of both hardware and duration to get results) of computational research has reduced significantly over the past decades, and this is set to continue, an increasing volume of thermo-hydraulic research today is conducted primarily through computational means. Numerical tools allow us to study complex geometries and perform parametric studies at a much higher pace than experiments would ever be able to do and at a fraction of the cost, particularly for high-temperature or reactive flows. When combined with advanced stochastic and parametric modeling, e.g., through clustering or neural networks, "computational experiments" can lead to novel designs with an improved efficiency (in whichever form the efficiency is defined for the considered application). It allows exploring areas/geometries to find new (local) optima intended for specific applications. With advances in additive manufacturing (e.g., 3D printing of metals), new previously infeasible designs can now become a reality and in fact can also be tested. Despite

advances in and strengths of numerical tools, the evidence of actual performance improvement is realized only through physical testing. This can
be done at various scales, often starting in the laboratory at a scaled model.
Additive manufacturing can now help laboratories generate test objects
quicker to allow these to be tested on, for example, [1–2]. This started in the
past decade opening up opportunities to improve energy efficiency or reduce
capital intensity of energy (heat or mass) transfer processes. This improvement process could bring engineering practices closer to considering fundamental limits, e.g., theoretical principles such as the constructal theory of
Bejan for heat exchangers [3], or drive application of novel materials such as
metal or polymer foams [4–5].

Care should be taken when using numerical tools for the design. First, despite
these advantages, computational tools have limits; for example, they will
consider typically simplified boundary conditions such as uniform velocity/
temperature fields or adiabatic walls. Second, at their very core, all computational approaches which are used today to model macroscale engineering problems rely on internal models to capture flow physics which drive heat transfer
and hydraulics, e.g., turbulence modeling (e.g., k-epsilon or large eddy simulation) and wall transfer functions. These models can be layered deep within
these tools, making it sometimes hard to reveal their details or the impact on
the modeled outcome. And at their core, these models are built on large-scale
experiments often in very different situations/configurations than those the
numerical tool is applied to as part of the study. As part of computational model
verification, sensitivity tests should be conducted by researchers on, for example, grid spacing and grid element count, temporal/spatial discretization, and
turbulence (sub-)model constants to confirm the predictions. Fundamentally,
the applicability of a certain numerical (turbulence) model should always be
evaluated ahead of the simulation, and the selection of a family of modeling
approaches, e.g., $k-\varepsilon$ or $k-\omega$, should be done based on the nature of the studied flow or region of flow of interest in the tested configuration.

Performing an experiment is the only way to fully validate a computational result, but it comes with its own set of challenges, just as numerical
modeling. Therefore, this evaluation step should be planned carefully to
ensure the right outcomes are achievable. When using computational tools to
optimize a design, "optimum" candidates should be selected to be tested in
actual/scaled conditions to verify the improved performance. Experiments
are always limited in the data they generate, be it in the spatial location of the
resulting data (typical point or planar data are measured vs. full-field information from simulations) or the frequency content (sampling rate—limited
experimental duration), and as such, the data acquisition requirements are
something to define in detail up ahead, thinking about the goal of the experiment while at the same time affirming the results are sound. This chapter is
focused on the latter aspect, looking at the experimental boundary conditions. In the next paragraph, several key principles are covered before several
case studies are explored as illustration.

1.2 Thinking Ahead

During the design of an experiment, ensuring the right boundary conditions are met is a key principle; one that at face value appears trivial, but this can be and in most cases is deceptive. Below are some key guiding principles to be applied when setting up experiments:

Scale appropriately: Almost all experiments require a level of scaling, based on a set of non-dimensional numbers. By use of dimensional analysis (Buckingham Π theorem, [6]), a physical problem can be cast in a minimum set of non-dimensional numbers. An example of a non-dimensional number is the Reynolds number, which physically links to the laminar or turbulent nature of a fluid flow. When designing an experiment, the studied geometry is in most cases scaled (up or down) for ease of access/to allow the experiment to take place within a reasonable space/time. Geometric similitude of the two samples will then be a given, resulting in scaling of other quantities such as the velocity to preserve a Reynolds number. Flexibility can be gained by switching fluids, but invariably, one will find that not all non-dimensional numbers can be kept constant in the experiment vs. reality. This is not a problem, but researchers should carefully assess this ahead and be aware of these choices. Also, the ratio of properties should be considered; for example, the ratio of heat flux to mass flux should be considered as this can be important, for example when studying heat transfer deterioration or natural convection. It is thus important to, ahead of the experimental design, understand the dimensionless properties which describe the physics of the problem (and in some cases ratios associated with it and the considered fluid property gradients/changes at the operating temperature). When scaling an experiment, explore the boundary conditions and the possible impact their non-ideal character has on the measurements that are intended. For "periodic structures" (illustrated in case study 1), it is key to ensure the tested configuration is indeed acting as a periodic one and avoid impact of the wall/channel on the results. This applies at various levels, e.g., the number of tube racks in a heat exchanger, fins rows in a tube fin bank, unit cells in a cooling tower, number of heat sinks cooling a CPU (see Ref. 7). This can be verified with computational tools or through expanded testing (i.e., increasing the count of the tested elements) and is critical to verify that the result is not affected by boundary effects.

Verify the global balances: In an experiment, thermal or hydraulic energy is transferred by or to the fluid; for example, in a heat exchanger a wall exchanges heat to a nearby (flowing) fluid. As the thermal measurements are taken, it is important that the experiment is set up to allow for an independent verification of the global heat or force balance for a hydraulic study. The term independent refers to using different instruments from those used to gather the main experimental data. With the use of independent sensors, there is no risk of confounding sensor errors, and it provides clarity on the heat flows or hydraulic losses within the system that is being studied. When setting up an experiment, ensure the overall balances can be recorded throughout the (transient) experiment with sufficient accuracy, focused on the key control volume that is being studied. This can be assessed ahead of the actual experiment, considering, for example, the accuracy of the transducers and the method to determine the fluid properties such as enthalpies. Researchers will find that to achieve a closure of <1%–2% can be (very) challenging related to non-uniform flow effects.

Check your ambient losses and boundary conditions: Adiabatic boundary conditions are often assumed initially during design; however, these do not exist in actual experiments. Even the thickest insulation layer will still result in heat loss, but it can be small enough to be ignored if well designed. During the design of an experiment, it is important to consider losses to ambient and establish measurement approaches which verify these at conditions as those present in the experiment. This can be done through artificial heating of the test sample (e.g., electrically) and recording the heat input required to reach a steady-state surface temperature, thus determining the heat lost to the environment at that condition. When performing such studies, it is important to consider the test object and how the measured losses will be used in the further analyses. Losses can be generated by the test setup itself; for example, heat ingress not only through piping feeding fluid to the test setup, but also through the instrumentation itself (heat loss/ingress through thermocouple wires and fittings) should be considered if relevant to the tested problem. Ensure sufficient margin is available to reach a target temperature by design as these heat losses incrementally can stack up to a large wattage lost to the environment if not properly accounted for. Fluid inflow, supplied by pumps or fans, can have a high turbulent intensity swirl and be non-uniform if the test sample is located close, which all may affect the measurements. Verifying the inflow profile during measurements and putting measures in place to smoothen the flow are important to ensure the right boundary conditions are present at the test sample.

Verify the steady or transient state: When extracting data and determining a measured quantity, an inherent assumption of steady state is often used (or for transient experiments that the change of the measured quantity is as per approximation). The use of thick insulation or large metal masses in the experiment can result in long transients, and varying ambient conditions (diurnal fluctuations) can result in difficulty achieving the desired experimental steady state. Considering the dynamics early in the design and determining (e.g., through computation) the timescale to reach the steady state are helpful to plan the measurement campaign and set up the data analysis tools to track the experimental quality. It may be better in fact not to insulate a transient experiment or to condition the insulation temperature such that limited heat transfer will take place (as illustrated in case study 2).

Real-time monitoring: Many researchers visualize data "quality" on a dashboard during the measurements; the state of the experiment (e.g., measured through time derivates of pressure or flow) and the calculated heat balance are good quality indicators to show continuously while measuring. As today, the instrumentation can be read continuously; it is straightforward to set up such a real-time visualization and this should always be done. It also provides the operators a means to assess the status and safe operation of the test rig.

Prepare to be uncertain: It is good practice to prepare the experimental data reduction up ahead of running the experiment to support the instrumentation design and subsequent uncertainty analysis. By analyzing the uncertainty propagation, key factors which impact the outcome will be found and this can drive the need to measure these properties more accurately (e.g., with more sensors or sensors with a higher accuracy), as illustrated in, for example, [8]. No fluid property is free of uncertainty—do not underestimate the impact of uncertainty on fluid properties (which can be derived from the equations of state) or on heat transfer coefficients determined from published correlations. Using these in calculations can result in a strong increase in the overall uncertainty. Adding uncertainty data to the real-time monitoring can be helpful to further monitor the data quality, though often it is quite complex for highly derived properties. Quality checking the uncertainty analysis through sensitivity studies is also effective to test data reduction methods for accuracy.

The principles described above form the foundation to design an experimental setup and will be reflected in the various case studies below. The sections below do not focus on the actual data and outcomes—the results can be found in the references—but zoom in on the practices applied in the experimental design and choices made to control the boundary conditions.

1.3 Case Studies

1.3.1 Air Heat Exchanger—Thermal Fin Design/Optimization

The first case study looks at thermal fin design for compact heat exchangers applied to indoor air conditioning units. Complex articulated patterns are used today, pressed in thin aluminum fins to promote heat transfer in such units, by acting on the air boundary layer formation (interruption), affecting the transition to turbulent flow and the fin tube wake area on the fin surface which is otherwise a poorly utilized heat transfer area. Fin designs continue to be studied today both numerically and experimentally, even for a technology which is arguably very mature. This is due to the need to continuously improve energy efficiency in space heating or industrial applications, which

can be realized through changes to the fin configuration (adding vortex generators has been studied for many years) or more fundamental changes to the
heat exchanger tube shape/layout. In this study (full details can be found in
[8]), various experimental setups were used; the focus of this section is on the
wind tunnel experiments with its scaled-up model of the fin configuration.

The intent of this experiment was to determine both the global averaged
heat transfer coefficient of the fin assembly, the local visualization of the
boundary layer on each of the louver elements and to explore the impact of
key fin parameters such as the fin spacing and louver angle. As the louvers
are normally in the order of 1 mm in length, the setup considered a scaled-
up version of the fin geometry to enable access to the louvers with thermocouples. The tube wall was not considered in this study, thus avoiding the
complexity of the wall-bounded vortices and their scale-up. In order to study
the tube effect on the louvered fins, the heat exchanger tubes would need to
be included in a sufficient number to ensure periodic effects, as illustrated,
for example, in [7].

Figure 1.1a schematically shows the wind tunnel, which consists of a fan
sending air through the scaled-up fin sample (as shown in Figure 1.1b). The
dimensions of the test setup were selected based on a trade-off between
scaling-up of the fin design (roughly 12/1 to place up to 10 thermocouples
on a single louver measuring wall temperatures on both sides of the louver) and the available space in the laboratory considering the air mass flow
metering requirements through an orifice plate. This measurement requires
sufficient length of undisturbed flow. The total test setup length exceeded
8 m. Uncertainty analysis indicated that mass flow (Reynolds number) was
a key property to get right, so rather than relying on the operating point
of the fan, the flow rate is measured with up to <2% accuracy through an
orifice plate (designed to ISO standard 5167). Computational studies were
performed to assess the flow behavior and deflection in the fin structure
to avoid the impact of the channel walls on the central fin structure where

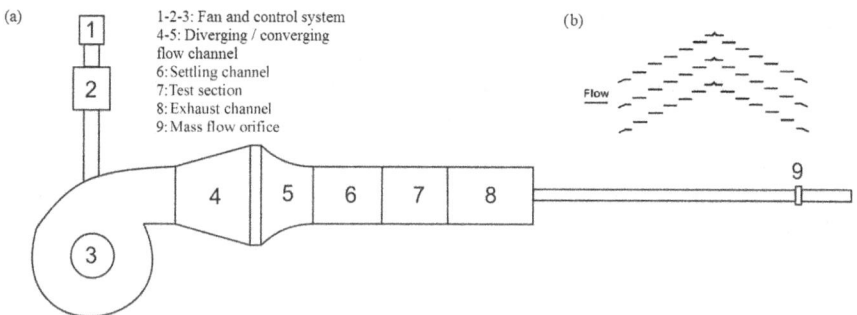

FIGURE 1.1
(a) Experimental test setup used to study a scaled-up fin configuration; (b) inclined louvered
fin geometry studied in the test section (7) showing three fin rows [9].

measurements would take place. The results showed that at least six fin rows would be needed to ensure periodic-like boundary conditions were present, which in turn constrained the maximum fin spacing that could be studied.

The wind tunnel was designed to provide flow with a low turbulent intensity by use of a diverging/converging section with, at its center, a 50-mm-thick honeycomb structure, with a cell size of ~10 mm. At the connection between the fan and the wind tunnel, a grate with circular drilled holes was inserted to ensure an even flow distribution. This was a common laboratory practice used to adjust for maldistribution in the fan exit plane. To even out the flow distribution, the holes were machined in such a manner to add more resistance locally where the flow rate was too high and to smoothen the velocity profile. In this specific case, the applied hole sizes were 10, 16, and 25 mm, set in vertical even spaced rows. The verification of the inflow conditions was done through 2D hot-wire anemometry. The results are presented in [9] for reference.

The local heat transfer coefficient h_{local} is defined in Eq. (1.1). The test setup was built to measure the local wall temperature on the louvers while imposing a constant surface heat flux q'' through electrical current heating. The local fluid bulk temperature was computed based on thermal balance considering the upstream heat input, but other reference values, e.g., the inlet temperature, can be used as well. The louvers were manufactured from balsa wood, which is lightweight and easy to handle. It also is electrically and thermally insulating, which is important considering that the temperatures at the top and bottom of a louver would differ due to flow phenomena such as impingement or local wakes. Figure 1.2 shows the cross section of the measurement louver, [10]. Thin (0.25-mm) K-type thermocouples were selected to measure the surface temperature. To mount these, the thermocouple junctions were curved upward before sliding them into the grooves cut into the balsa louver. The thermocouple end sections were then taped to the wood, resulting in the junctions curving upward and standing some 2 mm above the louver surface. This was done to ensure a good contact between the junctions and the metal copper foil (0.25 mm thick) which was placed

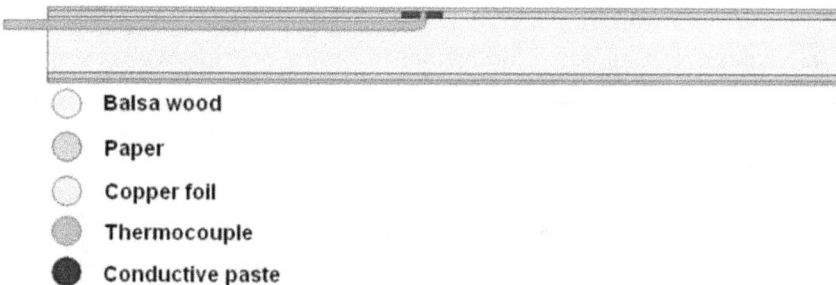

○ **Balsa wood**

◐ **Paper**

○ **Copper foil**

◐ **Thermocouple**

● **Conductive paste**

FIGURE 1.2
Cross section of the instrumented louver [9].

on top to provide electrical heating. Once the metal foil was glued to the balsa wood, the electrical connections were added on one side. Current was sent through the foil to set the heat flux, which is calculated based on the measured foil resistivity. To verify the heat flux was uniform, an IR image was recorded of the louver while being heated, which showed an even temperature profile apart from near the electrical connections at the edge of the louvers.

$$h_{\text{local}} = \frac{q''}{T_{\text{wall}} - T_{\text{bulk}}} \tag{1.1}$$

The heat balance over the louver bank was verified using temperature measurements; both upstream (4) and downstream (12–16) of the test setup, K-type thermocouples were spaced evenly across the channel cross section. The bulk averaged temperatures were calculated using the measured velocity profiles (hot-wire) as weighing factor. This provided sufficient data for a closure of the heat balance to within 5%. All thermocouples were mounted with their junctions pointing to the direction of the flow; no impingement heating occurred due to the low air velocity present in the wind tunnel, even at the highest Reynolds number. The thermocouples had to be supported by a metal grid structure to ensure they did not vibrate at certain velocities, improving data quality.

At the lowest air velocities, the temperature data recorded downstream of the test section showed higher temperatures at the top of the channel compared to the bottom. This showed how buoyancy started to affect the results, now operating in the mixed convective regime rather than forced convection, [10]. By scaling the heat flux down, this effect could be dampened; however, this cannot be completely mitigated. It is important to consider the implications of natural convection on test design and execution as it may drive the need to lower the surface heat flux significantly at lower Reynolds number, which affects the feasible temperature differences and thus the relative accuracy of the results. This is often overlooked early in the design phase of experiments.

1.3.2 Expanding Gas Jet—Impingement Cooling on Pipe Walls

This case study looks at impingement cooling of expanding gas flows (e.g., downstream a valve) in natural gas facilities. Due to Joule–Thomson cooling, when natural gas (mainly methane) expands from a high pressure, significant cooling will occur. These low temperatures get transferred to the pipe wall, which can then result in local embrittlement of the metal. This is a safety concern, as a brittle fracture will occur suddenly and result in a catastrophic loss of containment releasing gas into the environment. An experiment was set up to perform detailed measurements of the velocity and temperature field (planar in both cases) through particle image velocimetry (PIV) with smart tracers coated with a phosphor material. A dual-laser system was used, where one of the lasers provided the PIV data through local illumination of the target area (as per standard PIV practice taking two images in fast succession) and the other laser was attuned to the excitation frequency of the phosphor. By then taking high-speed images following the excitation laser pulse, the decay of the phosphorescence could be measured, which is a measure for the particle temperature. Through this dual-laser approach, both local temperature and velocity can be measured throughout the plane of interest. More details on the experimental setup can be found in [11]. Phosphor-based temperature measurements inside fluid flow is a developing field ([12]), and it offers a way to look closer at fluid and temperature dynamics in complex (reacting) flows. The experimental setup, however, becomes sensitive and complex, requiring optical access, controlled particle seeding, laser timing control, and well-defined environmental ambient/background conditions to allow for post-processing.

The test setup (Figure 1.3) was designed to mimic natural gas flow conditions and achieve fully turbulent flow inside the channel downstream of the gas expansion, while still providing a sizeable interrogation area for the PIV to record at high frequency (3 kHz). In addition, sufficient expansion cooling was required across the expansion to ensure the uncertainty on the temperature data was sufficiently low to calculate the local wall heat transfer coefficient. Various fluids were screened, and argon (stored in 300 bar cylinder) was selected as the working medium, expanding from 130 bar to atmosphere across a small round orifice opening (1.5 mm). The tube section was made square (50 mm) from aluminum, which allowed for optical access from three sides as required to generate the laser sheet and record the flow images. However, the square tube complicates the flow patterns, but this choice was accepted due to the challenge of light reflections/scatter through a circular glass wall which would have been the geometrically similar configuration. The experiment was conducted in transient mode, running for a

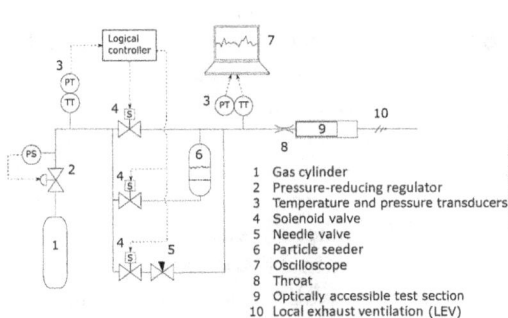

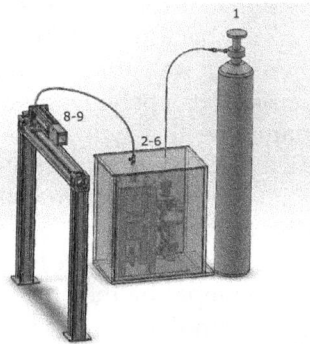

1 Gas cylinder
2 Pressure-reducing regulator
3 Temperature and pressure transducers
4 Solenoid valve
5 Needle valve
6 Particle seeder
7 Oscilloscope
8 Throat
9 Optically accessible test section
10 Local exhaust ventilation (LEV)

FIGURE 1.3
Jet impingement test setup [11].

period of 1–3 min triggering wall cooling. The transient mode was selected due to the seeding particles depositing inside the channel and causing high local brightness on the recorded data—hampering data quality. This effect was originally overseen in the design of the test setup requiring further modification. The working pressure level of 120 bar upstream of the orifice was established as a trade-off between experimental duration (1–3 min) and having sufficient cooling while avoiding depletion of multiple high-pressure gas bottles in a single run.

Analytical assessment of the gas expansion inside the cylinder revealed that as the gas expanded from 300 to 120 bar, it also cools down inside the cylinder. The expanding gas absorbs heat from the cylinder wall and the laboratory surroundings, which would result in temperatures of around –17°C upstream of the orifice for most of the duration of the experiment. However, the measurement setup and seeding equipment with associated valves (items 2–6 in Figure 1.3) are a large metal mass storing energy. This combined would result in a large thermal transient at the inlet of the orifice as the argon flow is cooling down in essence the rig, and this time is then "lost" due to highly non-uniform conditions. To counter this transient, the entire upstream end of the test setup was placed into a large refrigerator and kept at a temperature of ~–17°C and all piping connecting the channel was insulated. T-type thermocouples were used throughout this setup, measuring the temperature just upstream of the orifice as well as at various locations on the channel wall. Preconditioning of the unit in this manner enabled controlling the transients in short-duration experiments (<15 s to reach a near steady state) and provided a more constant inlet temperature as boundary condition. A fast-acting pressure regulator was selected to provide control of the mass flow entering the channel. No direct mass flow measurement was added to the system as the critical flow through the orifice provided a means to determine the rate based on the upstream pressure and temperature conditions which were measured.

To determine the local heat coefficient, the local surface temperatures were measured on the inner wall using a phosphorescent dye, which showed a uniform temperature distribution. To avoid the impact of axial conduction in the aluminum wall, the section where the inner wall temperature was measured consisted of a polymer insert which was insulated at the back. On the outer channel wall, T-type thermocouples were placed every few centimeters to track the wall cooling through the experiment (and indirectly provide an estimate of the through-wall heat flux). These thermocouples were connected using copper-based tape to ensure a good contact between the junction and the wall, and there were clear differences in the time response of these probes when comparing the polymer and aluminum wall section. Numerical work had shown that axial conduction would smoothen the local wall cooling. A surface averaged heat transfer coefficient was determined based on the rate of temperature decrease measured on the inner wall, and it was found to be significantly higher than that predicted by standard turbulent flow correlations due to the local impingement. Flow visualization including schlieren images further revealed that a strong back-flow area appears around the jet, which results in locally strong variations of the boundary layer and heat transfer, as illustrated in [11].

The thermal response of the seeding particles was calibrated through a dedicated setup, where small seeding rates were applied in a gas stream which impinged on a thin metal surface. The carrier gas was nitrogen, a stream taken from stored bottles and mixed with a second stream of boiling liquid nitrogen to provide a set temperature ranging from –120°C to 0°C. An exponential calibration curve was then established based on the measured phosphorescent response, as this was the physically expected behavior. The temperature of the back of the metal surface which was in the flow was measured through a thermocouple as direct input into the particle calibration curve.

Through careful design and calibration, it was possible to generate high-quality data on the jet flow in terms of both velocity and local temperature, which have been compared to numerical predictions. The closure of the heat balance was achieved to within 10% in this case through enthalpy-based calculations using the upstream conditions and transient heat lost by the wall during the experiment.

As the setup cooled down, condensation droplets started to appear on the optical access windows, which resulted in loss of signal. To resolve this, a continuous N2 purging was set up on the windows to keep it locally clear of droplets. This was originally not foreseen in the design and is an example of a challenge with optical access. At high temperatures, other challenges appear, such as local soot deposition/blackening. Seed particle deposition in walls and crevices resulted in high signal back to the detector; to reduce this, firstly all windows were installed flush mounted with the walls, avoiding a backward facing step which would otherwise accumulate particles, and secondly, after every run, the test setup was carefully cleaned with demineralized water and ethanol to reduce the tendency for particles to deposit. Metallic surfaces which could cause a local glare in the field of view were taped off with a matt black tape. This laboratory practice resulted also in less risk of scattering laser beams as sharp edges were taped off, which improved overall laboratory safety.

1.3.3 Longitudinal Fin Cooling of Electronics

Cooling of electronics remains an active subject in global research, looking to improve thermal efficiency or cooling performance. Various heat balance approaches can be applied to look at the overall efficiency, but understanding the local behavior in complex heat sink structures is a challenge due to limited access without disturbing the flow field. On a single fin, or part of a heat sink structure, it is possible to use infrared thermography to measure local temperatures as illustrated in [13], which applied an inverse problem statement to determine the local heat transfer coefficients. To do this, careful control of the heat flux boundary at the bottom of the fin was required, as well as understanding the accuracy of the measured temperatures using the IR camera. The latter was done through careful stepwise calibration following ASTM standards in situ to ensure the impacts of the wind tunnel surroundings were understood.

Guard heaters were applied at the base of the fin to limit the heat loss to the surroundings. The setup is illustrated in Figure 1.4. To provide heat to the fin, a heat foil was used at the base of the fin structure. To avoid heat loss to the bottom, firstly a layer of insulation that can resist high temperatures ("promatec") was applied. However, as there is a temperature gradient across this insulation layer, there is still a portion of the heat flux generated by the heat foil lost to the environment. To counter this, a second heat foil is applied underneath the insulation and both heat foils are instrumented with thermocouples. By having the power of the second heat foil controlled such that the temperature difference across the insulation layer is minimized, the heat flux lost is reduced and can be better estimated. The same principle was applied at the sides of the fin base (upper circular features - heating tape) and the sides of the insulation (lower circular features - second set of heating tapes). This resulted in a heat loss of <2% of the total heat input and a

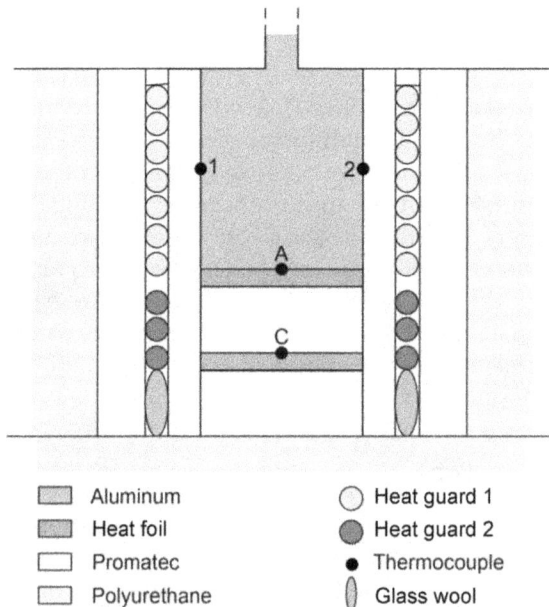

FIGURE 1.4
Guard heater configuration for a longitudinal fin [12].

well-known thermal boundary condition at the base of the fin structure. This then provided the input to the numerical model to assess the local heat flux distribution (inverse problem statement).

To ensure the guard heaters are effective, it is important to limit any thermal contact resistance between the thermocouples and the measured object. This was achieved by drilling small holes into the fin base and inserting the thermocouple junction together with thermal conductive paste before fixing it in place. Assessing local thermal resistance is challenging, and therefore, it is advised to design the setups up ahead to minimize the risk of these appearing.

1.3.4 Heated Pipe-in-Pipe Testing

Guard heaters are a useful technique to contain heat losses at end sections. In an experimental setup, a 30-m-long heated pipe-in-pipe section (manufactured from three 10-m-long samples) was tested to assess its thermal insulation performance as well as warm-up behavior. Full-scale pipelines (4″ inner pipe installed inside an 8″ outer pipe) were used in this test setup, which were mounted horizontally and filled with water, air, or a mixture. The thermal insulation was placed in the annulus space between the two pipes (hence the name "pipe-in-pipe"), and the pressure in the annulus volume was also lowered to <50 mbar, which further limits convection inside the annulus. Heating wires were mounted on the outside of the inner pipe which would receive current to heat up. The outer pipes were painted white to reduce the impact of solar radiation during the experiments as the setup was set in open air.

As the overall heat input was very low (<100 W) despite the good thermal insulation, the time to warm up the inner pipe fluid was still several days. Initial tests without guard heaters resulted in limited warm-up of the pipe, reaching only a few degrees above the ambient temperature, whereas it was designed to reach +20°C. Guard heaters were mounted by wrapping heating tape around a short 1-m end section at each end of the tested pipeline. This was then further insulated with glass wool and protected with aluminum coverings. This is illustrated in Figure 1.5. A thermocouple (black square) was mounted on the pipeline end wall as well as the short heating section where the power input to the heating tape was controlled such that the difference between the two temperatures is minimized. This effectively resulted in the pipeline mimicking an "infinitely" long section. As a result, the heat loss at the end sections was removed and the pipeline was able to warm up to its design value.

As the transient warm-up period also with guard heater was still long, the test setup experienced diurnal fluctuations in temperature as the ambient heat losses would vary between night and day. (The setup was set in open air.) This resulted in not reaching a pure steady state, but by having sufficient data during the temperature cycles, allowed to assess the thermal properties in a long-term averaged manner.

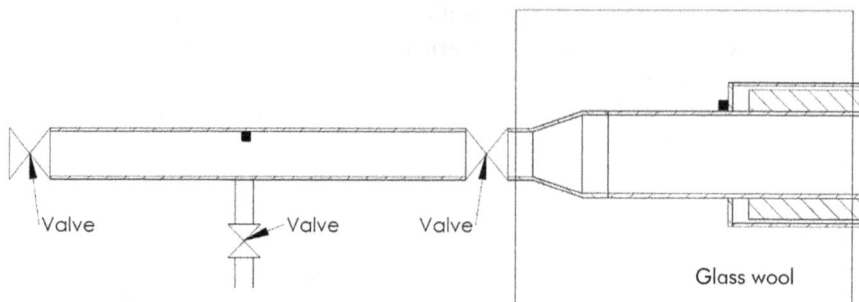

FIGURE 1.5
Guard heater principle applied to a pipeline.

Due to the long length of the pipeline, a fiber optic cable (distributed temperature sensing—DTS) was applied to measure the local temperatures inside the pipeline. This was supplemented by six internal thermocouple measurements. The DTS system provided a high resolution with data points every 0.5 m inside the pipe; however, each value corresponds to an averaged value over a length of 0.5 m, so at sharp temperature transitions, these data can be deceptive. Installing fiber optic cables in such setups is a delicate matter when splicing is required to connect the optical connectors and establish interfaces/repair fiber breaks. Careful calibration of the optical circuit and avoiding local losses is key to ensure data quality. When selecting novel techniques based on optical sensors, it is advised to double up on the instrumentation in case of failures and to verify data independently. Furthermore, by having access locally to the fiber, a verification can be done through adding heat and measuring the response.

During one of the experiments, the central 10-m section of the pipe was not heated. The results, however, showed that the temperature in the middle of the pipe was still well above the ambient temperature. This was due to natural convection inside the liquid, which resulted in an effective heat transport from inside the heated pipe sections into the central section. So, despite receiving only some 40 W of heat, the temperature was kept more than 7°C above the ambient temperature. Natural convection in liquids can result in strong thermal transport even with temperature differences of only a few degrees. This is due to the large density and resulting mass transport at the interfaces. Due to the large thermal insulation of the pipe-in-pipe configuration, limiting heat loss to ambient, this allowed the central pipe section to be warmed up. Without the guard heaters or the insulation, the natural convection acted as a strong heat sink, lowering the final temperature of the setup considerably. This illustrates the impact the unexpected convective terms can have on the outcome of a thermal experiment.

1.3.5 Multi-Pass Refrigerant Loop: Local Heat Transfer Coefficients

Multi-phase flow presents researchers with new challenges, as each phase has its own physical properties and characteristics to determine and build physical models for. Due to the large differences in basic fluid properties, the selection of the measurement technique is key to ensure good-quality results, with a preference for non-intrusive methods such as optical or electrical (resistive/capacitive). The control of the precise boundary conditions inside test setups is challenging due to the relatively large impact of uncertainty of standard measurement equipment (temperature/pressure) when expressing fluid properties such as the vapor pressure, the stochastic nature of the boiling process (and our limited understanding of its mechanisms), the nature of modern refrigerants which have a temperature glide as they are zeotropic mixtures.

To assess the local void fraction, flow patterns, and heat transfer coefficients, a test setup (Figure 1.6) was developed in [14] which consisted of several connected multi-pass counter-flow tube-in-tube heat exchangers. These could individually be controlled to examine void fractions ranging from 0 to 1 for various refrigerants. The refrigerant flowed through 3/8″ smooth bore copper tubes. Having a high flexibility in heat input, and associated control, is one of the main challenges experimental setups need to overcome to work effective in two-phase flow research. In this case, the length of the preheater sections varied between different passes to allow for more variation in setting the inlet quality. Optical access was provided through

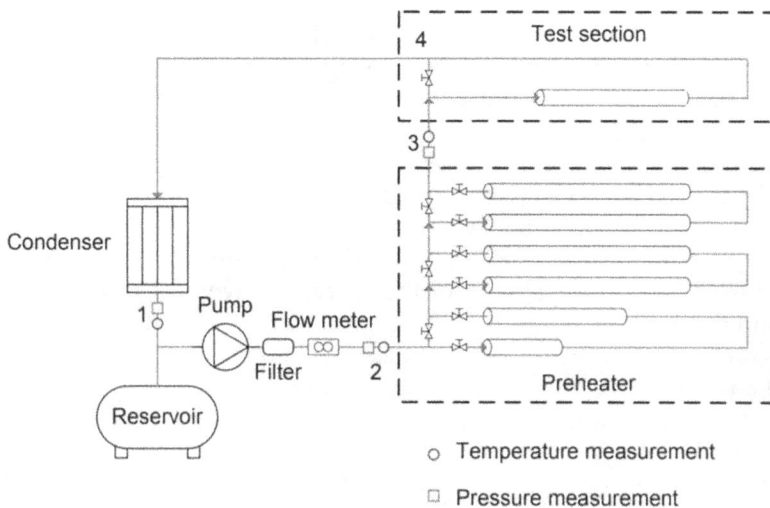

FIGURE 1.6
Schematic overview of the refrigerant multi-pass setup used in [13].

a sight-glass which enabled visual high-speed recordings to be made. The quartz sight-glass was annealed and hardened, to enable operation in high pressures. Local heat balances were computed on each of the pass segments to determine the vapor quality at the inlet of the test section, where a capacitance-based sensor was used to measure the void fraction and high-speed images were recorded of the flow.

To set the refrigerant pressure inside the loop, a reservoir was applied set in an open water bath. Inside the reservoir, a liquid–vapor interface is present which in effect sets the local vapor pressure at one location in the loop and is a key reference point. The water bath temperature is controlled through a heater/cold water loop. The saturation temperature of the loop could be controlled in this manner up to ±0.5°C.

To determine the local heat transfer coefficients inside the two-phase flow, the local wall temperature needs to be known. Cross-sectional averaged wall temperatures were measured by having three thermocouples mounted inside the 3/8″ copper tube wall at various locations. To mount these 0.25-mm K-type thermocouples, a small groove was made into the pipe wall, after which the junctions were carefully placed inside the groove and soldered into place using silver solder. This practice proved to be the most effective in ensuring no contact resistance was present when using these copper pipes. These thermocouples were then calibrated in situ to an accuracy of ±0.05°C against a precision PT-100 and using a triple point of water cell as cold junction reference temperature. The use of the triple point of water cell as cold junction temperature was required to achieve this high accuracy, as discussed in detail in [14]. More details on the experimental setup can also be found there.

To determine the local heat flux, the derivative was used of the water-side enthalpy variation as measured using 0.5-mm K-type thermocouples mounted inside the water annulus. By measuring this at four locations along the length of the tube together with a third-order polynomial fit, the heat flux was locally calculated. The water circulation rate was measured using a Coriolis-type flow meter. To ensure no temperature stratification took place inside the annulus, the flow rate was kept turbulent. As there are spacers present in the annulus to center the 3/8″ copper tube, the thermocouples were also positioned upstream of the spacers to avoid any impact on the measurements.

To control the refrigerant flow, a magnetic bearing pump was applied to avoid any oil contamination of the fluid, combined with Coriolis mass flow meter to provide sufficient accuracy. To reduce ambient heat losses, all heat exchanger sections were insulated with foam-type insulation, and by circulating only hot water in the setup, the heat loss to ambience was estimated. An integrated control system performed all the measurements of the system and contained all the data quality checks to verify the test loop status (heat balance and variation in flow rates/saturation pressure) and the data quality in the test section (local temperatures at each cross section and their variation over time and the last ten measurement points). This allowed the researcher to judge the data quality online while measuring and greatly helped reduce the duration of the experiments.

1.4 Conclusions

This chapter presents a set of guiding principles to experimentally verify boundary conditions in a test setup. These principles can be applied to any type of experiment; the case studies in this chapter have focused on measuring flow/heat transfer coefficients as most of these principles have to be applied in these cases to ensure good data quality. As computational tools continue to advance, it makes sense that more and more research will be done numerically. However, *the proof of the pudding is in the eating*, and therefore, experiments remain required to verify the numerical outcomes and to test whether realistic boundary conditions and actual equipment indeed result in the same behavior predicted through the often idealized computations. I encourage researchers to apply computational tools to help them design their setups and test conceptual ideas for sensor placement, the test setup dimensioning, and verifying whether the proposed scaling does not result in unexpected side effects (e.g., natural or mixed convection). By thinking ahead, and performing analytical pre-work, often it will become clear whether a desired boundary condition, be it an (a)diabatic wall or a fixed inlet temperature, can indeed be achieved throughout the duration of the experiment—and when combined with an uncertainty propagation will tell you where to best focus your attention to be more accurate. The devil is in the detail, and the number of devils to be put at rest scales perhaps exponentially with the complexity of a setup. But by being well prepared, setbacks can be avoided. And always appreciate the lessons you get from uncertainty propagation; understanding this will help you focus your experimental efforts at the right variable.

References

1. Arie, M.A., et al., 2018, Experimental characterization of an additively manufactured heat exchanger for dry cooling of power plants. *International Journal of Heat and Mass Transfer* 129, 187–198.
2. Salzman, D., et al., 2018, Design and evaluation of an additively manufactured aircraft heat exchanger. *Applied Thermal Engineering* 138, 254–263.

3. Bejan, A. and Errera, M.R., 2014, Technology evolution, from the constructal law: Heat transfer designs. *International Journal of Energy Research* 39, 919–928.

4. De Paepe, M., et al., 2011, The use of open cell metal foams in heat exchangers: Possibilities and limitations. *Proceedings of 8th International Conference on Heat Transfer, Fluid Mechanics and Thermodynamics*, Mauritius.

5. T'Joen, C., et al., 2009, A review of polymer heat exchangers for HVAC&R applications, *International Journal of Refrigeration* 32, 763–779.

6. Buckingham, E., 1914, On physically similar systems; illustrations of the use of dimensional equations. *Physical Review* 4, 345–376.

7. Ay, H., Jang, J.Y., and Yeh, J., 2002, Local heat transfer measurements of plate finned-tube heat exchangers by infrared thermography. *International Journal of Heat and Mass Transfer* 45, 4069–4078.

8. Moffat, R.J., 1985, Using uncertainty analysis in the planning of an experiment. *Journal of Fluids Engineering* 107(2), 173–178.

9. T'Joen, C., 2008, Thermo-hydraulic study of inclined louvered fins. PhD Dissertation, Gent University – Ugent (Belgium).

10. T'Joen, C., et al., 2011, Interaction between mean flow and thermo-hydraulic behaviour in inclined louvered fins. *International Journal of Heat and Mass Transfer* 54, 826–837.

11. Fond, B., et al., 2018, Investigation of a highly underexpanded jet with real gas effects confined in a channel: Flow field measurements. *Experiments in Fluids* 59, 160.

12. Abram, C., et al., 2018, Temperature measurement techniques for gas and liquid flows using thermographic phosphor tracer particles. *Progress in Energy and Combustion Science* 64, 93–156.

13. Willockx, A., 2010, Using the inverse heat conduction problem and thermography for the determination of local heat transfer coefficients and fin effectiveness for longitudinal fins. PhD Dissertation, Gent University – Ugent (Belgium).

14. Caniere, H., 2010, Flow pattern mapping of horizontal evaporating refrigerant flow based on capacitive void fraction measurements. PhD Dissertation, Gent University – Ugent (Belgium).

2

Measurement Error or Data Trend? How to Conduct Meaningful Experiments

Thorsten Klahm

GESMEX Exchangers GmbH

CONTENTS

2.1 Error Analysis in Experimental Investigations

Experimental studies consist of several parts that are usually experimental setup, observations/measurements, data reduction, and conclusions. Some studies compare their data with former investigations, some compare models with measurement data and propose some adjustments of existing equations, and some suggest new correlations. All these studies have something in common. They use measurement data with a limited accuracy. There are deviations in every experiment. Furthermore, they want to make a conclusion in the end. Does the set of data support the model or not? Are the deviations between the measurement and the model a physical effect or just a measurement error?

These questions can be answered if not only the measurement value but also the possible deviation or error that has to be accounted for is known. The following part of the chapter will give an introduction to error analysis and how it is done in the standardized "Guide to the expression of uncertainty in measurement" (GUM:1995). After the error is known, the set of data will reveal whether the deviation from the model is caused by a measurement error or a physical effect.

In the second part of this paragraph, it is shown how to calculate the expected error of the target quantities before the experimental rig is built and how this can improve the scientific conclusions that can be drawn.

2.1.1 Error Analysis

Error analysis is not rocket science anymore. You do not have to read several textbooks about statistics to perform the important steps of error analysis (however, you will have more insight in the fundamentals). In 1995, the first version of the "Guide to the expression of uncertainty in measurement" (GUM, 2008) was published by the International Bureau of Weights and Measures (BIPM) and the International Organization of Standardization (ISO). This guide was an attempt to standardize the way measurement errors were calculated. It was driven by the absence of any consensus between the countries and a need to make metrology measurements comparable. This ISO Guide is nowadays the standard that is used by calibration laboratories and industry.

The term measurement error is sometimes substituted by the term uncertainty (DIN 1319-1) to distinguish between wrong handling of equipment which may cause errors and the inevitable presence of doubt whether the measurement result corresponds to the true value of a quantity or not. If the word error is used in this chapter, it is a synonym for the deviation between the measured value and the true value in the absence of any handling mistakes or the like.

In general, one can distinguish between random errors that are recognized during the measurement as random fluctuations of the value and systematic errors. Random errors do not have a sign and can be handled by calculating the mean of the measurements over time. Systematic errors are not random, and they have a sign and a magnitude. For example, resistance thermometers such as Pt100 warm themselves up during the measurement because of the nature of the semiconductor material. This may cause a shift of the measured value in the positive direction. This shift is not easily recognized because systematic errors cannot be identified by repeated measurements. They are mostly constant, or they are a function of the measured value. They cannot be identified by repeated measurements. Calibration is one way to discover systematic errors or at least quantify the magnitude of the sum of systematic deviations.

Statistical analysis of the random error can only take place if systematic errors are removed (Moffat, 1988). Thus, first they must be identified and removed before a measurement can take place. Using this definition, life is easy, because all that is left is random fluctuations around a mean value that is already very close to the true value. Statistics about random deviation is all that is needed.

In metrology laboratories where sensors are calibrated, the equipment is optimized and systematic errors should not occur, but industrial equipment or laboratories at universities will not reach that precision and systematic errors have to be taken into account. A very recent example may illustrate that even very precise equipment may have some hidden systematic errors. During observations of the surface of the planet Mars with the Compact Reconnaissance orbiter and a special imaging spectrometer in 2015, scientists claimed to have identified concentrated regions of perchlorate, which suggested that water is present on Mars' surface. Three years later, they discovered that "noise in the data is smoothed in such a way that it mimics real mineral absorptions, falsely making it look as though certain minerals are present on Mars' surface" (Leask et al., 2018). Some pixels of the spectrometer were glared by the sun as the orbiter flew over the surface, and the afterglow of these pixels was misinterpreted as the perchlorate source. Usually, filtering and software corrections are done to compensate for these effects, but this systematic error was not discovered before.

It is clear that we cannot neglect that other errors than random ones may be present. Technical measurements are done using calibrated measurement devices. The deviation between the calibration normal and the probe consists of several effects, and there is only a calibration interval defined. This is not a pure systematic deviation with sign and value and not a statistical value. Contrary to some older textbooks, the GUM did not focus on systematic errors but generalized them even more.

Measurement uncertainties can be grouped into category A, which includes all uncertainties that are evaluated by statistical methods, and category B, which is evaluated by other means.

Uncertainties of category A can be characterized by the estimated variances or estimated standard deviations. Uncertainties of category B are not directly accessible by statistical tools. Think about the shift of temperature in the thermometer example. This deviation is not a statistical fluctuation. If the systematic shift is not known and cannot be removed or compensated, we can only define the estimated interval in which the shift probably could be found. In the methodology of the GUM, an approximated variance and standard deviation for that uncertainty are defined based on that interval. The aim is to treat both uncertainties A and B equally. Although this approach is subject to criticism (Grabe, 2007), the GUM is the standard way how uncertainties of measurement should be calculated today.

2.1.2 Uncertainty of Category A

Random fluctuations of a measurement may develop because of many small disturbances of the measurement process which are independent of each other and have positive or negative sign. The process should be in a steady state. The values may fluctuate around a mean value that should not increase or decrease continuously.

How much effort is needed to achieve steady-state conditions depends on its definition. You will have to define how long a steady-state measurement should be taken and which deviation is reasonably tolerable. The values of these limits depend on the application, but in general, the duration time should reflect the inertia of the system. Imagine a large container that is filled with water and equipped with an inlet pipe and an outlet pipe. Water flows in and out at the same flow rate, and inlet and outlet temperatures are measured. A change in the inlet temperature will change the outlet temperature with a delay that depends on the transfer function of the system. The response time of the transfer function is related to the mass and mass flow rate of the water in the container. You need to measure the inlet and outlet temperatures at least during that time period that a change of temperature needs to travel through the container; otherwise, you cannot be sure that you recognize changes in the temperature correctly.

As an example, imagine that at $t = 0$, the inlet temperature changes from 50°C to 55°C and the outlet temperature is 20°C. There is some kind of heat removal that should be constant with time and may cool the water in the steady-state case by 30 K. Let us assume that it takes half an hour until temperature changes of the inlet are recognized at the outlet. Saving measurement values between $t = 5$ min and $t = 15$ min will show that the inlet temperature is 55°C and the outlet temperature is still 20°C. The 5 K change is not recognized at the outlet, and the steady-state heat flow rate calculated using inlet and outlet temperature will be wrong. It is 35 K, but in steady state, it should be 30 K. The duration time of the steady-state measurement should be estimated according to the transfer function of the system (Liptak & Moore, 2006).

The maximum deviation of the measurements in that time should be, ideally, in the range of the measurement accuracy of the instrument. Nevertheless, process control of complex technical systems is often characterized by periodic fluctuations that come from the combination of higher order transfer functions and control loops. In this case, the researcher might relax the definition of the allowed deviation at steady state and increase the number N of measurements taken.

The N measurements are averaged over time by calculating the mean value $\bar{x}$.

$$\bar{x} = \frac{1}{N} \sum_{k=1}^{N} x_k \qquad (2.1)$$

Deviations from the mean value are quantified by the standard deviation s that is calculated by a special way of averaging the squared difference between the mean value and the measurement value.

$$s = \sqrt{\frac{1}{(N-1)} \sum_{k=1}^{N} (\bar{x} - x_k)^2} \qquad (2.2)$$

This is closely related to the normal distribution function that C.F. Gauss had described. In Figure 2.1, the curve with light gray color shows an idealized distribution of measured values with the mean value $\bar{x}_S$ and standard

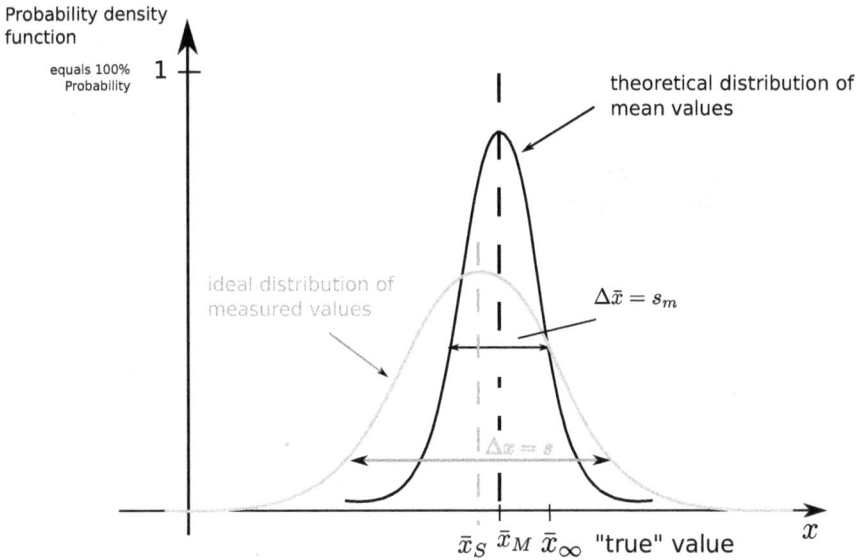

FIGURE 2.1
Statistical distribution function, mean, and standard deviation.

deviation s. Imagine that you carry out further measurements with the same amount of measured values. All these measurements can be described by a distribution curve and a mean value.

The mean value of these repetitions will all be different and scattered, which can be described by a distribution itself. This distribution of the mean values $\bar{x}_S$ is represented by the solid black curve in the figure. The mean value of the mean values $\bar{x}_S$ is $\bar{x}_M$, and the standard deviation is called s_m. The standard deviation is smaller, and it is closer to the true value $\bar{x}_\infty$. The true value will be reached in the limiting case of an infinite number of measurements.

The deviation of $\bar{x}_M$ from $\bar{x}_\infty$ is not known because $\bar{x}_\infty$ is not known, but an estimate of that deviation is s_m.

$$\Delta\bar{x} = s_m = \frac{s}{\sqrt{N}} = \sqrt{\frac{1}{N(N-1)}\sum_{k=1}^{N}(\bar{x}-x_k)^2} \tag{2.3}$$

In the absence of systematic errors, the uncertainty of a measurement is characterized by the deviation of the calculated mean value of N measurements from the mean value of infinite measurements s_m.

This deviation can be reduced by just repeating the measurements and increasing N. This may lead to the assumption that measurement uncertainty does not have to be included in the planning phase of an experiment because the uncertainty can be reduced afterward by just taking a lot of measurements. Usually, measurements are carried out using commercial sensor equipment such as pressure transducers, thermocouples or Pt100 resistance thermometers, and flow meters. All these sensors have been calibrated, and the accuracy is given. This accuracy includes effects attributed not only to the sensor element but also to electrical connection, electronics to turn the sensor signal into a voltage that can be readily measured, or even an additional analog–digital converter that creates a computer-readable signal. This accuracy is the sum of several effects, and we only know that the deviation caused by these effects should not be outside the given accuracy interval.

This is nothing random but a possible shift of the measured value that should be treated as an uncertainty of category B.

2.1.3 Uncertainty of Category B

The GUM does not distinguish between random uncertainty and systematic uncertainty, but systematic uncertainties are included in category B. The measurement x_k is shifted by a constant deviation of magnitude f from the true value that is not known. The only known value is the interval in which the true value may fall.

Figure 2.2 shows that this can be treated as a rectangular distribution function with the width of $2f_s$. A statistical treatment of this distribution leads to the standard deviation

$$\Delta x_s = \frac{f_s}{\sqrt{3}}. \tag{2.4}$$

The reason why the deviation f_s is not used directly is that the definition of a standard deviation makes both uncertainty categories A and B comparable. They can be added, and a combined uncertainty U can be defined by

$$U = \sqrt{\Delta \bar{x}^2 + \Delta x_s^2} \tag{2.5}$$

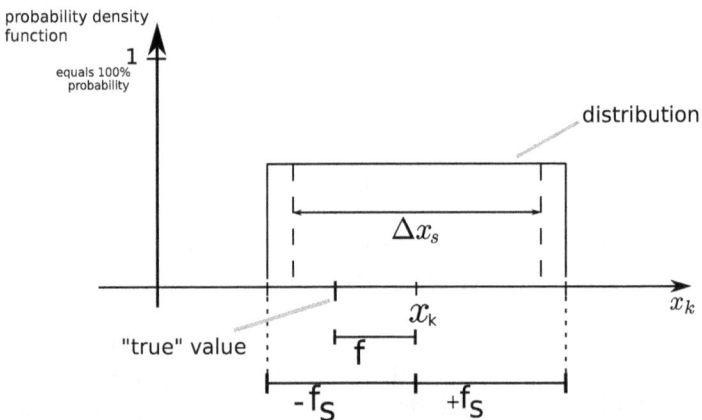

FIGURE 2.2
Systematic deviation.

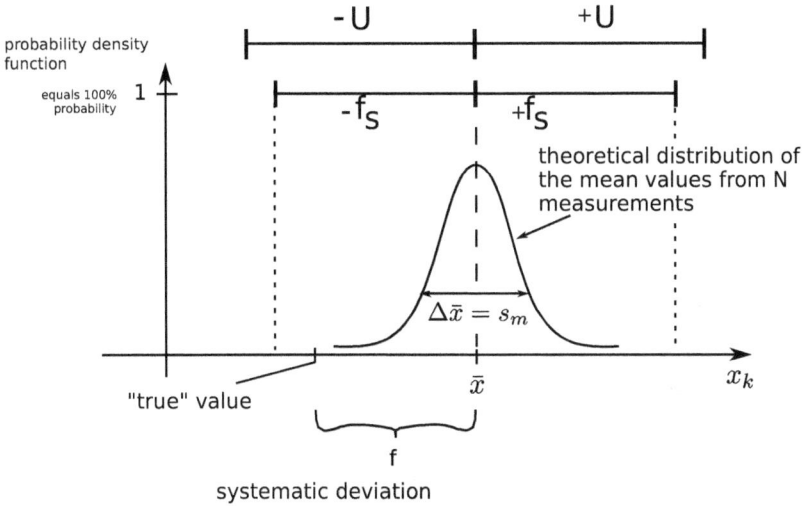

FIGURE 2.3
Random and systematic uncertainties.

This is illustrated in Figure 2.3. In this picture, the random deviation is smaller than the systematic shift, which might often be the case in thermal measurement of industrial equipment but has to be avoided in calibration laboratories.

The combined uncertainty is a combination of two standard deviations. Assuming that U could be treated similarly as the standard deviation of a Gaussian normal distribution, this would mean that 68% of all measurements are covered by this uncertainty. If 95% of the measurements should be covered, U must be doubled. The GUM recommends to use the extended uncertainty $U_E = 2U$ in order to be confident about 95% of the measurements. Although statistically several assumptions must be used to validate this, a factor of 2 could also be considered as a safety factor to be sure that the deviations are covered. The GUM does not state that this method guarantees that the true value is inside the boundary of U_E (Grabe, 2007).

2.1.4 Error Propagation

The previous sections showed how to calculate the uncertainty of a measured mean value. This can be a temperature measurement. Often the knowledge of the temperature itself is not the purpose of the experiment. The target may be the heat flow rate or the heat transfer coefficient, which is not measured directly. The interesting question is: How will the uncertainties of each individual measurement x_i influence the uncertainty of the target quantity y? The equations that are used for data reduction propagate each uncertainty into the result.

There are two ways of error propagation calculation described in the GUM:

1. Taylor series calculation
2. Monte Carlo simulation

Taylor series are useful if the model equations are known and explicit.

$$y = f(x_1, x_2, \ldots, x_i) \tag{2.6}$$

In that case, differential variations of y with differential variations of measured quantities x_i can be expressed by the total differential or a Taylor series, which is given by

$$\Delta y = \sum_{i=1}^{N} \frac{\partial f}{\partial x_i} \Delta x_i \tag{2.7}$$

Higher order terms are assumed to be negligible. The uncertainty of y must be expressed in terms of a statistical variance or standard deviation. That means that not the difference Δy is needed but the squared value which is expected to be the variance of y

$$U_y^2 = \left(\sum_{i=1}^{N} \frac{\partial f}{\partial x_i} U_i \right)^2 \tag{2.8}$$

This can be rewritten as

$$U_y^2 = \sum_{i=1}^{N} \left(\frac{\partial f}{\partial x_i} \right)^2 U_i^2 + 2 \sum_{i=1}^{N-1} \sum_{j=i+1}^{N} \frac{\partial f}{\partial x_i} \frac{\partial f}{\partial x_j} U_i U_j C_{ij} \tag{2.9}$$

with C_{ij} as the correlation coefficient. The second term in this equation is needed if the individual uncertainties of the measurements x_i or the measurements itself depend on each other. If this is not the case, only the first term remains. It may be advantageous to express the uncertainty as relative uncertainty ($u_i = U_i/x_i$ percentage of the measured value):

$$u_y^2 = \sum_{i=1}^{N} \frac{x_i^2}{y^2} \left(\frac{\partial f}{\partial x_i} \right)^2 u_i^2. \tag{2.10}$$

To compute the desired uncertainty of the target variable needs a lot of calculation work, so why do we need this? Is it not enough to keep the uncertainty of the measurement as low as possible?

After data reduction, you will need to compare your data to an existing model or invent a new way of correlating some quantities. Graphical representations are a good way to have a look at the data and judge whether there is a significant trend, a proportionality, or a change in slope. A lot of papers deal with arguing whether the data support the model or not. Having no indication of data accuracy makes it impossible to judge whether the distance between model curve and data points is within tolerance or not. Even the decision what kind of curve fitting is necessary depends on measurement uncertainty. A cloud of data may look like a curve, but is it really a curve or does the measurement uncertainty increase and cause the curve to decline?

Knowing the uncertainty of the target variables is necessary because it helps to interpret the results and draw conclusions. Comparisons with models and correlations will always lead to the question whether it is a physical phenomenon or a measurement error. As a researcher, you want to be able to exclude the measurement errors and identify the physics behind the measurements. This is only possible by defining the uncertainty of the results. A simple example should illustrate this.

A common way of calculating the heat flow rate is using the following equation:

$$\dot{Q} = \dot{M}cp(T_{in} - T_{out}) \tag{2.11}$$

where $\dot{Q}$ is the heat flow rate (W), $\dot{M}$ is the mass flow (kg/s), cp is the heat capacity (kJ/kg K), and T is the temperature (°C).

We need three measurements, $\dot{M}$, T_{in}, and T_{out}, to calculate the target quantity $\dot{Q}$. We assume the accuracy of the fluid property data to be very good and neglect it. Each of these measurements is not perfectly accurate, and the calculation procedure propagates the uncertainties into the result $\dot{Q}$. Assuming that all measurements and uncertainties are independent, the first term of Eq. (2.9) is sufficient. The derivative $\dfrac{\partial f}{\partial x_i}$ is called sensitivity factor. If the sensitivity factor is high, the uncertainty is amplified and increases the uncertainty of the target value. Looking at different forms of equations, some rules can be deduced on how the propagation of uncertainties can be estimated.

Products of several variables: The relative uncertainty of a product of several variables is the sum of the relative uncertainties of these variables. If the mass flow measurement has a relative uncertainty of 2%, cp of 0.01%, and temperature difference of 4%, the relative uncertainty of the heat flow is 6.01%.

The error propagation of sums and differences, expressed with relative uncertainties, includes the measured quantity and depends on that value. Using Eq. (2.10) to express the relative uncertainty of the temperature difference $(T_{in} - T_{out})$, the partial derivative of the temperatures equals 1 and the term x_i/y does not vanish:

$$u_{\Delta T}^2 = \left(\frac{T_{in}}{(T_{in} - T_{out})} \right)^2 u_{T_{in}}^2 + \left(\frac{T_{out}}{(T_{in} - T_{out})} \right)^2 u_{T_{out}}^2 \qquad (2.12)$$

This equation shows that the relative uncertainty of the temperature difference depends on the level of T_{in} and T_{out} itself. Assuming that the temperature difference in the denominator and the relative uncertainty are constant, the relative uncertainty of measuring T_{in} is multiplied by a higher value if T_{in} is 300°C than it is if T_{in} is 30°C. The temperature difference plays also an important role. The uncertainty is bigger the smaller the temperature difference is.

The error propagation calculation is not just a nasty calculation effort that results in some uncertainty numbers. It shows the researcher where the setup is sensitive to small deviations of the measurements and points to the critical range of values that are prone to uncertainty and, in the end, might not even produce meaningful results.

Above, we considered products of variables and sums/differences. Powers are also very common in engineering correlations and also have an impact on error propagation.

$$y = ax_1^p x_2^q \qquad (2.13)$$

$$u_y^2 = (p \cdot u_1)^2 + (q \cdot u_2)^2. \qquad (2.14)$$

Equation (2.13) is an example of a correlation. a, p, and q are correlation parameters. Equation (2.14) shows that powers turn to factors due to the calculation of the derivative in Eq. (2.10). That means that powers that are greater than 1 will increase the uncertainty contribution and powers that are smaller than 1 will decrease the uncertainty contribution of the measurements of x_1 and x_2.

These propagation rules are useful if the impact of uncertainties of the measurements on the target quantity needs to be estimated. The outcome of this analysis is the information which measurement range causes higher uncertainties than the other and which measured quantity has the highest contribution to the uncertainty of the target quantity. Both data are needed in the planning phase of an experiment.

Imagine you are planning to analyze the heat flow during experiments with an uncertainty of 1 kW and you are measuring temperatures and mass flow as in the example above. What if the temperature difference is only 5 K and the measured inlet temperature is 200°C? The factor $T_{in}/(T_{in} - T_{out})$ in Eq. (2.12) will be 40, and even if the relative uncertainty of the measurement of 200°C is only 0.1%, the contribution to the uncertainty of the heat flow will be 4%. The second term in Eq. (2.12) adds another 3.9%. The relative uncertainty of the temperature difference $u_{\Delta T}$ according to Eq. (2.12) sums up to 5.6%. Even if the uncertainty of the mass flow measurement is negligible, a heat flow rate of 50 kW will have an uncertainty of 2.8 kW (use Eq. (2.10) with (2.11)). This means that the desired accuracy cannot be achieved in this temperature range.

The evaluation of the different contributions to the overall uncertainty of the target is especially useful if you want to know how accurate the measuring equipment should be. Usually, the decision how accurate the measurements must be is a matter of scientific common sense or best practice of the laboratory or institute. Using uncertainty calculations, you know in advance which measurement must be highly precise and which is of minor importance. Imagine the contribution of the temperature measurement is about 20% and the contribution of the mass flow measurement is only 2%. It is not necessary to spend a lot of money to buy very accurate flow meters because the source of uncertainty is the temperature measurement. First, these sensors must be improved, and if the level is decreased to the level of the flow meter, it is worthwhile to think about increasing the accuracy of the flow meter.

Nevertheless, calculation of the whole error analysis might be much more complicated than in the simple example above. Analytical calculation of the Taylor series method leads to very long equations especially if a cross-correlation has to be considered. One might be tempted to avoid the effort and to set up a spreadsheet or a computer program instead. This way, one can change the input values of the data reduction equations by small portions and monitor the sensitivity of the results which will indicate the propagation of uncertainty based on datasets. Actually, this is not far away from the other method of error propagation analysis, the Monte Carlo method (MCM).

This numerical method simply uses the data reduction equation to test the impact of little changes around a fixed value of each input variable. These changes cause changes in the output variable, and all uncertainty contributions are already included in these output values, even if the uncertainties are correlated. First of all, a set of input values are fixed. These can be considered as real values, whereas these little changes we introduce can be considered as arbitrary errors and the fluctuating values as "measurements." The way of introducing the arbitrary error should follow a statistical approach, for example, using random generated values that form a Gaussian curve. The output values will also show a certain statistical distribution, probably Gaussian. This distribution can be used to calculate the standard deviation of the result. Repeating this procedure with different datasets spread over the whole range of applicable values will cover the whole domain of the experiment. Although it seems that, again, a lot of calculation is involved, this numerical calculation will be done with the help of computer programs, whereas the TSM may be carried out by hand. Powerful scripting languages such as Python, R, or MATLAB® are available. The advantage of this method is that it does not linearize or simplify the system of equations. As long as the user is able to calculate the result, this method works. It is also very useful for identifying potential cross-correlations that are not obvious. If you have already done a simplified TSM analysis, you can calculate some MCM datasets and compare the results with the analytical results. If there are large deviations, it is very likely that there is a hidden cross-correlation. Further reading can be found in Stockton & Wilson (2012).

Most of the time, a combination of TSM and MCM will yield the best results. TSM gives a good overview about uncertainties and lets you choose the accuracy of your measuring devices, whereas MCM is a way to test the propagation of probability distributions.

2.2 Design of Experiments for Validation of CFD Calculations

Computational fluid dynamics (CFD) calculations are a very popular way to model the behavior of a certain system. It is possible to include a lot of complexity without the need to solve a set of differential equations analytically. In fact, approaching real-world systems makes the direct use of analytical solutions mostly impossible. The differential equations are often strongly coupled and must be solved in three geometrical dimensions and sometimes additionally in time. Using only the important equations that obviously dominate the problem reduces the complexity; however, the loss in the accuracy of the description is then unknown.

CFD uses a numerical approach to estimate the solution of the coupled system of differential equations. This needs further assumptions concerning the boundary conditions that are used, the design of the mesh, and other assumptions inside the used equations. For example, it is necessary to model turbulent flow with the help of empirical assumptions concerning the boundary layer, which results in different model categories such as $k - \varepsilon$, $k - \omega$, or SST model (Kanaris et al., 2005). All these assumptions build up the model error of the CFD system, Δ_{Model}.

Numerical calculations are estimations of the analytical solution. The differential equations are discretized along the spatial dimensions and time and solved with repeated iteration. Whether the solution of the discretized system converges toward the analytical solution must be verified by solving equations that have known analytical solutions. The accuracy of the numerical estimate can be determined that way and gives the error of the numerical algorithm, Δ_{Num}.

The third error that needs to be considered is the error of input values. All CFD calculations depend on a given set of input data that are usually taken from the validation experiments. Among the set of data are values for fluid properties, geometrical dimensions, and physical measurements such as temperature, pressure, or velocity and even calculated values from measurements such as heat flux and wall shear stress. All these values include a certain variance and can be taken as the deviation or error of the input values, Δ_{Input}.

We can conclude that the overall error of the CFD calculation consists of the error of the CFD model, the error of the calculation algorithm itself, and the error of the input data.

$$\Delta_{\text{CFD}} = \Delta_{\text{Model}} + \Delta_{\text{Num}} + \Delta_{\text{Input}}. \qquad (2.15)$$

The comparison with experimental data is a way of quantifying the error of the CFD calculation. It is called validation of the CFD simulation, whereas the pure comparison of the numerical results of the modeling with known analytical solutions is called verification. The process of validation includes the verification of the model and also the experimental deviation Δ_D which is addressed in the first part of this chapter. The answer whether the experiment can validate the CFD calculation depends on both the experimental deviation and error of the CFD calculation.

A straightforward approach is looking at the difference between the experimental result D and the CFD result S of the target quantity. The result is the comparison error E.

$$E = S - D \tag{2.16}$$

Knowing nothing about uncertainty analysis, we would expect E to be close to zero to have a successful validation. Often research is done in teams that consist of scientists that are responsible for the simulation and others that are focused on the experimental work. Team meetings will spend a lot of time arguing about the reasons why E is bigger than expected or what could be the reasons for E. The experiments are done to validate the CFD solution, so experimental results are taken as the benchmark that the CFD calculation has to reach. As a first approach, E will be attributed to the numerical result. The CFD team will argue that the deviation from calculation could also have something to do with model assumptions that are not reflected by the experiment or that measurement errors are the cause of the deviation.

With the knowledge of this chapter and the preparation of the right numbers, the discussion would be much shorter:

Considering that there is a true value T but this is not known, we can at least state that the deviation from this value can be defined by

$$\Delta_S = S - T \tag{2.17}$$

and

$$\Delta_S = D - T \tag{2.18}$$

Using Eqs. (2.17) and (2.18), we get

$$E = S - D = T + \Delta_S - (T + \Delta_D) = \Delta_S - \Delta_D. \tag{2.19}$$

The validation comparison error E is thus the combination of the errors of the simulation result and the experimental result.

Considering that the simulation error was defined in Eq. (2.13), we can write

$$E = \Delta_{\text{Model}} + \Delta_{\text{Num}} + \Delta_{\text{Input}} - \Delta_D \tag{2.20}$$

Although this equation defines E, only the effect of deviations of the numerical code Δ_{Num}, the input data Δ_{Input}, and the experiments Δ_D can be estimated. There is no method to estimate the effect of Δ_{Model} (Coleman, 2009).

The standard uncertainty of the experiments U_D is explained in the first part of the chapter and can be used to express the deviation U_D. Verification of the numerical code will result in an uncertainty of the calculation U_{Num}, and the estimation of the uncertainty of input values U_{Input} is also covered by the above explanation because inputs are often measured values. The deviation of a model with and without a certain assumption cannot be determined in most cases because if we could compute the result without the assumption, we would immediately do so. Only in cases where there are several possible assumptions that seem equally applicable, we could estimate the effect of the different assumptions and thus had an indication how much the result will be affected by the choice of assumption. An example is the choice of turbulence model. Consecutive simulations with different turbulence models may indicate how the target quantity is affected by this assumption.

Returning to the team meeting, it is possible to shorten the discussion just by presenting the different uncertainties together with the result E. The CFD team cannot question the validity of the experimental results outside the range of uncertainty. Numerical and input uncertainties are also given. The discussion must focus on the model assumptions and their applicability. Therefore, Eq. (2.20) can be written as

$$\Delta_{\text{Model}} = E - \left(\Delta_{\text{Num}} + \Delta_{\text{Input}} - \Delta_D \right) \tag{2.21}$$

Using standard uncertainties to express the known deviations, we can write

$$U_{\text{val}} = \sqrt{U_{U_{\text{Num}}}^2 + U_{\text{Input}}^2 + U_D^2} \tag{2.22}$$

This is done assuming that the deviations are independent of each other (other examples can be found in Coleman & Glenn Steele, 2018).

The sign of this standard uncertainty is not known; thus, only an interval can be used.

$$\Delta_{\text{Model}} = E - U_{\text{val}} \tag{2.23}$$

The discussion of the validation error should focus on the uncertainty of the model which can be calculated with the equation above. E can be interpreted as an estimate of the model error Δ_{Model}, and U_{val} is the standard uncertainty of that estimate. Conclusions that can be drawn at this stage are as follows: If U_{val} is small enough, all components of U_{val} are negligible, and E is a good estimate of the model error. This means that simulation and experiment deviate by E. The model must be adjusted in order to better represent the experimental findings. If this is not possible or if it adds too much complexity to the model, which leads to very long computing time or a big investment in hardware, it is also possible to adjust the experiment. We will come back to this later.

On the other hand, if U_{val} is of the magnitude of E, any attempts to reduce the model error are questionable because validation has been achieved at the U_{val} level. This means that the model may be part of the problem but not necessarily. Reducing other uncertainties already reduces the validation error.

If the model error dominates the validation error and the error is too big, the assumptions of the model must be proven. Boundary conditions, for example, are simplified assumptions. A constant pressure at the entrance of a pipe is traditionally the first choice, but in the experiment, a velocity profile exists in the pipe flow and an assumption of constant pressure may lead to a deviation between the model and the experiment. Unfortunately, the choice of boundary conditions may influence the simulation result in a way that can only be predicted with a lot of experience. The deviations are not known in advance. Test simulations of a smaller part of the domain are a good way to judge the magnitude of imposed deviations.

In general, model assumptions are not solely a matter of the simulation. One way to reduce the model error may be to adjust the simulation in order to meet the conditions of the experiment. That may lead to extra computational time or is sometimes simply not possible because the commercial simulation package of choice does not allow this adjustment. The other way is to adjust the experimental setup to meet the assumptions. In laboratory experiments that are designed for the purpose of validation, this is obviously easier than in the case of measurements at engineering equipment built for a process plant. In laboratory experiments, long entrance pipes for fully developed flow or conditioning of the fluid flow in a pipe to form the desired velocity field is possible.

This means that CFD and laboratory people in a research team must work toward elimination of assumptions through adjustment of the simulation and/or the experiment.

Again, this should be done in the planning phase of the project to avoid costly adjustments of the experimental setup and/or the simulation later in the project. If the team cannot or is not willing to remove an assumption or compensate it through adjustments of the experiment, the team should try to estimate the possible error or uncertainty that will be introduced into the system. Besides the already-mentioned test simulations that can help to estimate the influence of an assumption, preliminary experiments can be built with the aim to test how close the measurement can meet the model assumption.

It is important to take into account that a validation error may show up because of an effect that nobody has anticipated before. In that case, the analysis of this effect is much easier if all other sources of validation error are as good as possible eliminated.

References

BIPM, IEC, IFCC, ILAC, ISO, IUPAC, IUPAP and OIML 2008. Evaluation of measurement data: Supplement 1 to the "Guide to the expression of uncertainty in measurement", Joint Committee for Guides in Metrology, JCGM 101.

Coleman, H. & Glenn Steele, W. 2018. *Experimentation, Validation, and Uncertainty Analysis for Engineers*, Fourth Edition. Wiley, New York.

DIN 1319-1:1995-01, Fundamentals of metrology – Part 1: Basic terminology.

Grabe, M. 2007. Ten theses for a new GUM, *Poster Paper, PTB-BIPM Workshop on the Impact of Information Technology in Metrology*, Berlin, 4–8 June.

ISO/IEC Guide 98-3:2008 (GUM:1995). Guide to the expression of uncertainty in measurement, Joint Committee for Guides in Metrology (JCGM/WG1/100,2008). https://www.bipm.org/utils/common/documents/jcgm/JCGM_100_2008_E.pdf.

Kanaris, A.G., Mouza, A.A., & Paras, S.V. 2005. Flow and heat transfer in narrow channels with Corrugated Walls a CFD Code Application, Institution of Chemical Engineers Tans IChemE Part A.

Leask, E.K., Ehlmann, B.L., Dundar, M.M., Murchie, S.L., & Seelos, F.P. 2018. Challenges in the search for perchlorate and other hydrated minerals with 2.1-μm absorptions on Mars. *Geophysical Research Letters*, 45, 12180–12189. doi: 10.1029/2018GL080077.

Liptak, B.G. & Moore, C.F. 2006. *Control Basics, IN: Instrument Engineers' Handbook, Volume Two: Process Control and Optimization*. CRC Press/Taylor & Francis, Boca Raton, FL.

Moffat, R.J. 1988. Describing the uncertainties in experimental results. *Experimental Thermal and Fluid Science*, 1(1), 3–17.

Stockton, P. & Wilson, A. 2012. Allocation uncertainty: Tips, tricks and Pitfalls. *In Proceedings: 30th International North Sea Flow Measurement Workshop*, 23–26 October, St. Andrews, UK.

3

Uncertainty Analysis of Indirect Measurements in Thermal Science

X. Cui
Xi'an Jiaotong University

L.C. Wei
Xi'an Jiaotong University
Shenzhen Envicool Technology Co., Ltd.

C.F. Ma and X.Z. Meng
Xi'an Jiaotong University

J.C. Chai
University of Huddersfield

L.W. Jin
Xi'an Jiaotong University

CONTENTS

3.1 Introduction 38
    3.1.1 Importance of Uncertainty Analysis..38
    3.1.2 Direct and Indirect Measurements 39
    3.1.3 Objectives...40
3.2 Measurement Uncertainty...40
    3.2.1 Sources and Types of Uncertainties 40
    3.2.2 Normal Distribution and Standard Deviation 42
3.3 Dealing with Indirect Measurement..43
    3.3.1 Uncertainty Determination of Individual Variables 43
    3.3.2 Uncertainty Determination of a Multivariable Function..........45
3.4 Case Study 47
    3.4.1 Determination of Uncertainty of Direct Measurement.............48
        3.4.1.1 Determination of Type A Uncertainty u_A......................48
        3.4.1.2 Determination of Type B Uncertainty u_B......................50
        3.4.1.3 Combined Standard Uncertainty u_c 50
    3.4.2 Evaluation of Uncertainty of Indirect Measurements................51
3.5 Remarks...52
References..53

3.1 Introduction

3.1.1 Importance of Uncertainty Analysis

In thermal science, experimental investigation could enhance the understanding for thermodynamics, heat and mass transfer, fluid mechanics, and relevant applications. Advanced methodologies and techniques have been widely applied to the experimental investigations in multiphase flow, turbulence combustion, micro- and nanoscale heat transfer, etc. However, no matter how advanced the experimental methods are, experimental errors always exist in the experimental data due to unavoidable uncertainties (Labudová and Vozárová 2002, Hay et al. 2005). Strictly speaking, an experimental result must be associated with an indication of its uncertainty to deliver a scientific meaning of the measurement.

In theory, experimental errors are the differences between the "true" values of the parameter being measured and the practically measured value. The error of a measurement could be either positive or negative and is always existent because the "true" value is never exactly known. An essential feature of every scientific measurement is the measured value accompanied by an uncertainty (Gupta 2012). Being able to assess measurement uncertainties is an important measure in any type of experiments in thermal science. For example, thermocouples are commonly used in heat transfer experiments to measure temperature. One may measure the temperature of an objective through multiple measurements and use the averaged value as the result. The averaged value alone does not deliver sufficient information for the measurement, because errors always exist, and the actual value may be far from the averaged value, due to the systematic and random errors. An average value of multiple measurements associated with an indication of error limit defines the range that the actual value falls into. That is to say, the "true" value of the objective locates in the defined range with taking various errors into consideration.

In probability and statistics, the concept of error does not express the connotation of mistake, but the unavoidable uncertainty in experimental data acquisition. Errors are not mistakes, and one cannot eliminate them by careful operation. The best expectation is to ensure that the experimental errors or uncertainties are as small as practically possible and to make some reliable estimates for them (Analytical Methods Committee 1995). Error analysis or experimental uncertainty estimation is the study and evaluation of uncertainty in measurement. In many scenarios, the words error and uncertainty are not explicitly distinguished and are interchangeable. In this chapter, for the sake of clarity, errors are used to describe the basic assessment of measurement, such as relative errors and absolute errors, while uncertainties are used to estimate the interval to which the "true" value belongs.

In practical operation, we generally assume that the average values of the variables obtained from multiple runs represent the attainable "true" values, and the uncertainties are used to estimate the degree of unsureness of these "true" values, due to the unavoidable measurement errors mentioned above. The smaller the uncertainty is, the closer the average value is to the "true" value of the measurement (Mills and Chang 2004). In other words, the uncertainties are indicators to represent the reliability of the average values.

3.1.2 Direct and Indirect Measurements

A variety of parameters, such as flow rate, temperature, pressure, and humidity, are required to be measured in various thermal processes, especially in experimental heat transfer. It is not a complex problem to determine the uncertainty of a single measurement, which is acquired directly by the corresponding sensors or devices.

If only a single parameter is considered, multiple equal precision runs on this direct measurement can be conducted to reduce the measurement uncertainty. Due to the existence of the random errors, the multiple measurements of the single parameter normally deliver different values if the sensor and data acquisition system are sensitive enough. In this way, the calculation of the statistical standard deviation associated with random distribution is then applicable to characterize the extent to which the average values of the multiple measurements approach the true value. The normal distribution of the direct measurement with ~99% of the results falling within ±3 times standard deviations of the mean value can be regarded as the best estimation of the true value of the quantity being measured (White 2008).

On the other hand, in many cases, the final result of an experimental study is not a direct measured value, but a calculated one based on some direct measured parameters. Normally, the uncertainties of these measured parameters lead to an uncertainty in the derived final result, which is referred to as the uncertainty of the indirect measured parameter. The individual uncertainty of each direct measurement obviously affects the uncertainty of the indirect measurement (Duta et al. 2007). When a final result has to be determined from other direct measurements, the combined uncertainty of each measurement can be used to represent the estimated standard deviation of the final result based on the law of propagation of uncertainty.

The law of propagation of uncertainty allows the combined standard uncertainty of one result to be readily incorporated in the evaluation of the combined standard uncertainty of another result in which the first is used (Betta, Liguori, and Pietrosanto 2000). For example, the heat transfer rate of a fluid through a heated pipe can be obtained by measuring the mass flow rate and the temperature difference between inlet and outlet, considering the specific heat capacity is a constant. The individual uncertainties arising from the measurement of mass flow rate and temperature difference are then combined to estimate the standard deviation of the heat transfer rate, which is equal to the positive square root of the variance obtained from all variance and covariance components (Rad, Afshin, and Farhanieh 2015).

3.1.3 Objectives

Based on the above introduction, the determination of uncertainty of the indirect measurement is more complicated than that of the direct measurement. This chapter interprets briefly the uncertainty of direct measurement and focuses on the uncertainty estimation of the indirect measurement. The rules in the uncertainty propagation for dealing with functions involving addition or subtraction, multiplication or division, and exponent will be interpreted in detail. It is expected that the chapter delivers a quick and handy guide for estimating the uncertainty of indirect measurement by scientific and standard procedures.

3.2 Measurement Uncertainty

3.2.1 Sources and Types of Uncertainties

Uncertainty is an indispensable component of the measured value that indicates the range of value in which the "true" value is ascertained to fall. An experimental uncertainty is generally caused by either random errors or systematic errors. Random errors are statistical fluctuations in the process of repeated measurements with equal precision, while systematic errors are reproducible inaccuracies that exist consistently in the measurement system (Moffat 1988).

Random errors, also known as accidental errors, are produced by unknown and unpredictable variations in the experimental process. These random variations may occur in the measuring equipment, the environmental conditions, and the operation instability. Examples of origins of random errors are as follows: room temperature fluctuation, changes in relative humidity and air pressure, electronic noise in the circuit of an electrical instrument, and irregular changes in the heat loss rate. The size and positive and negative of random errors are not fixed as shown in Figure 3.1. Most measurements could achieve that the probability of positive and negative random errors with the same absolute value is roughly equal, indicating that they are mutually compensatory errors and can often offset each other (Bevington, Robinson, and Bunce 1993). Thus, random errors can be reduced by increasing the number of parallel measurements to take the average.

Systematic errors are often introduced by an inaccuracy inherent in a measuring system or operation process, which are unable to be diminished by the multiple parallel measurements. In other words, systematic errors yield results systematically in repeated measurements, either greater than or less than the true value, in the same direction (Fornasini 2009). Sources of systematic error in the thermal measurements are possibly due to imperfect calibration of thermocouple, resistance temperature detector, flow meter,

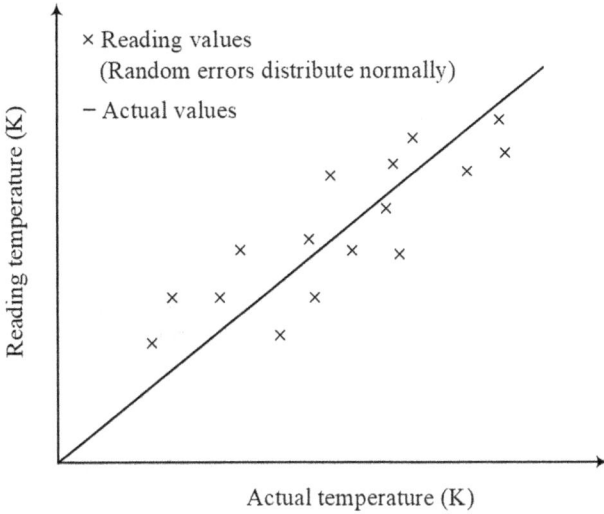

FIGURE 3.1
Characteristics of random errors.

pressure sensor, and measurement instruments. Figure 3.2 shows a system error caused by a wrong calibration which yields a fixed bias from the actual value. Systematic errors can also be caused by improper measurement methods and data reduction methods, which may be either zero error or percentage error. Therefore, it is conceivable that the systematic errors can be neglected, provided the careful calibrations are conducted and the proper measuring methods are strictly examined.

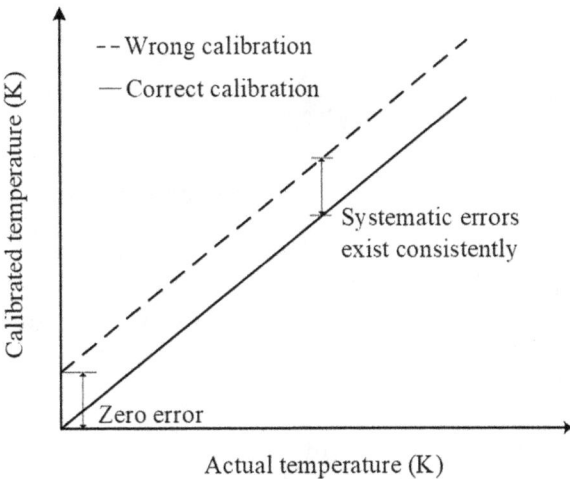

FIGURE 3.2
Characteristics of systematic errors.

According to the Guide to the Expression of Uncertainty in Measurement (GUM), measurement uncertainty is classified into two types. Type A uncertainty is the evaluation of a component of measurement uncertainty by a statistical analysis of measured quantity values obtained under defined measurement conditions. Essentially, Type A uncertainty is data collected from a series of observations and evaluated using statistical methods associated with the analysis of variance. Among the available statistical algorithms, standard deviation is the most commonly used method for evaluating the uncertainty rather than arithmetic mean and degrees of freedom methods. Type B evaluation of uncertainty is defined as the method of evaluation of uncertainty by means other than the statistical analysis of series of observations. Compared with Type A uncertainty, Type B uncertainty is to determine the errors for which the necessary statistical data are absence or impossible to obtain (BIPM et al. 2008).

In thermal experiments, obviously in most cases, Type A uncertainty should be applied mainly to express the statistical dispersion of the measurement from its "true" value for evaluating the measurement quantity. Every measurement should be subject to an uncertainty, and a measured result possesses the scientific meaning, provided it is associated with a statement of the uncertainty.

3.2.2 Normal Distribution and Standard Deviation

The measurement uncertainty is often taken as the standard deviation of a probability distribution over the possible values that could be attributed to a measured quantity. The mean value of a number of measurements of the same quantity is the best estimate of that quantity, and the standard deviation σ of the measurements shows the accuracy of the estimate.

$$\sigma = \sqrt{\frac{1}{n}\sum_{i=1}^{n}(x_i - \bar{x})^2} \tag{3.1}$$

The standard deviation is a statistic that measures the dispersion of a dataset relative to its mean. The precision is limited by the random errors and could be evaluated through repeated tests. As shown in Figure 3.3, systematic errors e_s yield results consistently in repeated measurements with a constant bias from the expected mean value. Comparatively, random errors e_r exhibit positive and negative variations with the same absolute value. Taking careful calibration and proper experimental design, the systematic errors are often assumed to be negligible.

In Type A evaluations of measurement uncertainty, the normal distribution is usually used to justify random errors. It indicates that averages of measured random variables independently derived from independent distributions converge in distribution to the normal. The averages become

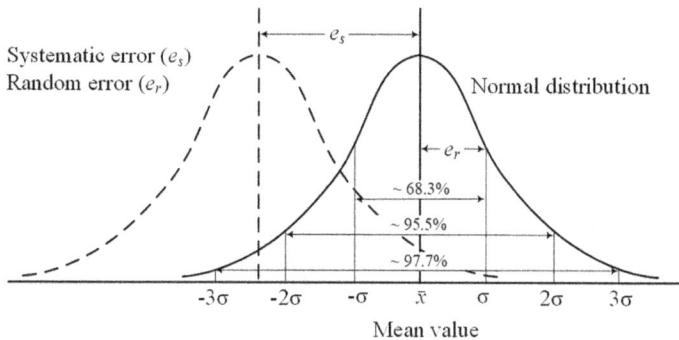

FIGURE 3.3
Normal distribution associated with standard deviations.

normally distributed if the number of measurements is adequately large. Physical quantities that are expected to be the sum of measurement errors often have distributions that are nearly normal (Stephanie 2001). The variable x then has expectation value equal to the average measured value and standard deviation equal to the standard deviation of the average. The normal distribution of the direct measurement with ~99.7% of the results falling within ±3 times standard deviations of the mean value can be regarded as the best estimation of the true value of the quantity being measured.

3.3 Dealing with Indirect Measurement

Many parameters cannot be directly measured. For example, the velocity (v) of an object is determined by measuring both the distance (l) and the time (t), followed by the calculation using a function ($v = l/t$). Therefore, the estimation of uncertainties of velocity involves two steps. The first step is to determine the uncertainties of parameters measured directly. The second step is to estimate the propagation of uncertainty through the calculation (Hughes and Hase 2010). This section aims to illustrate how the uncertainties "propagate" for indirect measured parameters.

3.3.1 Uncertainty Determination of Individual Variables

Suppose a variable x is measured in the standard form $\bar{x} \pm \delta x$, and a parameter A is calculated as a function of $f(x)$. The best estimation of A is $\bar{A} = f(\bar{x})$. To facilitate the analysis, a graph of $f(x)$ is assumed in Figure 3.4.
    The uncertainty of A for the measured variable x is given as:

$$\delta A = f(\bar{x} + \delta x) - f(\bar{x}) \tag{3.2}$$

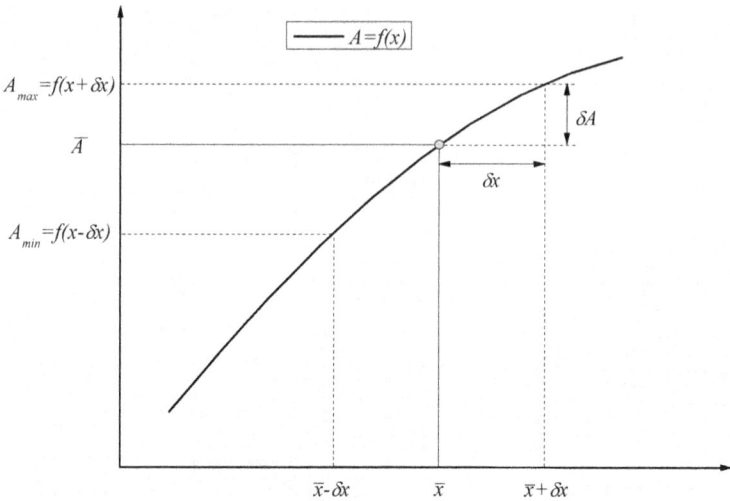

FIGURE 3.4
Uncertainty determination of individual variables with a function of $A = f(x)$.

Provided the increment δx is small, the error can be obtained using the gradient of the tangent $\left(\dfrac{df(x)}{dx} \right)$:

$$f(\bar{x} + \delta x) - f(\bar{x}) = \frac{df(x)}{dx} \delta x \tag{3.3}$$

Therefore, the uncertainty of A can be calculated by considering the derivative $\left(\dfrac{df(x)}{dx} \right)$ and the uncertainty of the measured variable (x):

$$\delta A = \frac{df(x)}{dx} \delta x \tag{3.4}$$

Equation (3.4) is derived for a function with a positive slope. Thus, for a function with a negative slope, the uncertainty of A can be obtained similarly:

$$\delta A = -\frac{df(x)}{dx} \delta x \tag{3.5}$$

As a result, a general rule for calculating the uncertainty δA is expressed as:

$$\delta A = \left| \frac{dA}{dx} \right| \delta x = \left| \frac{df(x)}{dx} \right| \delta x \tag{3.6}$$

Equation (3.6) requires the calculation of derivative of the function. Table 3.1 presents the examples of uncertainty propagation for several commonly used functions.

3.3.2 Uncertainty Determination of a Multivariable Function

In this section, we will illustrate the uncertainty propagation scheme when a function involves several measured variables. For example, the parameter A is assumed to be a sum of two variables ($A = B + C$). The uncertainty δA for the calculated parameter A is impacted by the uncertainties of δB and δC. The best estimation for A is:

$$\bar{A} = \bar{B} + \bar{C} \tag{3.7}$$

It can be inferred that the highest probable value of A is:

$$\bar{B} + \bar{C} + (\delta B + \delta C) \tag{3.8}$$

The lowest probable value of A is:

$$\bar{B} + \bar{C} - (\delta B + \delta C) \tag{3.9}$$

Therefore, the uncertainty of A can be derived as:

$$\delta A \approx \delta B + \delta C \tag{3.10}$$

In other words, for the function of $A = B + C$, the uncertainty of the parameter A is the sum of the uncertainties of each variable in the function.

Next, we suppose a function of two quantities with a general form of $A = f(B,C)$, in which B and C are independent variables. The best estimation of A is expressed as:

TABLE 3.1

Examples of Uncertainty Propagation for Functions of Individual Variables

Function $A = f(x)$	Uncertainty of A
$f(x) = \sin x$	$\lvert \cos x \rvert \delta x$
$f(x) = \cos x$	$\lvert \sin x \rvert \delta x$
$f(x) = \dfrac{1}{x}$	$\dfrac{1}{x^2} \delta x$
$f(x) = kx$	$\lvert k \rvert \delta x$
$f(x) = x^n$	$\lvert nx^{n-1} \rvert \delta x$
$f(x) = e^x$	$e^x \delta x$
$f(x) = 10^x$	$10^x \ln(10) \delta x$
$f(x) = \ln x$	$\dfrac{1}{x} \delta x$
$f(x) = \log(x)$	$\dfrac{1}{\ln(10)x} \delta x$

$$\bar{A} = f(\bar{B}, \bar{C}) \tag{3.11}$$

Similar to the analysis in Eq. (3.2), for a small increment in B and C, the error can be approximated using the gradient of the tangent:

$$f(\bar{B} + \delta B, \bar{C}) - f(\bar{B}, \bar{C}) = \frac{\partial f}{\partial B} \delta B \tag{3.12}$$

$$f(\bar{B}, \bar{C} + \delta C) - f(\bar{B}, \bar{C}) = \frac{\partial f}{\partial C} \delta C \tag{3.13}$$

The extreme probable value for A can be estimated as:

$$f(\bar{B} \pm \delta B, \bar{C} \pm \delta C) \approx f(\bar{B}, \bar{C}) \pm \left(\left| \frac{\partial f}{\partial B} \right| \delta B + \left| \frac{\partial f}{\partial C} \right| \delta C \right) \tag{3.14}$$

Thus, the uncertainty of A is written as:

$$\delta A \approx \left| \frac{\partial f}{\partial B} \right| \delta B + \left| \frac{\partial f}{\partial C} \right| \delta C \tag{3.15}$$

Provided the uncertainties of B and C are independent and random, the uncertainty of A is given as:

$$\delta A = \sqrt{ \left(\frac{\partial f}{\partial B} \delta B \right)^2 + \left(\frac{\partial f}{\partial C} \delta C \right)^2 } \tag{3.16}$$

Therefore, a general rule for multivariable functions can be further obtained. Considering a function of independent variables $A = f(B, C, ..., Z)$, the uncertainty for each variable is expressed using Taylor series expansion as:

$$\left\{ \begin{array}{l} f(\bar{B} \pm \delta B, \bar{C}, ... \bar{Z}) = f(\bar{B}, \bar{C}, ..., \bar{Z}) + \frac{\partial f}{\partial B} \delta B + \frac{1}{2} \frac{\partial^2 f}{\partial B^2} \delta B^2 + \cdots + \frac{1}{n!} \frac{\partial^n f}{\partial B^n} \delta B^n + R_n(B) \\[2mm] f(\bar{B}, \bar{C} \pm \delta C, ..., \bar{Z}) = f(\bar{B}, \bar{C}, ..., \bar{Z}) + \frac{\partial f}{\partial C} \delta C + \frac{1}{2} \frac{\partial^2 f}{\partial C^2} \delta C^2 + \cdots + \frac{1}{n!} \frac{\partial^n f}{\partial C^n} \delta C^n + R_n(C) \\[2mm] \qquad\qquad\qquad\qquad\qquad \cdots \\[1mm] \qquad\qquad\qquad\qquad\qquad \cdots \\[2mm] f(\bar{B}, \bar{C}, ..., \bar{Z} \pm \delta Z) = f(\bar{B}, \bar{C}, ..., \bar{Z}) + \frac{\partial f}{\partial Z} \delta Z + \frac{1}{2} \frac{\partial^2 f}{\partial Z^2} \delta Z^2 + \cdots + \frac{1}{n!} \frac{\partial^n f}{\partial B^n} \delta B^n + R_n(Z) \end{array} \right.$$

$$\tag{3.17}$$

TABLE 3.2

Examples of Uncertainty Propagation for Some Common Functions

Function $A = f(B, C, \ldots)$	Uncertainty of A
$A = B + C$	$\delta A = \sqrt{(\delta B)^2 + (\delta C)^2}$
$A = B - C$	$\delta A = \sqrt{(\delta B)^2 + (\delta C)^2}$
$A = B \times C$	$\dfrac{\delta A}{A} = \sqrt{\left(\dfrac{\delta B}{B}\right)^2 + \left(\dfrac{\delta C}{C}\right)^2}$
$A = \dfrac{B}{C}$	$\dfrac{\delta A}{A} = \sqrt{\left(\dfrac{\delta B}{B}\right)^2 + \left(\dfrac{\delta C}{C}\right)^2}$

In these equations, the increments in all the variables are small so that the higher order derivatives are negligible compared with the gradient term. As a result, the uncertainty of the multivariable function is approximated as:

$$\delta A = \sqrt{\left(\frac{\partial f}{\partial B}\delta B\right)^2 + \left(\frac{\partial f}{\partial C}\delta C\right)^2 + \cdots + \left(\frac{\partial f}{\partial Z}\delta Z\right)^2} \qquad (3.18)$$

By using the general rule for uncertainty propagation, we can easily achieve the expressions of calculating the errors for some common functions as illustrated in Table 3.2.

3.4 Case Study

In this case study, the evaluation of uncertainty of direct and indirect measurements is performed based on the above discussion.

A detailed experimental study of forced convection in a metal foam sample subjected to a constant heat flux is conducted in the case study. The systematic diagram of the experimental system is shown in Figure 3.5. The objective of the investigation is to explore the heat transfer enhancement and hydraulic characteristics of metal foams with different pore densities. As shown in Figure 3.6, the average temperature on the heating surface and the temperatures of inlet and outlet flow were measured. The heating power was controlled by a programmable DC power supply.

To minimize the data reduction uncertainty, the time-averaged method was employed to deal with the experimental data, for each group of repeated measurements with equal precision. For uncertainties of the measured parameters, they can be classified into two groups. The random uncertainties are treated statistically, and the systematic uncertainties that cannot be treated statistically are minimized with careful calibrations and experimental operations.

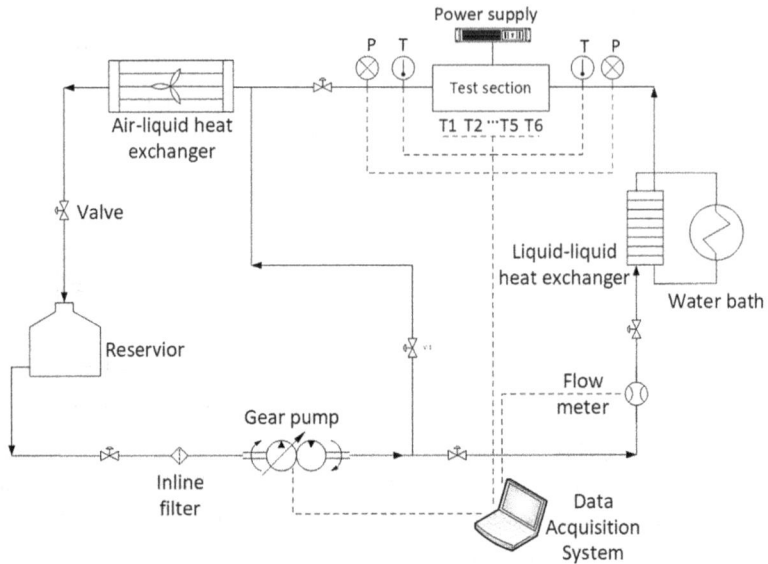

FIGURE 3.5
Schematic flowchart of experimental setup.

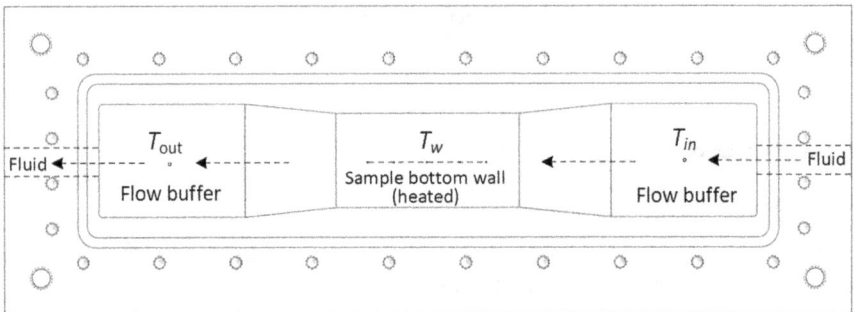

FIGURE 3.6
Top view of the test section.

3.4.1 Determination of Uncertainty of Direct Measurement

For direct measurement and multiple metering, the inlet water temperature (T_{in}) and the wall temperature (T_w) are taken as examples to demonstrate the determination of uncertainty of direct measurements.

3.4.1.1 Determination of Type A Uncertainty u_A

Performing eight independent repeated measurements, the measured values of T_{in} are listed in Table 3.3.

TABLE 3.3

Measured Inlet Water Temperature T_{in}

	T_{in} (°C)							
Variable	x_1	x_2	x_3	x_4	x_5	x_6	x_7	x_8
Values	26.9	26.1	25.5	26.0	26.3	25.4	26.3	25.7

The arithmetical mean $\overline{T_{in}}$ of the eight measurements is:

$$\overline{T_{in}} = 26.0 \tag{3.19}$$

The standard deviation $s(T_{in})$ of the measured inlet water temperature can be calculated by Eq. (3.1) as:

$$s(T_{in}) = 0.49 \tag{3.20}$$

Then, the Type A uncertainty of T_{in} is:

$$u_A(T_{in}) = s(\overline{T_{in}}) = \frac{s(T_{in})}{\sqrt{n}} = 0.17 \tag{3.21}$$

where n is the repeated number of the independent measurements. In addition, the Type A uncertainty of T_w can be calculated similarly as follows. Table 3.4 presents the examples of 12 independent repeated measurements of the wall temperature T_w.

The arithmetical mean $\overline{T_w}$ of the multiple independent measurements on the wall temperature is:

$$\overline{T_w} = 65.0 \tag{3.22}$$

TABLE 3.4

Measured Wall Temperature T_w

T_w (°C)	Values
x_1	64.4
x_2	64.6
x_3	66.3
x_4	67.2
x_5	67.5
x_6	66.2
x_7	63.8
x_8	65.2
x_9	65.6
x_{10}	62.7
x_{11}	62.9
x_{12}	63.5

The standard deviation of the measured wall temperature T_w is:

$$s(T_w) = 1.60 \qquad\qquad (3.23)$$

Similar to the treatment of T_{in}, the Type A uncertainty of T_w can be obtained:

$$u_A(T_w) = s(\overline{T_w}) = \frac{s(T_w)}{\sqrt{n}} = 0.46 \qquad\qquad (3.24)$$

3.4.1.2 Determination of Type B Uncertainty u_B

In most cases, Type B uncertainty cannot be assessed by statistical methods within the measurement range, which is generally estimated based on the experimental facilities or other relevant information. In actual applications, only the errors caused by instrumental errors are considered in the evaluation of Type B uncertainty.

In some cases, Type B uncertainty is based on the instrument specification or verification certificate, while in other cases, it is estimated by the accuracy level of the instrument. That information related to the instrumental precision can be used to represent the limit of system error Δ (also indicated as the allowable error or the indication error). In the absence of the statistical information, it is supposed that the probability distribution function of general instrumental errors obeys the uniform distribution for most experimental measurements; that is, $c = \sqrt{3}$, where c is the confidence coefficient for the confidence probability at 0.683. Then, Type B uncertainty can be determined by $\Delta/\sqrt{3}$.

Applying the evaluation of Type B uncertainty discussed above, the Type B uncertainties of T_{in} and T_w can be determined as:

$$u_B(T_{in}) = \frac{\Delta}{\sqrt{3}} = \frac{0.5}{\sqrt{3}} = 0.29 \qquad\qquad (3.25)$$

$$u_B(T_w) = \frac{\Delta}{\sqrt{3}} = \frac{0.5}{\sqrt{3}} = 0.29 \qquad\qquad (3.26)$$

where the accuracy of the used thermocouples is 0.5°C. It is seen that Type B uncertainties for the instruments with the same accuracy are the same.

3.4.1.3 Combined Standard Uncertainty u_c

After determining Type A and Type B uncertainties separately, the combined standard uncertainties of T_{in} and T_w can be easily obtained by calculating the square root of the sum of u_A and u_B squares, respectively:

$$u_c(T_{in}) = \sqrt{0.17^2 + 0.29^2} = 0.34 \qquad\qquad (3.27)$$

$$u_c(T_w) = \sqrt{0.46^2 + 0.29^2} = 0.54 \qquad\qquad (3.28)$$

TABLE 3.5

Uncertainty of Individual Variables for a Multivariable Function

Variables	Uncertainties
$u_c(q)$	9424.3 W/m²
$u_c(T_{in})$	0.34°C
$u_c(T_{out})$	0.34°C
$u_c(T_w)$	0.54°C

The uncertainties of other variables can be obtained similarly. Table 3.5 summarizes the uncertainties of individual variables for a multivariable function, which are involved in the determination of the indirect measurement of heat transfer coefficient h.

3.4.2 Evaluation of Uncertainty of Indirect Measurements

In order to evaluate the heat transfer performance of metal foam practically, the convective heat transfer coefficient is introduced, as given in Eq. (3.29). The convective heat transfer coefficient synthetically reflects the parameters that influence the heat transfer performance of the metal foam. This equation is a useful benchmark for evaluating experimental methods used to measure the convective heat transfer coefficient.

$$h = \frac{q}{\Delta T} = \frac{q}{\left(T_w - \dfrac{T_{in} + T_{out}}{2}\right)}$$ (3.29)

where q is the heat flux and ΔT is the temperature difference, which is determined by T_{in}, T_{out}, and T_w. In the current case, the heating power is provided by a high-accuracy DC power supply. After conducting the careful calibration, the precision error of instrument and bias error in the meters are expected to be <0.5%, which is neglected for a sake of delivering a straightforward instruction. According to the method introduced in Section 3.2, the uncertainty transfer coefficients are obtained as follows:

$$\frac{\partial h}{\partial q} = \frac{1}{\left(T_w - \dfrac{T_{in} + T_{out}}{2}\right)} = \frac{1}{\left(65.0 - \dfrac{26.0 + 29.1}{2}\right)} = 0.03$$ (3.30)

$$\frac{\partial h}{\partial T_w} = -\frac{q}{\left(T_w - \dfrac{T_{in} + T_{out}}{2}\right)^2} = \frac{167,100}{\left(65.0 - \dfrac{26.0 + 29.1}{2}\right)^2} = -119.2$$ (3.31)

$$\frac{\partial h}{\partial T_{in}} = \frac{q}{2\left(T_w - \frac{T_{in}+T_{out}}{2}\right)^2} = \frac{167,100}{2\left(65.0 - \frac{26.0+29.1}{2}\right)^2} = 59.6 \tag{3.32}$$

$$\frac{\partial h}{\partial T_{out}} = \frac{q}{2\left(T_w - \frac{T_{in}+T_{out}}{2}\right)^2} = \frac{167,100}{2\left(65.0 - \frac{26.0+29.1}{2}\right)^2} = 59.6 \tag{3.33}$$

where $\partial h/\partial q$, $\partial h/\partial T_w$, $\partial h/\partial q$, and $\partial h/\partial q$ are the uncertainty transfer coefficients that are affected by the direct measurement parameters q, T_w, T_{in}, and T_{out}, respectively.

Thus, applying Eq. (3.18), the uncertainty of the indirect measured h can be written as:

$$u(h) = \sqrt{\left(\frac{\partial h}{\partial q}\right)^2 u_c^2(q) + \left(\frac{\partial h}{\partial T_w}\right)^2 u_c^2(T_w) + \left(\frac{\partial h}{\partial T_{in}}\right)^2 u_c^2(T_{in}) + \left(\frac{\partial h}{\partial T_{out}}\right)^2 u_c^2(T_{out})}$$

$$= 261.5 \tag{3.34}$$

And, the relative uncertainty of heat transfer coefficient h is:

$$\frac{u(h)}{h} \times 100\% = 5.9\% \tag{3.35}$$

The relative uncertainty could provide the basis for improving the overall measurement accuracy of a composite test system. For a measurement system with multiple subsystems, the overall accuracy of the composite system can be improved at a low cost by promoting the measurement of the subsystem with the highest relative uncertainty.

3.5 Remarks

This chapter illustrates the importance of attempting to conduct the uncertainty analysis. The sources of uncertainties include random errors and systematic errors. The uncertainty of direct measurement was illustrated by considering the standard deviation and normal distribution. The uncertainty of indirect measurement involves the propagation of uncertainty. The uncertainty of the multivariable function can be calculated by:

$$\delta A = \sqrt{\left(\frac{\partial f}{\partial B}\delta B\right)^2 + \left(\frac{\partial f}{\partial C}\delta C\right)^2 + \cdots + \left(\frac{\partial f}{\partial Z}\delta Z\right)^2}$$

The general rule for uncertainty propagation was employed to obtain the expressions of calculating the errors for several common functions such as addition, subtraction, multiplication, and division. A case study was reported to illustrate the procedure for calculating the uncertainty of indirect measurement, which aimed to provide a quick example for estimating the uncertainty of indirect measurement by scientific and standard procedures.

References

Analytical Methods Committee. 1995. "Uncertainty of measurement: Implications of its use in analytical science." *Analyst* 120(9): 2303–8.

Betta, G., C. Liguori, and A. Pietrosanto. 2000. "Propagation of uncertainty in a discrete Fourier Transform algorithm." *Measurement* 27(4): 231–39.

Bevington, P.R., D. Keith Robinson, and G. Bunce. 1993. "Data reduction and error analysis for the physical sciences, 2nd Ed." *American Journal of Physics* 61(8): 766–67.

BIPM, IEC, IFCC, ISO, IUPAC, IUPAP, and OIML. 2008. "Evaluation of Measurement Data—Guide to the Expression of Uncertainty in Measurement JCGM 100: 2008." *Citado En Las* 5–7.

Duta, S., P. Robouch, L. Barbu, and P. Taylor. 2007. "Practical aspects of the uncertainty and traceability of spectrochemical measurement results by electrothermal atomic absorption spectrometry." *Spectrochimica Acta: Part B Atomic Spectroscopy* 62(4): 337–43.

Fornasini, P. 2009. *The Uncertainty in Physical Measurements: An Introduction to Data Analysis in the Physics Laboratory. Radiation Protection Dosimetry.* Springer Science & Business Media, Berlin, Heidelberg.

Gupta, S.V. 2012. *Measurement Uncertainties: Physical Parameters and Calibration of Instruments.* Springer Science & Business Media, Berlin, Heidelberg.

Hay, B., J.R. Filtz, J. Hameury, and L. Rongione. 2005. "Uncertainty of thermal diffusivity measurements using the laser flash method." *International Journal of Thermophysics* 26(6): 1883–98.

Hughes, I.G. and T.P.A. Hase. 2010. *Measurements and their Uncertainties: A Practical Guide to Modern Error Analysis.* Oxford University Press, Oxford.

Labudová, G. and V. Vozárová. 2002. "Uncertainty of the thermal conductivity measurement using the transient hot wire method." *Journal of Thermal Analysis and Calorimetry* 67(1): 257–65.

Mills, A.F. and B.H. Chang. 2004. "Error Analysis of Experiments–a Manual for Engineering Students."

Moffat, R.J. 1988. "Describing the uncertainties in experimental results." *Experimental Thermal and Fluid Science* 1(1): 3–17.

Rad, S.E., H. Afshin, and B. Farhanieh. 2015. "Heat transfer enhancement in shell-and-tube heat exchangers using porous media." *Heat Transfer Engineering* 36(3): 262–77.

Stephanie, B. 2001. *A Beginner's Guide to Uncertainty of Measurement.* National Physical Laboratory Teddington, Teddington.

White, G.H. 2008. "Basics of estimating measurement uncertainty." *The Clinical Biochemistry Reviews/Australian Association of Clinical Biochemists* 29 Suppl 1(August): S53–60.

4

Modern Error Analysis in Indirect Measurements in Thermal Science

Denis Maillet

Université de Lorraine, CNRS, LEMTA, F-54000 Nancy, France

CONTENTS

4.1 Introduction

Sound measurements in thermal sciences rely on a proper data reduction technique together with a credible assessment of the errors made, taking into account the characteristics of the digital measurement equipment used.

Getting the raw signal provided by a sensor, after its conversion into a thermal quantity, a temperature or more seldom a flux or a rate of heat flow, through an appropriate calibration, is not always the objective of the people running an experiment and making measurements. The indirect measurement of a quantity present in the model they use to simulate heat transfer in the experiment may be their final motivation. This quantity is called a parameter, which can be the thermal conductivity or diffusivity of a material in a thermal characterization experiment [1], or a heat transfer coefficient in a thermal convection experiment. The methodology for dealing with these indirect measurement problems in a systematic way already exists: The parameter estimation techniques that mix statistical approaches (least squares) and linear algebra [2–4] are devoted to this task. They constitute a subdomain of the wide field of inverse problems of experimental nature that comprises also the estimation of functions, the problem that is not considered here.

4.2 Model, Direct Measurements, and Definition of the Parameters That Are Looked For

We consider here an unsteady experiment in heat transfer where there is a unique thermal excitation $u(t)$, also called the input in the terminology of system dynamics, and a unique thermal measurement $y(t)$ at a given point in the system, which is also called experimental output or "direct" measurement here.

In parallel, we can design a thermal model M that gives the theoretical response $y_{mo}(t)$ at the same location, also called output, and which depends on time t, on the input u, and on a given number n of structural parameters β_j, for $j = 1$ to n, that can be gathered in a set of n parameters β (see Figure 4.1).

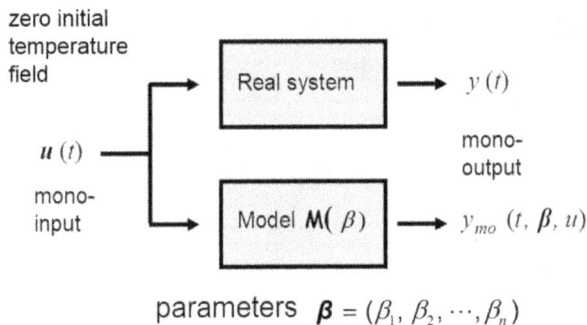

FIGURE 4.1
Real system and its representation by a model (the case of a zero-initial-temperature field).

Each parameter β_j can be any quantity linked to the structure of the model (conductivities, volumetric heat capacities, heat transfer coefficients, etc.) or to the instrumentation (location of thermocouples, intrusive character or response time of the sensor, etc.).

This model $\mathbf{M}(\beta)$ is derived here from the heat equation and from basic physical laws that govern heat transfer in the considered material system.

We also assume that both u and y_{mo} are observable quantities, which means that they can be measured through an ad hoc instrumentation, starting from an initial time $t_0 = 0$.

There are two ways the output y can be considered: It can be either an electrical signal delivered by the sensor, such as a tension V in the case of a thermocouple, or directly a temperature if the calibration law of the sensor is available.

We consider the second approach here, which requires the parameters β_{cal} of the calibration law to be estimated first. For a thermocouple, this calibration "model," within a temperature range where the output tension-to-temperature relationship can be linearized, writes out:

$$V = kT + T_{cold} \tag{4.1}$$

where T is the temperature of the hot junction, T_{cold} is the temperature of the reference cold junction, and k is the thermoelectric power of the couple. We have here $\beta_{cal} = (T_{cold}, k)$.

All the equations for calculating the output of the model constitute the structure of the model. Its solution at a given time t can be written as a functional relationship as follows:

$$y_{mo} = \eta(t, x) \tag{4.2}$$

where x is a *list* of explanatory quantities $x = \{\beta, u\}$ composed of input u or of structural parameters, the β_j's.

Figure 4.2 defines the terminology used in a model of the system dynamic type: Variable t is not necessarily a time but is used for the independent variable here.

When the output can be observed at several times, or more generally for several values of the independent variable, one deals with an output vector (not a scalar y_{mo} anymore), which requires the use of a vector function $\eta(\cdot)$ whose arguments are the vector of the measurement times t and the column-vector version x of the x list:

$$y_{mo} = \eta(t, x) \tag{4.3}$$

The preceding analysis shows that any variation in the data present in the x vector (including structural and position parameters β_{struct} and β_{pos}) will

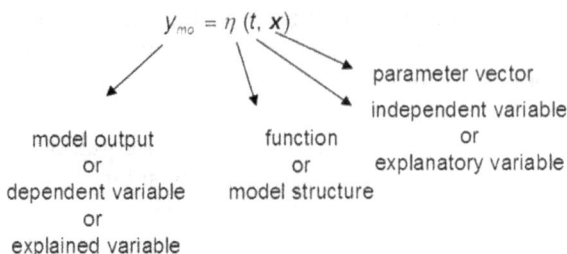

$$y_{mo} = \eta\,(t,\,\mathbf{x})$$

model output	function	parameter vector
or	or	independent variable
dependent variable	model structure	or
or		explanatory variable
explained variable		

FIGURE 4.2
Terminology of a model of the system dynamic type.

produce a variation of the y_{mo} output. Conversely, any variation of this output y_{mo} is necessarily caused by variation of some data inside $\mathbf{x}$.

The parameter estimation approach is based on this principle. However, solving the inverse problem to find all the parameters present in $\mathbf{x}$ constitutes very often an impossible task: They are too many and their effect on the output may be undistinguishable. This feature can be studied through a study of the sensitivity matrix (see Section 4.5 further down).

So, when knowledge of part of the variables that are necessary to solve the direct problem is lacking, data vector $\mathbf{x}$ of this problem can be split into two vectors in the following way:

$$x = \begin{bmatrix} x_r \\ x_c \end{bmatrix} \tag{4.4}$$

where only x_r, that is, the (column) vector gathering the unknown part of the data, is *sought (researched)*, while its *complementary* part x_c contains data that are *supposed to be known*. These last data may stem from the literature and may differ from their exact values in the model of the considered experiment.

So, any process aimed at finding $\mathbf{x}$ requires some *additional information* about the values of x_c (see Figure 4.3).

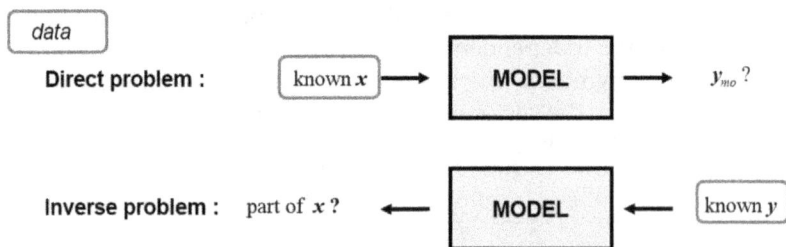

FIGURE 4.3
Direct and estimation (inverse) problems.

Problems whose objective is to find a value for x, starting from additional information and experimental measurements, are called *inverse experimental problems*.

Any inverse problem consists in making the model work in the "backward" way: If outputs y as well as model structure $\eta(\cdot)$ are known, together with part x_c of x, its complementary part x_r will be sought (see Figure 4.3). In this chapter, we only consider inverse problems of the parameter estimation type, and therefore exclude the function estimation problems, such as recovering the input function $u(t)$, for example.

4.3 Statistical Characterization of Noise and Bias for a Parameter Estimation Problem

4.3.1 Static Model and Estimation of the Parameters of a Probability Law

We assume here that the model output (Eq. (4.3)) is a constant μ (it does not depend on the independent variable, called "times" here), which means that $y_{mo} = \mu$ while its unbiased measured value $y = y_{mo} + e_y$ is a realization of a random variable noted $Y = y_{mo} + \varepsilon$. This means that its output is "static." Repetition of measurements y of y_{mo} corresponds to different observed values e_y of the (random) noise ε. This repetition constitutes a simple random sampling process, without replacement here, in order to construct a sample $(y_1, y_2, \ldots, y_m)$ that is a realization of a random "vector" $(Y_1, Y_2, \ldots, Y_m)$.

This very classical procedure is used to get a statistical estimation of the stochastic mean μ and/or of the variance σ^2 of the distribution of Y (the infinite number of "potential" measurements) starting from the real discrete measurements of y in the sample.

With these assumptions, it can be shown that, as soon as the number of measurements m is high enough, higher than 30 in practice, the sample mean $\bar{Y} = \dfrac{1}{m}\sum_{i=1}^{m} Y_i$ tends to follow a normal law, noted $\mathcal{N}$ here (central limit theorem), and the sample variance, noted $S_s^2 = \dfrac{1}{m}\sum_{i=1}^{m}\left(Y_i - \bar{Y}\right)^2 = \dfrac{1}{m}\sum_{i=1}^{m} Y_i^2 - \bar{Y}^2$, after a scaling by σ^2/m, follows a chi-square law, noted χ^2 here, with $(m-1)$ degrees of freedom:

$$\bar{Y} : \mathcal{N}\left(y_{mo}, \sigma^2/m\right) \quad \text{and} \quad mS_s^2/\sigma^2 : \chi^2(m-1) \tag{4.5}$$

This allows to find non-biased estimations of μ and σ^2, using the "hat" notation ($\wedge$) to designate an estimated value and the variance of the corresponding parameter:

$$\hat{y}_{mo} = E(\bar{Y}) = \bar{y} \quad \text{and} \quad \hat{\sigma}^2 = E(S_s^2) = \frac{m}{m-1}s^2 \qquad (4.6a,b)$$

where $\bar{y}$ and s^2 are the observed values of $\bar{Y}$ and of S_s^2, and replacing random variables Y_i by the corresponding measured values y_i, $E(\cdot)$ is the mathematical expectation of a random variable (see its definition in Table 4.1).

So, the error for μ is in fact a random quantity and the true value for this parameter can be found within a confidence interval centered on $\bar{y}$:

$$\text{Prob}\left\{|\mu - \bar{y}| \le z_\alpha \sqrt{\frac{m}{m-1}}s\right\} = \alpha \quad \text{with} \quad \alpha = 2\int_0^{z_\alpha} \frac{1}{\sqrt{2\pi}}\exp\left(-\frac{z^2}{2}\right)dz \quad (4.7)$$

with Prob$\{\cdot\}$ designing the probability of an event.

This shows that when a noise is present in an unbiased signal, the absolute error for the parameter, which is commonly noted $\Delta\mu$ in the present case, is not a deterministic quantity. However, it can be quantified by its standard deviation, and one can take, as a definition:

$$\Delta\mu = \hat{\sigma} \qquad (4.8)$$

We will now focus on estimation of parameters of dynamic models, whose output is not constant anymore.

4.3.2 Measurement of the Output of a Perfect Dynamical Model and Its Vector Form

We consider the measurements y_i of $y_{mo} = \eta(t_i, x)$ at m times t_i (for $i = 1$ to m). They derive from a sampling operation, but now the expectation of the y_i measurements varies with time. This means that this sampling is made for a *dynamical* population.

The first natural idea is to consider each measurement as the realization of a scalar random variable Y_i, of expectation $E(Y_i) = y_i^{\text{perfect}} = \eta_{\text{exact}}(t_i, x^{\text{exact}})$ that is the "perfect" (unbiased and noiseless) output of the model whose structure $\eta_{\text{exact}}(t_i, \cdot)$ is exact as well as the set of parameters gathered in a column vector x^{exact}:

$$y_i = y_i^{\text{perfect}} + \varepsilon_i \qquad (4.9)$$

where ε_i is the measurement noise, that is, the difference between the measured signal and the perfect output of the model that corresponds exactly to this measurement.

TABLE 4.1

Definitions and Quantities Associated with Two Random Variables

- Expectation

$$E(Y_k) = \int_{-\infty}^{\infty} y_k f_{Y_k}(y_k) \, dY_k = y_{mo}(t_k, x) \quad \text{for} \quad k = 1 \quad \text{or} \quad 2$$

- Expectation of a random vector

$$E(Y) = \begin{bmatrix} E(Y_1) \\ E(Y_2) \end{bmatrix} = \begin{bmatrix} y_{mo}(t_1, x) \\ y_{mo}(t_2, x) \end{bmatrix}$$

- Marginal distributions (probability density functions)

$$f_{Y_1}(y_1) = \text{Prob}(y_1 \le Y_1 < y_1 + dy_1) = \int_{-\infty}^{\infty} f_Y(y_1, y_2) \, dy_2$$

$$f_{Y_2}(y_2) = \text{Prob}(y_2 \le Y_2 < y_2 + dy_2) = \int_{-\infty}^{\infty} f_Y(y_1, y_2) \, dy_1$$

- Specific case of the binormal law:

$$f_{Y_k}(y_k) = \frac{1}{\sigma_k \sqrt{2\pi}} \exp\left(-\frac{1}{2}\left(\frac{y_k - E(Y_k)}{\sigma_k}\right)^2\right) \quad \text{for} \quad k = 1 \quad \text{or} \quad 2$$

- Variance

$$\text{var}(Y_k) = \sigma_k^2 = \int_{-\infty}^{\infty} (y_k - E(Y_k))^2 \, f_{Y_k}(y_1, y_2) \, dy_k \quad \text{for} \quad k = 1 \quad \text{or} \quad 2$$

- Covariance

$$\text{cov}(Y_1, Y_2) = \int_{-\infty}^{\infty}\int_{-\infty}^{\infty} (y_1 - E(Y_1))(y_2 - E(Y_2)) f_Y(y_1, y_2) \, dY_1 \, dY_2$$

$$= E\left([Y_1 - E(Y_1)][Y_2 - E(Y_2)]\right) = \text{cov}(\varepsilon_1, \varepsilon_2)$$

- Correlation coefficient

$$\rho_{12} = \frac{\text{cov}(Y_1, Y_2)}{\sigma_1 \sigma_2}$$

- Variance-covariance matrix

$$\text{cov}(Y) = \begin{bmatrix} \text{var}(Y_1) & \text{cov}(Y_1, Y_2) \\ \text{cov}(Y_1, Y_2) & \text{var}(Y_2) \end{bmatrix} = \begin{bmatrix} \sigma_1^2 & \rho_{12}\,\sigma_1\,\sigma_2 \\ \rho_{12}\,\sigma_1\,\sigma_2 & \sigma_2^2 \end{bmatrix}$$

Let us note that it is impossible to discriminate in a measurement y_i between the contributions of the perfect model output y_i^{perfect} and of the noise ε_i.

That is why the measurement noise is considered as a random variable. The expectation of the noise is equal to zero for a good measurement chain, which means that the (direct) measurement y_i of y_i^{perfect} is unbiased.

So, because of Eq. (4.9), the experimental signal y_i (of expectation equal to y_i^{perfect}) is also a random variable.

One calls $\mathcal{L}$ the probability law followed by noise ε_i. One good measuring instrument provides a noise whose standard deviation σ is constant. This probability law is characterized by several parameters: its expectation (i.e., its stochastic mean, equal to zero here), its variance σ^2, and possibly other parameters required to define the probability density function (pdf) $f_{\varepsilon i}$ of this random variable. If one limits oneself to the mean and variance (the case of a normal law by example), one notes:

$$\varepsilon_i : \mathcal{L}(0, \sigma^2) \quad \Rightarrow \quad Y_i : \mathcal{L}\left(y_i^{\text{perfect}}, \sigma^2\right) \tag{4.10}$$

A capital character has been used here to designate the random variable Y_i whose realization at time t_i is the measured signal y_i.

If the expectation of y_i is different from y_i^{perfect}, either the model is wrong (the case of a model bias, but we have excluded this assumption above), or the measurement is biased, which means that in average (i.e., with repetitions), the sensor (with its acquisition chain and its calibration law) does not yield the exact value it is supposed to measure.

The second idea is to consider the whole set of measurements (a multi-dimensional sample) as a column vector $\mathbf{y} = [y_1, y_2, \ldots, y_m]^T$, that is, the realization of a random vector $\mathbf{Y} = [Y_1, Y_2, \ldots, Y_m]^T$, where the superscript T denotes the transpose of a vector, or of a matrix, using bold characters for them here.

So, Eq. (4.9) is given a vector form:

$$\mathbf{y} = \mathbf{y}^{\text{perfect}} + \boldsymbol{\varepsilon} \tag{4.11}$$

where the noise vector $\boldsymbol{\varepsilon}$ is also a random vector.

In the general case, the noise vector $\boldsymbol{\varepsilon}$ defined in Eq. (4.11) has a probability law noted $\mathcal{L}$:

$$\boldsymbol{\varepsilon} : \mathcal{L}\big(E(\boldsymbol{\varepsilon}) = 0, \operatorname{cov}(\boldsymbol{\varepsilon}), \ldots\big) \tag{4.12}$$

where the (symmetrical) variance–covariance matrix $\operatorname{cov}(\boldsymbol{\varepsilon})$ is defined by:

$$\operatorname{cov}(\boldsymbol{\varepsilon}) = \operatorname{cov}(\mathbf{Y}) = \begin{bmatrix} \operatorname{var}(\varepsilon_1) & \operatorname{cov}(\varepsilon_1, \varepsilon_2) & \cdots & \operatorname{cov}(\varepsilon_1, \varepsilon_m) \\ & \operatorname{var}(\varepsilon_2) & \ddots & \operatorname{cov}(\varepsilon_2, \varepsilon_m) \\ & & \ddots & \vdots \\ \text{symmetric} & & & \operatorname{var}(\varepsilon_m) \end{bmatrix} \tag{4.13}$$

With this second point of view, the joint probability density function f_Y of $\mathbf{Y}$ is defined as follows:

$$\text{Prob}\left(y_1 \le Y_1 < y_1 + dy_1, y_2 \le Y_2 < y_2 + dy_2, \dots, y_n \le Y_n < y_n + dy_n\right)$$
$$= f_Y\left(y_1, y_2, \dots, y_n\right) dy_1 dy_2, \dots, dy_n = f_Y(y) dy_1 dy_2, \dots, dy_n \tag{4.14}$$

The stochastic properties of a two-dimensional random vector are summed up in Table 4.1. It can be easily generalized for higher dimensions (see Section 4.3.5; the specific case of a binormal joint distribution is presented in Section 4.3.4 further down).

Let us note that, by definition, the expectation of a real noise should be equal to zero ($E(\varepsilon) = 0$). This vector noise is:

- non-correlated (or independent), if its variance–covariance matrix is diagonal: $\text{cov}\left(\varepsilon_i, \varepsilon_k\right) = \sigma_i^2 \delta_{ik}$, where δ_{ik} is the Kronecker symbol (null if $i \ne k$, and equal to 1 if $i = k$) and σ_i^2 is the variance of ε_i.
- independent and identically distributed (i.i.d.) if this noise is furthermore uncorrelated with a constant variance: $\text{var}(\varepsilon_i) = \sigma^2$.

In this last case, its variance–covariance matrix is spherical:

$$\text{cov}(\varepsilon) = \sigma^2 I_m \tag{4.15}$$

where I_m is the identity matrix of dimensions $m \times m$. In this case, each component of ε is independent and follows the same probability law.

4.3.3 Model Bias

Modern data acquisition techniques allow to sample the temperature signal y with a given time period Δt for a duration $t_f = m\Delta t$. This can be written as:

$$y_i = y_{\text{mo}}\left(t_i, \beta, u\right) + e_i \quad \text{with} \quad e_i = b_i + \varepsilon_i \quad \text{for} \quad i = 1 \text{ to } m \quad \text{and} \quad t_i = i\Delta t \tag{4.16}$$

where e_i is the error in the measured output; ε_i is the noise of the sensor at time $t = t_i$, with a random quantity of zero mean; and b_i is the bias possibly present in the model or in the sensor at the same time. The bias b_i is an a priori unknown deterministic quantity. A nonzero bias means that one of the four following assumptions is not met:

a. There is no error in the structure $\eta(\cdot)$ of the model, which respects the physical assumptions that correspond to the experiment;
b. There is no error in the values of all its parameters x;
c. The measuring equipment fulfills the specifications and there is no error in the parameters β_{cal} of the calibration law of the sensors, whose presence does not modify heat transfer (non-intrusive character);
d. There is no error in the calculation of the output of the model, for given values of the x parameters. If this output results from the

solution of a partial differential equation such as the heat equation, the numerical solution is exact, within the numerical resolution of the solver.

Once the parameters x_r to be estimated (see Eq. (4.4)) are defined, the output of the system is given a column-vector form, in the case where assumptions a), c), and d) are met:

$$y = y_{mo}\left(x_r^{exact}, x_c^{sk}\right) + e_y \quad \text{with} \quad e_y = \varepsilon - b_y\left(x_c^{sk}\right)$$

$$\text{with } y = \begin{bmatrix} y_1 \\ y_2 \\ \vdots \\ y_m \end{bmatrix}; y_{mo} = \begin{bmatrix} y_{mo,1} \\ y_{mo,2} \\ \vdots \\ y_{mo,m} \end{bmatrix}; b_y = \begin{bmatrix} b_1\left(x_c^{sk}\right) \\ b_2\left(x_c^{sk}\right) \\ \vdots \\ b_m\left(x_c^{sk}\right) \end{bmatrix}; \varepsilon = \begin{bmatrix} \varepsilon_1 \\ \varepsilon_2 \\ \vdots \\ \varepsilon_m \end{bmatrix} \quad (4.17)$$

$$\text{and } \begin{cases} y_{mo,i} = \eta\left(t_i, x_r^{exact}, x_c^{sk}\right) \\ \quad \text{for } i = 1 \text{ to } m \end{cases}$$

The error e_y in the output is composed of two terms: a random quantity, the measurement noise ε, whose stochastic mean is equal to zero; and a bias, which corresponds to an error in the parameters "supposed to be known" whose value x_c^{sk} may differ from their exact value x_c^{exact}.

So, the output bias is defined as follows:

$$b_y\left(x_c^{sk}\right) = y_{mo}\left(x_r^{exact}, x_c^{sk}\right) - y_{mo}\left(x_r^{exact}, x_c^{exact}\right) \quad (4.18)$$

4.3.4 Example of a Binormal Joint Distribution

We consider here the particular case of a Gaussian distribution. Figure 4.4 shows the graphical plot of the joint probability distribution (pdf) f_Y whose expression is given below:

$$f_Y(y_1, y_2) = \frac{1}{2\pi\sigma_1\sigma_2\sqrt{1-\rho_{12}^2}} \exp\left[-\frac{1}{2}Q(y_1, y_2)\right]$$

with:

$$Q(y_1, y_2) = \frac{1}{\left(1-\rho_{12}^2\right)}\left(\frac{y_1 - y_{mo}(t_1, x)}{\sigma_1}\right)^2 + \left(\frac{y_2 - y_{mo}(t_2, x)}{\sigma_2}\right)^2 \qquad (4.19)$$

$$-2\rho_{12}\left(\frac{y_1 - y_{mo}(t_1, x)}{\sigma_1}\right)\left(\frac{y_2 - y_{mo}(t_2, x)}{\sigma_2}\right)$$

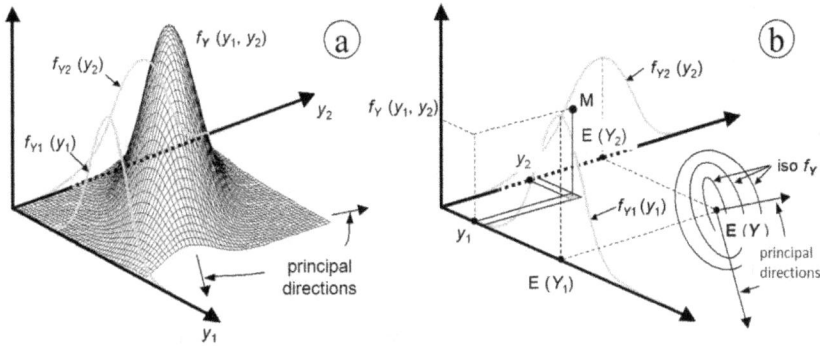

FIGURE 4.4
Binormal law: the case of two correlated measurements.
f_Y: joint probability density function of Y
f_{Y1}, f_{Y2}: marginal probability density functions of Y_1 and Y_2

Three parameters are present in this law: the variances σ_1^2 and σ_2^2 of the two individual random variables Y_1 and Y_2 and their correlation coefficient ρ_{12} ($-1 \le \rho_{12} \le 1$). The variance–covariance matrix of Y is defined in Table 4.1 in the very general case of a random vector Y of size 2 and of joint probability density function $f_Y(y_1, y_2)$. This can be easily generalized in the case $m > 2$.

4.3.5 Properties of Random Vectors

The notion of a random vector has been introduced above. In the type of applications concerned here, it is the column vector of measurements Y, of dimensions $m \times 1$.

This vector has an expectation, noted $E(Y)$, that is, a vector of same size, whose coefficients are the expectations of the corresponding coefficients of Y. It has also a variance–covariance matrix, noted cov(Y), of dimensions $m \times m$ whose coefficients are the covariances of the coefficients of Y:

$$[E(Y)]_i = E(Y_i) \quad \text{and} \quad [\text{cov}(Y)]_{ik} = \text{cov}(Y_i, Y_k) \tag{4.20}$$

It is always possible to linearly transform signal Y using a linear transformation whose deterministic coefficients are set in a matrix G of size $(p \times m)$, in order to get a transformed signal $Z = G \cdot Y$ of size $(p \times 1)$. Since Y is a random vector, such is also the case for vector Z. The expectation and the variance–covariance matrix of Z depend on the same properties of Y:

$$Z = GY \quad \Rightarrow \quad E(Z) = G\,E(Y) \quad \text{and} \quad \text{cov}(Z) = G\,\text{cov}(Y)\,G^T \tag{4.21}$$

4.4 Example 1: Estimation of a Heat Transfer Coefficient in a One-State Experiment

We consider here the case of forced convection over a wall with an external flow, and our objective is to estimate the heat transfer coefficient h, using the following equation:

$$q = h(T_w - T_\infty) \qquad (4.22)$$

as well as the measurements of the heat flux q, of the wall temperature T_w, and of the fluid temperature T_∞. Let us note that Eq. (4.22), which defines h, is not really a model, as considered in Section 4.2 above, but only constitutes the definition of this coefficient as a by-product of the heat flux and of the two temperatures.

The measured values of these quantities are:

$$q^{\exp} = q + e_q; \quad T_w^{\exp} = T_w + e_w; \quad T_\infty^{\exp} = T_\infty + e_\infty \qquad (4.23)$$

Here, e_x, for $x = q, w$ or ∞, designates the measurement error for x (wall flux, wall temperature, or temperature of the flow). Each of them can be decomposed into a systematic error b_x and a measurement noise ε_x:

$$e_x = b_x + \varepsilon_x \qquad (4.24)$$

ε_x is a continuous random variable, which follows a probability law that is not necessarily a Gaussian one and which is noted $\mathcal{L}$. Its stochastic mean is equal to zero, and its variance is noted as σ_x^2:

$$\varepsilon_x : \mathcal{L}\left(E(\varepsilon_x) = 0, \text{var}(\varepsilon_x) = \sigma_x^2, \ldots\right) \qquad (4.25)$$

where $E(\cdot)$ designates the mathematical expectation and var($\cdot$) the variance.

This noise has an electronic component: Digitization of an analogical signal yields a noise whose probability density function is a uniform one between two adjacent binary levels, with an additional contribution caused by thermal fluctuations in the electronics of the measuring equipment.

So, the measurement noise stems mainly from the acquisition system of the different signals, that is, for example, the tension output of two thermocouples and the tension V and the electrical current I of an electrical heater. These allow to reconstitute the dissipated power by the Joule effect at the origin of the flux q.

The standard deviation σ_∞ in the noise ε_∞ in the wall temperature $T_\infty^{\exp}$ may also stem from turbulence in the fluid, with a higher value if the fluid is a gas compared to the liquid case (higher heat transfer coefficient between junction and fluid in the latter case). Let us also note that the standard deviations σ_∞ and σ_w of the measured temperatures can be estimated through measurements of their fluctuations before the start of the power stimulation, that is, for $q = 0$ (see Eq. (4.6b) in Section 4.3.1).

NB: We assume here that the flux is measured by an "active" (traditional) method, that is, using an electrical heater. An alternative is to measure it through solution of a basic inverse heat conduction problem (IHCP) (see Ref. [5]), where two thermocouple junctions are set in the system: The first one is localized at the "cold" face of the wall, while the second one is inserted at an internal point within the thickness of the wall. This requires a 1D steady-state assumption, where an extrapolation of the temperature field allows to recover, with some regularization, the hot wall temperature and the corresponding flux q [6].

The systematic error is a statistical bias that is observed if no noise is present in the signal:

$$b_x = E(e_x) \tag{4.26}$$

In the present case, for q, this bias may originate from an improper equalization between the dissipated power and the wall heat flux: This one does not take into account the intrusive character of the heater, its localization that is not always the wall surface (the case of a heat cartridge, which is a volume heater unlike a foil heater), and the heat losses through the current supply cables. As a consequence, the flux bias is:

$$b_q = q - VI/A \tag{4.27}$$

where A is the concerned area of the wall or of the heater, for the foil heater configuration here.

The temperature biases may originate from cold junction temperatures that are not strictly equal for each thermocouple: This is the case if they are located in a computer housing where temperature gradients are present.

Another possible bias is the intrusive character of thermocouples that is the consequence of a connection by cables that are perpendicular and not parallel to the isotherms within the volume of the solid wall. This introduces a fin effect in the cables with tension differences that do not correspond anymore to the temperature of the non-instrumented system it is supposed to measure. So, eventually, each signal is in fact a random variable:

$$e_x: \mathcal{L}\big(E(\varepsilon_x) = b_x, \operatorname{var}(\varepsilon_x) = \sigma_x^2, ...\big) \;\Rightarrow\; x^{\exp}: \mathcal{L}\big(E(\varepsilon_x) = x + b_x, \operatorname{var}(\varepsilon_x) = \sigma_x^2, ...\big) \tag{4.28}$$

This writing goes against the traditional, and a bit simplistic, way of defining the absolute uncertainty of a measured quantity $x^{\exp}$:

$$x^{\exp} = x \pm \Delta x \tag{4.29}$$

A confidence interval can obviously be used to replace the absolute uncertainty (see Section 4.3.1 above), with a chosen confidence level $1 - \alpha$ (99%, 95%, ...), where α is the risk level associated with the corresponding interval, but it does not present much interest in practice since:

1. It will be constructed around a biased value $x + b_x$ and, moreover, the bias is not generally known;
2. The distribution law $\mathcal{L}$ of the noise is unknown.

However, if the measurement noise is small, not in the absolute, but with respect to the measured quantity x^{exp}, because both quantities are expressed with physical units, the error becomes deterministic and one writes:

$$x^{exp} = x + b_x \tag{4.30}$$

Let us note a + sign that appears in front of b_x and not the initial ± sign in front of Δx in Eq. (4.29).

In the general case, it is more interesting to replace Δx by the quadratic mean (the root mean square) of the bias of the signal and of the standard deviation of the noise:

$$e_x^{rms} = \left(E\left(x^{exp} - x \right)^2 \right)^{1/2} = \left(E(b_x + \varepsilon_x)^2 \right)^{1/2} = \left(b_x^2 + \sigma_x^2 \right)^{1/2} \tag{4.31}$$

Return to the initial problem of calculation of the h coefficient from measured values of quantities present in its definition (Eq. (4.22)) provides its estimation under the following form:

$$\hat{h} = \frac{q^{exp}}{T_w^{exp} - T_\infty^{exp}} = \frac{q + e_q}{\Delta T + e_{\Delta T}} = \frac{q + e_q}{\Delta T \left(1 + \dfrac{e_{\Delta T}}{\Delta T} \right)}$$

$$\approx \frac{q}{\Delta T} \left(1 + \frac{e_q}{q} \right)\left(1 - \frac{e_{\Delta T}}{\Delta T} \right) \approx h \left(1 + \frac{e_q}{q} - \frac{e_{\Delta T}}{\Delta T} \right) \tag{4.32}$$

with $\Delta T = T_w - T_\infty$ and $e_{\Delta T} = e_w - e_\infty$

Two sequential assumptions have been made in the construction of the above equation:

1. The relative error on the temperature difference $e_{\Delta T}/\Delta T$ is small, which has allowed the linearization of the inverse of the denominator of $\hat{h}$;
2. The relative error on flux e_q/q is also small, which allows to neglect the second-order term $e_q e_{\Delta T}/q\Delta T$ with respect to the first-order terms.

So, the estimation error of $\hat{h}$, which is also a random quantity, can be expressed as:

$$e_h = \hat{h} - h = h\left(\frac{e_q}{q} - \frac{e_{\Delta T}}{\Delta T}\right) \tag{4.33}$$

Its mathematical expectation, which is the estimation bias, is:

$$b_h = E(e_h) = h\left(\frac{b_q}{q} - \frac{b_{\Delta T}}{\Delta T}\right) = h\left(\frac{b_q}{q} - \frac{b_w - b_\infty}{\Delta T}\right) \Rightarrow \frac{b_h}{h} = \frac{b_q}{q} - \frac{b_w - b_\infty}{T_w - T_\infty} \tag{4.34}$$

Its variance is:

$$\sigma_h^2 = \mathrm{var}(e_h) = h^2 E\left[\left(\frac{\varepsilon_q}{q} - \frac{\varepsilon_w - \varepsilon_\infty}{\Delta T}\right)^2\right]$$

$$= h^2\left[\frac{\sigma_q^2}{q^2} + \frac{E\left((\varepsilon_w - \varepsilon_\infty)^2\right)}{(\Delta T)^2} - \frac{2E\left(\varepsilon_q \varepsilon_w - \varepsilon_q \varepsilon_\infty\right)}{q\Delta T}\right]$$

$$= h^2\left[\frac{\sigma_q^2}{q^2} + \frac{\sigma_w^2 + \sigma_\infty^2 - 2\mathrm{cov}\left(\varepsilon_w, \varepsilon_\infty\right)}{(\Delta T)^2} - \frac{2\left(\mathrm{cov}\left(\varepsilon_q, \varepsilon_w\right) - \mathrm{cov}\left(\varepsilon_q, \varepsilon_\infty\right)\right)}{q\Delta T}\right] \tag{4.35}$$

Three covariances are present in Eq. (4.35): They are equal to zero since the errors on q, T_w, and T_∞ are independent here. They would be different from zero if the flux were calculated by an inverse heat conduction method (see Ref. [6]).

One finds here:

$$\left(\frac{\sigma_h}{h}\right)^2 = \frac{\sigma_q^2}{q^2} + \frac{\sigma_w^2 + \sigma_\infty^2}{(T_w - T_\infty)^2} \tag{4.36}$$

Classically, the calculation of relative error $\Delta h / h$ allows to assess the quality of the estimation of the heat transfer coefficient. As a proposal, in this situation where a bias and a random component are involved, a good corresponding indicator could be the root mean square of the relative bias and of the relative standard deviation, replacing the exact values of q, T_w, and T_∞ by their measured values in their calculation:

$$\frac{\Delta h}{h} = \frac{e_h^{\mathrm{rms}}}{h} = \left(\frac{b_h^2}{h^2} + \frac{\sigma_h^2}{h^2}\right)^{1/2} \approx \left[\left(\frac{b_q}{q^{\exp}} - \frac{b_w - b_\infty}{T_w^{\exp} - T_\infty^{\exp}}\right)^2 + \frac{\sigma_q^2}{q^{\exp 2}} + \frac{\sigma_w^2 + \sigma_\infty^2}{\left(T_w^{\exp} - T_\infty^{\exp}\right)^2}\right]^{1/2} \tag{4.37}$$

If the bias on each temperature and the noise in the measurement of the flux are neglected, one gets:

$$\frac{\Delta h}{h} = \frac{e_h^{rms}}{h} \approx \left[\left(\frac{b_q}{q^{exp}} \right)^2 + \frac{\sigma_w^2 + \sigma_\infty^2}{\left(T_w^{exp} - T_\infty^{exp} \right)^2} \right]^{1/2} \tag{4.38}$$

The relative error is calculated here for the following values:

$$b_q/q^{exp} = 0.05; \quad T_w^{exp} - T_\infty^{exp} = 4°C; \quad \sigma_w = \sigma_\infty = 0.1°C \Rightarrow \Delta h/h = 6.1\%$$

If the temperature difference is reduced to 1°C, without any change in the other quantities, one gets $\Delta h/h = 15\%$. This means that the impact of the noise on the error on h cannot be neglected when small temperature differences between wall and fluid are present.

In the above example, only measurements for one state of the system have been used to estimate the h coefficient. We will see in Section 4.6 further ahead that several points corresponding to different values of the flux can be used to estimate the heat transfer coefficient. So, we focus now on the estimation of the parameters of dynamic models whose output is not constant anymore.

4.5 Least Squares and Estimation Residuals

4.5.1 Least Square Sum and Residuals

The more general "indirect" measurement technique consists in minimizing the Euclidian distance between the measurements and the model outputs. If one calls x the parameters whose estimation is looked for and if the measurement bias b_y is equal to zero, the square of this distance, the ordinary least square (OLS) sum, writes out:

$$J_{OLS}(x) = \|y - \eta(t,x)\|^2 = \sum_{i=1}^{m} \left[y_i - \eta(t_i,x) \right]^2 = (y - \eta(t,x))^T (y - \eta(t,x)) = \|r(x)\|^2 \tag{4.39}$$

For a given value of x, this sum is also the square of the length of the vector of residuals $r(x) = y - \eta(t,x)$, that is, the square of its norm $\|r(x)\|$ in a vector space of dimensions $m \times 1$, where m is the number of measurements of the output y_i and t is the vector of the measurement "times " y_i, of the same dimensions, and the term "times " designates the independent variable here (see Figure 4.2 in Section 4.2).

The value of the OLS estimator results in the minimization of $J_{OLS}(x)$ with respect to each of the n parameters:

$$\left.\frac{\partial J_{OLS}}{\partial x_1}\right|_{t,x_k \text{ for } k \neq 1} = \left.\frac{\partial J_{OLS}}{\partial x_2}\right|_{t,x_k \text{ for } k \neq 2} = \cdots = \left.\frac{\partial J_{OLS}}{\partial x_n}\right|_{t,x_k \text{ for } k \neq n} = 0 \quad (4.40)$$

In order to get this minimum, it is very useful to implement a sensitivity analysis, that is, to calculate the n sensitivity functions:

$$S_j(t,x) = \left.\frac{\partial y_{mo}}{\partial \beta_j}\right|_{t,\beta_k \text{ for } k \neq j} \quad (4.41)$$

These sensitivity functions are sampled on the time grid corresponding to the measurements, to yield n sensitivity column vectors that, once concatenated, form an $m \times n$ sensitivity matrix S:

$$S(x) = \begin{bmatrix} S_1 & S_2 & \cdots & S_j & \cdots & S_n \end{bmatrix}$$

$$\xrightarrow{\quad j: \text{ subscript of } x_j \quad}$$

$$= \begin{bmatrix} S_{11} & \cdots & & \cdots & S_{1n} \\ \vdots & \ddots & & & \vdots \\ \vdots & & S_{ij} = \frac{\partial y_{mo}}{\partial x_j}(t_i, x) & & \vdots \\ \vdots & & & & \vdots \\ S_{m1} & \cdots & & \cdots & S_{mn} \end{bmatrix} \begin{matrix} \\ \\ \downarrow i: \text{ subscript of } t_j \\ \\ \end{matrix} \quad (4.42)$$

4.5.2 Linear Least Squares

When a model is linear, with respect to its parameters, the sensitivity functions $S_j(t,x)$ do not depend on their values. Consequently, the model output can be put under the form:

$$y_{mo} = \eta(t,x) = Sx \quad (4.43)$$

One shows that the least square solution, the minimum of the criterion $J_{OLS}(x)$, the OLS estimator $\hat{x}_{OLS}$ of the parameter vector x, has an explicit value:

$$\hat{x}_{OLS} = (S^T S)^{-1} S^T y \quad (4.44)$$

Of course, this result is valid only if the number of measurements m is equal to or greater than the number n of parameters: $m \geq n$.

Furthermore, if the noise ε is independent and identically distributed (i.i.d.) over the m measurement times, its variance–covariance matrix becomes spherical, with a standard deviation σ, and thus the variance–covariance matrix of $\hat{x}_{\text{OLS}}$ can be easily calculated using Eq. (4.44) and the covariance theorem given in Eq. (4.21):

$$\text{cov}(\varepsilon) = \sigma^2 I_m \Rightarrow \text{cov}(\hat{x}_{\text{OLS}}) = \sigma^2 \left(S^T S\right)^{-1} \tag{4.45a,b}$$

Equation (4.45b) allows to get the standard deviation of the estimate of any of the n parameters $\hat{x}_j$ that compose the parameter vector $\hat{x}_{\text{OLS}}$. This one is simply the product of the standard deviation of the noise by the square root of the corresponding diagonal term of the inverse of the information matrix $A = S^T S$:

$$\sigma_{\hat{x}_j} = \sigma \sqrt{B_{jj}} \quad \text{with} \quad B = \left(S^T S\right)^{-1} \tag{4.46}$$

It is important to notice that the results given by Eqs. (4.45a) and (4.46) do not depend on the precise stochastic law followed by noise ε (the normality assumption is not required). It simply stems from the covariance theorem (Eq. (4.21)).

When the relative standard deviation $\sigma_{\hat{x}_j}/x_j$ or $\sigma_{\hat{x}_j}/\hat{x}_j$ of one of the parameters that is looked for is too large, this means that the system $\{S_1, S_2, \ldots, S_j, \ldots, S_n\}$ formed by the n sensitivity vectors is close to being dependent, in the linear algebra sense, which means that the sensitivity matrix is nearly rank-deficient.

So, in this case, at least one parameter of the original parameter vector has to be removed from the unknowns and given a value "supposed to be known," with the decomposition of x already given by Eq. (4.4) in Section 4.2. This results in the partitioning of the sensitivity matrix into two matrices, composed of the sensitivity vectors linked to either x_r or x_c, with the following expression for the model and for the OLS estimate:

$$y_{\text{mo}} = \begin{bmatrix} S_r & S_c \end{bmatrix} \begin{bmatrix} x_r \\ x_c \end{bmatrix} \Rightarrow \hat{x}_{r,\text{OLS}} = \left(S_r^T S_r\right)^{-1} S_r^T \left(y - S_c x_c^{\text{sk}}\right) \tag{4.47}$$

The error e_r of $\hat{x}_{r,\text{OLS}}$ is composed of a zero average random quantity caused by the presence of the noise ε in the measurement, and of a (deterministic) bias b_r produced by the error in the parameters supposed to be known:

$$e_r = \hat{x}_{r,\text{OLS}} - x_r^{\text{exact}} = \left(S_r^T S_r\right)^{-1} S_r^T \varepsilon - b_r \tag{4.48a}$$

with

$$\text{cov}(\hat{x}_{r,\text{OLS}}) = \text{cov}(e_r) = \sigma^2 \left(S_r^T S_r\right)^{-1} \quad \text{and}$$

$$E(e_r) = -b_r = -\left(S_r^T S_r\right)^{-1} S_r^T S_c \left(x_c^{\text{exact}} - x_c^{\text{sk}}\right) \tag{4.48b}$$

4.5.3 Nonlinear Least Squares

If the output y_{mo} of the model is not linear with respect to its parameters x, its n sensitivity vectors S_j defined in Eq. (4.42) depend on x. The minimum of the ordinary least square criterion $J_{OLS}(x)$ is not explicit anymore but can be found through an iterative Gauss–Newton algorithm, where the initial value $\hat{x}^{(0)}$ has to be given:

$$\hat{x}^{(k+1)} = \hat{x}^{(k)} + \left(S^{(k)T}S^{(k)}\right)^{-1} S^{(k)T}\left(y - y_{mo}\left(\hat{x}^{(k)}\right)\right) \quad \text{with} \quad S^{(k)} = S\left(\hat{x}^{(k)}\right) \quad (4.49)$$

The corresponding value of $J_{OLS}(x)$ is calculated at each iteration, and the algorithm is stopped as soon as its reaches its minimum. If the number n of parameters is large, one has to prevent the residuals to go below the level of the measurement noise, which would mean an over-fitting of the experimental points by the model:

$$\hat{x}_{OLS} = x^{(k_{max})} \quad \text{with} \quad J_{OLS}\left(x^{(k_{max})}\right) \geq m\sigma^2 \quad (4.50)$$

This criterion is called the "discrepancy principle" or "Morozov's principle" [7] in function estimation.

More stable minimization algorithms can be used, such as the Levenberg–Marquardt minimization (e.g., see Ref. [8]).

Once the minimum $\hat{x}_{OLS} = x^{(k_{max})}$, given by Eq. (4.50), is reached, the error analysis is exactly the same as in the linear case, since Eq. (4.49) can be written by linearization around this minimum:

$$e_x = \hat{x}_{OLS} - x^{exact} = \left(S^T\left(\hat{x}_{OLS}\right)S\left(\hat{x}_{OLS}\right)\right)^{-1} S^T\left(\hat{x}_{OLS}\right)\left(y - y_{mo}\left(\hat{x}_{OLS}\right)\right) \quad (4.51)$$

So, calculation of the variance–covariance matrix of e_x and therefore of $\hat{x}_{OLS}$ is also given by Eq. (4.45b), replacing sensitivity matrix S by $S\left(\hat{x}_{OLS}\right)$.

4.6 Example 2: Estimation of a Heat Transfer Coefficient through Multiple-Steady State Measurements

Returning to Example 1, about estimating a heat transfer coefficient over a wall, we now assume that m measurements of q, T_w and T_∞ are now available, each for a different level q_i of flux q (for $i = 1$ to m). So, Eq. (4.19) has to be given a vector form. We decide to call $y_{mo} = T_w - T_\infty$ the dependent variable (model output) and q the independent variable. So, the vector-form equation (Eq. (4.43)) of the model is:

$$y_{mo} = \eta(q, R) = SR \quad \text{with} \quad S = \begin{bmatrix} q_1 & q_2 & \cdots & q_m \end{bmatrix}^T \quad (4.52)$$

where $R = 1/h$ is the thermal resistance for a unit area, which has replaced h in the model under the form of Eq. (4.46); and S is the corresponding sensitivity matrix, which is here a single column vector corresponding to the sensitivity vector of the output to this unique parameter.

We assume that both measured temperature vectors $T_w^{\exp}$ and $T_\infty^{\exp}$ are unbiased, independent, and identically distributed (i.i.d.), with variances σ_w^2 and σ_∞^2, and that they are independent of each other. As a consequence, the experimental output $y = T_w^{\exp} - T_\infty^{\exp}$ is also unbiased and i.i.d., with a variance $\sigma^2 = \sigma_w^2 + \sigma_\infty^2$.

Two cases are considered here:

- Case 1: There is no error in the flux measurement: $q^{\exp} = q$. So, the parameter estimation problem is linear, and the estimate of parameter R is given by Eq. (4.40) and its variance is deduced from Eq. (4.41b):

$$\hat{R}_{\mathrm{OLS}} = \left(S^T S\right)^{-1} S^T y = \frac{\sum\limits_{i=1}^{m} q_i y_i}{\sum\limits_{i=1}^{m} q_i^2} \; ; \quad \sigma_{\hat{R}}^2 = \sigma^2 \left(S^T S\right)^{-1} = \frac{\sigma_w^2 + \sigma_\infty^2}{\sum\limits_{i=1}^{m} q_i^2} \qquad (4.53a,b)$$

Return to the estimation of h yields:

$$\hat{h} = 1/\hat{R}_{\mathrm{OLS}} = \frac{\sum\limits_{i=1}^{m} q_i^2}{\sum\limits_{i=1}^{m} q_i y_i} \; ; \quad \frac{\sigma_{\hat{h}}}{\hat{h}} = \frac{\sigma_{\hat{R}_{\mathrm{OLS}}}}{\hat{R}_{\mathrm{OLS}}} = \frac{\left(\left(\sigma_w^2 + \sigma_\infty^2\right) \sum\limits_{i=1}^{m} q_i^2\right)^{1/2}}{\sum\limits_{i=1}^{m} q_i y_i} \qquad (4.54a,b)$$

For $m = 1$ (estimation from one unique state), Eq. (4.54a) reduces to Eq. (4.32) and Eq. (4.54b) reduces to Eq. (4.34)—for $b_q = 0$. Comparison of the quality of the estimation, that is, checking the level of $\sigma_{\hat{h}}/\hat{h}$, depends on the values of the fluxes that are measured in both the single- and multiple-state experiments. The main advantage of this technique is to check whether the residuals, for the different values of q, have a zero average without any signature. If this is the case, model (4.52) is validated.

- Case 2: There is an error in the flux measurement: $q^{\exp} = q + e_q$
  This case corresponds to a parameter estimation problem where there is an error in the independent variable. So, the parameter vector has to be replaced by an extended parameter vector, with a new sensitivity matrix S that is a concatenation of the sensitivity matrices to R and q:

$$x = \begin{bmatrix} R \\ q \end{bmatrix} \Rightarrow S = \begin{bmatrix} S_R & S_q \end{bmatrix} \quad \text{with} \quad S_R = \begin{bmatrix} q_1 & q_2 & \cdots & q_m \end{bmatrix}^T \text{ and } S_q = R I_m$$

$$(4.55)$$

In that case, this new sensitivity matrix depends on R and on q. This last quantity is both a signal (it is measured) and a parameter. This type of parameter estimation problem is called a total least square problem [9] or an error-in-variable (EIV) or non-orthogonal regression in statistics [10]. It can be solved through minimization of a sum that can be given the following form in the present case, where the model is a straight line passing through the origin in the (q, y_{mo}) plane (see Refs. [11–13]):

$$J_{\text{Bayes}}(x, Q) = J_{\text{OLS}}(x) + Q \left\| q^{\exp} - q \right\|^2 \quad \text{with} \quad J_{\text{OLS}}(x) = \left\| y - h(q, R) \right\|^2$$

$$\text{and } Q = (\sigma_q / \sigma)^2 \qquad (4.56)$$

Both y and $q^{\exp}$ are supposed to be unbiased, i.i.d., and independent of each other, with standard deviations noted σ and σ_q, respectively.

NB: Let us note that minimization of the sum (4.56) corresponds to a Bayesian estimation if the distributions of the noise and of the error on the independent variable q are both multinormal.

The estimates are as follows:

$$\hat{R} = Z + \frac{s_{qy}}{|s_{qy}|}\left(Z^2 + \frac{1}{Q} \right) \quad \text{and} \quad \hat{q}_i = \frac{q_i^{\exp} + Q y_i \hat{R}}{1 + Q\hat{R}^2} \quad \text{for} \quad i = 1 \text{ to } m$$

$$(4.57)$$

$$\text{with } Z = \frac{Q s_{yy} - s_{qq}}{2 Q s_{qy}}; \; s_{qy} = \text{cov}\left(q^{\exp}, y\right); s_{yy} = \text{cov}(y, y); s_{qq} = \text{cov}\left(q^{\exp}, q^{\exp}\right)$$

Here, the covariances are statistical covariances between measured column vectors:

$$\text{cov}(a, b) = \frac{1}{m}\sum_{i=1}^{m} a_i b_i - \overline{ab} \quad \text{with} \quad \overline{c} = \frac{1}{m}\sum_{i=1}^{m} a_i b_i \text{ for } c = a \text{ or } b \qquad (4.58)$$

Since the ratio Q of standard deviations is usually unknown, it is adjusted in such a way that the ordinary least square component of the Bayesian criterion is of the same order of magnitude as the noise:

$$J_{\text{OLS}}\left(\hat{x}_{\text{OLS}}(Q)\right) \approx m\sigma^2 \qquad (4.59)$$

The bias and the standard deviation of parameter R can be given approximate values:

$$b_{\hat{R}} \approx \frac{\hat{R}/m}{\left(s_{qq}/\sigma_q\right)^2} \text{ and } \sigma_{\hat{R}} \approx \frac{1}{\sqrt{m}}\frac{\left(\hat{R}^2+1/Q\right)^{1/2}}{s_{qq}/\sigma_q} \text{ if } \frac{\sigma}{s_{yy}} \ll 1 \text{ and } \frac{\sigma_q}{s_{qq}} \ll 1 \quad (4.60)$$

The preceding equations show that an increase in the number m of measurements makes the bias for R decrease as $1/m$ and its standard deviation decrease as $1/\sqrt{m}$. So, if a high enough number of measurements is available, the ratio of bias to standard deviation, $b_{\hat{R}}/\sigma_{\hat{R}}$, remains low and the relative error in the estimation of R, and therefore of h, is dominated by the noise.

4.7 Conclusions

We have shown above that indirect measurement of a quantity, using "direct" measurements (temperature and flux) and a model, requires revisiting the notion of absolute and relative errors for a given measurement. These notions deserve to be interpreted in terms of a random quantity, the noise, and of a bias, using a statistical approach. Once these quantities are defined, inversion of a model, either static or dynamic, allows not only to estimate the parameters that are looked for, using least square techniques, but also to assess their errors. The four causes of bias one can meet in a parameter estimation problem—see their labels (a) to (d) in Section 4.3.3 as well as the cause of the dispersion of the estimate, that is, the measurement noise—have been reviewed. Vectorization of the different quantities present in both the model and the measurements allows to handle estimation problems with several unknowns in a powerful way. Two examples about estimation of a heat transfer coefficient have been used to illustrate some of these concepts.

References

1. Jannot, Y. and A. Degiovanni. 2018. *Thermal Properties Measurement of Materials.* London: Wiley/ISTE.
2. Beck, J.V. and K.J. Arnold. 1977. *Parameter Estimation in Engineering and in Science.* Hoboken, NJ: Wiley.
3. Le Masson, P., O. Fudym, J.-L. Gardarein, and D. Maillet. 2015. Getting started with problematic inversions with three basic examples. *Lecture 1, Metti Advanced Spring School, Thermal Measurements and Inverse Techniques*, Biarritz, March 1–6, 2015.

4. Rigollet, F., O. Fudym, and D. Maillet. 2015. Basics for linear estimation, the white box case. *Lecture 3, Metti Advanced Spring School, Thermal Measurements and Inverse Techniques*, Biarritz, March 1–6, 2015. http://www.sft.asso.fr/metti-6.html (accessed November 29, 2018).

5. Beck, J.V., B. Blackwell, and C.R. St Clair. 1985. *Inverse Heat Conduction-Ill-Posed Problems*. New York: Wiley Interscience.

6. Maillet, D. 2014. Experimental inverse problems: Potentials and limitations. *Keynote Lecture in Proceedings of the 15th International Heat Transfer Conference*, Kyoto, Japan, August 10–15, 2014.

7. Morozov, V. 1984. *Methods of Solving Incorrectly Posed Problems*. New York: Springer Verlag.

8. Gill, P.E. and W. Murray. 1978. Algorithms for the solution of the nonlinear least-squares problem. *SIAM Journal on Numerical Analysis* 15(5): 977–992.

9. Van Huffel, S. and P. Lemmerling. *Total Least Squares and Errors-in-Variable Modelling: Analysis, Algorithms and Applications*. Dordrecht: Kluwer Academic Publishers.

10. Cheng, C.L. and J.W. Van Ness. 1999. *Statistical Regression with Measurement Errors*. London: Arnold.

11. Metzger, T. and D. Maillet. 2011. Multisignal least squares: Dispersion, bias and regularization. In: *Thermal Measurements and Inverse Techniques*, eds. H.RB. Orlande, O. Fudym, D. Maillet, and R. Cotta, 599–618. Boca Raton, FL: CRC Press/Taylor & Francis.

12. Maillet, D., T. Metzger, and S. Didierjean. 2003. Integrating the error in the independent variable for optimal parameter estimation. Part I: Different estimation strategies on academic cases. *Inverse Problems in Engineering* 11(3): 599–618.

13. Testu, A., D. Maillet, C. Moyne, T. Metzger, and T. Niass 2007. Thermal dispersion for water or air flow through a bed of glass beads. *International Journal of Heat and Mass Transfer* 50(7–8): 1469–1484.

Part B

Convection Challenges and Energy Balances

5

Wilson Plots and Measurement Accuracy

Jaime Sieres

University of Vigo

CONTENTS

5.1 Introduction

Convection is a heat transfer mechanism of utmost importance in many thermal engineering applications and equipment. The physics that describe convection heat transfer are complex, and most convection heat transfer problems are actually solved by virtue of the Newton law of cooling, which relies on the knowledge of a convection heat transfer coefficient (HTC). In order to obtain the HTC, carefully designed experiments are generally performed.

Frequently, the desired outcome of this experimental work is a built-in correlation based on dimensional analysis.

The HTC is a local parameter; however, in heat exchange devices we are often interested on an average value for the HTC. According to Newton's law of cooling, the heat transfer rate ($\dot{Q}$) exchanged by convection between a solid surface at temperature (T_s) and a fluid with temperature (T_f) is:

$$\dot{Q} = h \cdot A \cdot \left(T_s - T_f\right) \tag{5.1}$$

where A is the area of the surface and h is the average HTC for the entire surface.

Obtaining the HTC from the previous equation requires knowing the surface temperature (T_s). In many applications, it is not feasible to measure this temperature because the surface area is not accessible. In others, measuring this temperature might not be recommended since the use of a temperature sensor might affect the heat transfer rate or the fluid flow near the surface, which will also affect the measured value of the HTC. One alternative simple method to determine the HTC from experimental data is the Wilson plot method (Wilson 1915) or any of its subsequent modifications.

The Wilson plot method dates back from 1915 when Wilson proposed a technique to obtain HTC for turbulent flow inside a circular tube. The HTC correlation involved two unknown parameters that were determined by a linear regression analysis of the experimental data. Since then, multitude improvements and modifications of the original method have been proposed in order to improve its accuracy and applications. A literature review (Fernández-Seara et al. 2007) shows that most works are related to the application of the Wilson plot method in order to obtain HTC correlations that involve two parameters. However, multiple applications that involve three or more parameters have been studied through modifications of the original method. In these last cases, a simple linear regression analysis can no longer be used and iterative regression schemes are used to determine the HTC correlation parameters.

As the HTCs obtained from the Wilson method (or its subsequent modifications) are based on experimental data, they are sensitive to measurement errors. Then, good and accurate measured data with known uncertainties are needed for reliable correlations of HTCs.

Little attention has been given in the open literature to the effect of the uncertainty of the measured data in the determination of the HTC model parameters and their corresponding uncertainties (Khartabil and Christensen 1992, Uhía et al. 2013, Sieres and Campo 2018). As pointed out by van Rooyen et al. (2012), simple statistics such as the coefficient of determination of the fit R^2 (i.e., the proportion of the variance in the dependent variable that is predictable from the independent variable) or the mean absolute error are used as metrics to measure the accuracy of the outcome HTC correlation. However, this approach does not give any information about what is the confidence of the correlation in predicting future HTC values. Neither does this approach give information of the effect of the uncertainty of the measured data on the reliability of the HTC correlation. As a result, the experimenter has no information of which measured data variable's accuracies are critical, and this might lead to a non-optimal decision on the investment or selection of the measurement instrumentation or procedures.

In this chapter, the basics of the original and modified versions of the Wilson plot method are explained with a special focus on addressing correctly how the experimental measured data uncertainties affect the HTC model parameters. Examples will be used to explain and compare HTC results obtained with simple linear regression models extensively used in the open literature (which do not account for the uncertainty of the measured data), with more advanced methods based on unweighted and weighted nonlinear regression models. It is not the author intention to present HTC correlations that can be applied to a specific problem, but rather to explain how the different methods can be applied to determine HTC correlations.

5.2 Wilson Plot Method

5.2.1 Estimation of HTC

Consider a heat exchanger employed as a test section for determining HTC. Two fluids ("h" and "c") flow on each side of the heat exchanger. A set of experiments is designed with the final goal of obtaining a HTC correlation, for which the inlet and outlet conditions of both fluids are measured. From the measured data, the heat transfer rate and the mean (or average) temperature difference (ΔT_m) in the heat exchanger can be determined. Thus, the product of the surface area and the overall heat transfer coefficient (UA), or the overall thermal resistance (R_{ov}), can be determined from the following equation:

$$\frac{1}{\text{UA}} = R_{ov} = \frac{\Delta T_m}{\dot{Q}} \tag{5.2}$$

The overall thermal resistance can be expressed as the sum of the thermal resistances corresponding to the hot and cold stream convections ($R_{c,h}, R_{c,c}$), the hot and cold side fouling films ($R_{f,h}, R_{f,c}$), and the heat exchanger wall (R_w):

$$R_{ov} = R_{c,h} + R_{f,h} + R_w + R_{f,c} + R_{c,c} = \frac{1}{(\eta hA)_h} + R_{f,h} + R_w + R_{f,c} + \frac{1}{(\eta hA)_c} \tag{5.3}$$

where η is the finned surface efficiency (equal to 1 if a surface side has no fins).

Wilson applied his method on a shell-and-tube condenser in order to obtain the HTC for water (fluid "c") flowing in turbulent flow inside plain circular tubes ($\eta_c = 1$ and $\eta_h = 1$) with condensing steam on the shell side (fluid "h"). For this problem, Eq. (5.3) can be rewritten as:

$$R_{ov} = \left(R_{c,h} + R_{f,h} + R_w + R_{f,c} \right) + \frac{1}{h_c A_c} \tag{5.4}$$

For fully developed turbulent flow inside a pipe, the following Nusselt number correlation for fluid "c" may be considered:

$$Nu_c = C_c \; Re_c^n \; Pr_c^{1/3} \left(\frac{\mu}{\mu_w} \right)_c^{0.14} \tag{5.5}$$

where $\left(\mu/\mu_w \right)_c^{0.14}$ accounts for the viscosity changes due to temperature differences between the bulk liquid and the wall; for gases, an absolute temperature function is used. By substitution of the Reynolds, Prandtl, and Nusselt numbers:

$$\left(\frac{hD}{k} \right)_c = C_c \left(\frac{\rho VD}{\mu} \right)_c^n \left(\frac{c_p \mu}{k} \right)_c^{1/3} \left(\frac{\mu}{\mu_w} \right)_c^{0.14} \tag{5.6}$$

where k is the fluid thermal conductivity, ρ is the fluid density, D is the pipe diameter, and V is the fluid velocity inside the pipe. In Eqs. (5.5) and (5.6), all properties are evaluated at the fluid bulk temperature, except μ_w which is evaluated at the wall temperature.

Originally, Wilson considered a number of assumptions when applying his method in order to obtain a HTC correlation for the tube side: (H1) negligible variations of fluid "c" properties during the different experiments; (H2) negligible effect of the $\left(\mu/\mu_w \right)_c^{0.14}$ correction term; (H3) constant thermal resistances in the wall (R_w) and in the shell side ($R_{c,h}$); (H4) negligible fouling resistances ($R_{f,h} = R_{f,c} = 0$); and (H5) a presumed value of $n = 0.82$, which was proposed after a trial and error procedure. By considering assumptions H1, H2, and H5, the HTC on fluid "c" can be expressed as:

$$h_c A_c = C_c \left(\frac{Ak}{D} \right)_c \left(\frac{\rho VD}{\mu} \right)_c^{0.82} \left(\frac{c_p \mu}{k} \right)_c^{1/3} = \frac{1}{m} V^{0.82} \tag{5.7}$$

After substitution into Eq. (5.2) and under assumptions H3 and H4, he obtained the following:

$$R_{ov} = \frac{m}{V^{0.82}} + b \tag{5.8}$$

This equation represents the equation of a straight line (in the form $y = mx + b$) of R_{ov} (equivalent to y) in terms of $1/V^{0.82}$ (equivalent to x). If R_{ov} versus $1/V^{0.82}$ is represented on a 2D plot, the slope (*m*) and intercept (*b*) can be determined, as shown in Figure 5.1. Once these constants are obtained, Eq. (5.7) can be used to determine the HTC correlation for fluid "*c*" as a function of velocity. Additionally, since $b = (R_w + R_{c,h})$, the value of $R_{c,h}$ (or h_h) can be determined if R_w is known.

The experimental data required to apply the original Wilson plot method may be troublesome to obtain because of the different assumptions of the method. In particular, it requires a constant bulk temperature in fluid "*c*" during the different experiments (related to the constant properties assumption H1). Wilson was aware of this issue and found convenient to apply correction factors in order to express results to a standard viscosity and a standard diameter. However, during the original work of Wilson, other property variations (density, conductivity, or specific heat) were negligible and were not taken into account.

In order to adequately handle transport bulk property variations during different experiments, all terms in Eq. (5.7) can be allowed to vary, so Eq. (5.8) is replaced by:

$$R_{ov} = \frac{m}{\left(kRe^n Pr^{1/3} A/D\right)_c} + b \tag{5.9}$$

If the value of *n* is presumed to be known (such as $n = 0.8$ for fully developed turbulent flow inside a circular tube), then Eq. (5.9) represents the equation of a straight line (in the form $y = mx + b$) of R_{ov} (equivalent to *y*) in terms of $(D/A)_c/(kRe^n Pr^{1/3})_c$ (equivalent to *x*). The two constants *m* (slope) and *b* (intercept) can be easily extracted by performing a simple regression analysis of the experimental data, so the method can be applied to determine the fluid "*c*" HTC.

For a general model of the form $y = f(a, x)$ that contains *v* parameters $a = (a_1, a_2, ..., a_v)$ to be determined from a set of *N* observations of *n* input variables $x = (x_1, x_2, ..., x_n)$, the parameters $a = (a_1, a_2, ..., a_v)$ can be obtained by minimizing the merit function:

$$\chi^2 = \sum_{i=1}^{N} \left(\frac{y_i - f(a, x)}{\sigma_i} \right)^2 \tag{5.10}$$

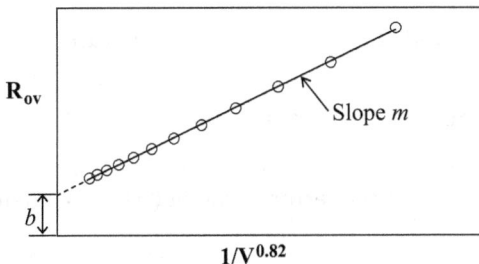

FIGURE 5.1
Original Wilson plot of Eq. (5.8).

where $1/\sigma_i^2$ are weighting factors that depend on the results of the fit and the uncertainties of the measured data.

The numerator of Eq. (5.10) is a measure of the spread of the model predictions with the measured data, and the denominator is a measure of the expected spread. For a moderately good fit, both values should be similar, i.e., $\chi^2 \sim N - v$ or $\chi_v^2 = \chi^2/(N-v) \sim 1$, since v parameters are in fact estimated from the data. If χ_v^2 is much larger than 1, then the discrepancies in the numerator of Eq. (5.10) are unlikely to be accidental fluctuations but probably indicate that the model is not appropriate or that the uncertainties have been underestimated. On the other hand, χ_v^2 values much lower than 1 might indicate an error in the assignment of the uncertainties of the measured data (overestimated).

In the case of a linear model $f = mx + b$, the previous equation takes the following form:

$$\chi^2 = \sum_{i=1}^{N} \frac{(y_i - mx_i - b)^2}{\sigma_i^2} \tag{5.11}$$

If the weighting factors are assumed to be all equal, $\sigma^2 = \sigma_i^2$, then minimizing Eq. (5.11) reduces to fitting data to a straight line by the least square method. In this case, the following expressions for m and b are obtained (Bevington and Robinson 2003, p. 105):

$$m = \frac{1}{\Delta}\left(N \sum_{i=1}^{N} x_i y_i - \sum_{i=1}^{N} x_i \sum_{i=1}^{N} y_i \right) \tag{5.12}$$

$$b = \frac{1}{\Delta}\left(\sum_{i=1}^{N} x_i^2 \sum_{i=1}^{N} y_i - \sum_{i=1}^{N} x_i \sum_{i=1}^{N} x_i y_i \right) \tag{5.13}$$

where $\Delta = \sum_{i=1}^{N} x_i^2 - \left(\sum_{i=1}^{N} x_i\right)^2$. Once the two constants are calculated, the parameter $C_c = 1/m$ of the fluid "c" side HTC correlation can be determined.

5.2.2 Standard Errors of the HTC Parameters

Once the parameters $a = (a_1, a_2, ..., a_v)$ that minimize the χ^2 given by Eq. (5.10) are obtained, the mean square error of the fit (σ^2) can be determined:

$$\sigma^2 = \frac{1}{N-v} \sum_{i=1}^{N} (y_i - f(a,x))^2 \tag{5.14}$$

According to Bevington and Robinson (2003) or Orear (1982), the uncertainties in the estimates of the parameters $a = (a_1, a_2, \ldots, a_v)$ are given by the following expression:

$$\left[u^2_{a_j a_k} \right] = \left(\alpha^{-1} \right)_{jk} \tag{5.15}$$

where the matrix elements α_{jk} are given by:

$$\alpha_{jk} = \sum_{i=1}^{N} \left[\frac{1}{\sigma_i^2} \left(\frac{\partial f}{\partial a_j} \right)_i \left(\frac{\partial f}{\partial a_k} \right)_i \right] \tag{5.16}$$

In the previous section, the merit function given by Eq. (5.11) was minimized considering that the uncertainties σ_i^2 for the experimental data points y_i were not used or were unknown. As pointed out by Bevington and Robinson (2003), in these cases it may be assumed that all measurements have equal standard uncertainties ($\sigma_i^2 = \sigma^2$) and are estimated from the experimental data and the results of the fit. The assumption may be satisfied if the uncertainties are instrumental and all the data are recorded with the same instrument and on the same scale. For the model given by Eq. (5.9), the parameters are $(a_1, a_2) = (b, m)$, so assuming $\sigma_i^2 = \sigma^2$, Eqs. (5.14)–(5.16) become:

$$\sigma^2 = \sigma_i^2 = \frac{1}{N-v} \sum_{i=1}^{N} (y_i - b - mx_i)^2 = \frac{1}{N-2} \sum_{i=1}^{N} \left(R_{ov_i} - b - \frac{m}{\left(k_i Re_i^n Pr_i^{1/3} A_i / D_i \right)_c} \right)^2 \tag{5.17}$$

$$\alpha = \frac{1}{\sigma^2} \begin{bmatrix} N & \sum_{i=1}^{N} x_i \\ \sum_{i=1}^{N} x_i & \sum_{i=1}^{N} x_i^2 \end{bmatrix} \tag{5.18}$$

$$u_b^2 = \frac{\sigma^2 \sum_{i=1}^{N} x_i^2}{N \sum_{i=1}^{N} x_i^2 - \left(\sum_{i=1}^{N} x_i \right)^2} \tag{5.19}$$

$$u_m^2 = \frac{\sigma^2 N}{N \sum_{i=1}^{N} x_i^2 - \left(\sum_{i=1}^{N} x_i \right)^2} \tag{5.20}$$

$$u_{b,m}^2 = u_{m,b}^2 = -\frac{\sigma^2 \sum_{i=1}^{N} x_i}{N \sum_{i=1}^{N} x_i^2 - \left(\sum_{i=1}^{N} x_i \right)^2} \tag{5.21}$$

The coefficient of correlation between the uncertainty in b and the uncertainty in m is given by:

$$r_{b,m} = r_{m,b} = \frac{u_{b,m}^2}{\sqrt{u_b^2 u_m^2}} \tag{5.22}$$

Finally, the uncertainty for C_c can be determined using the uncertainty propagation formula (JCGM 2008):

$$u^2(C_c) = u^2(m)\left(\frac{\partial(m)}{\partial(C_c)}\right)^2 = u^2(m)\left(-\frac{1}{C_c^2}\right)^2 \tag{5.23}$$

5.2.3 Experimental Procedure and Limitations

A set of experiments should be designed with the final goal of obtaining a HTC correlation. Consider that fluid "c" is the test fluid for which a HTC correlation is being determined. For fluid "h," the HTC may or may not be known. During these experiments, the mass flow rate of fluid "c" is changed systematically, but the operating conditions of fluid "h" (flow rate and average temperature) should remain nearly constant in order to have a constant thermal resistance on the fluid "h" side (in compliance with the Wilson plot assumption). From the measured data, the HTC in fluid "c" can be determined following the guidelines given in the previous section. In order to guarantee that the results are reliable, it would be appropriate to perform sets of experiments with different constant thermal resistance values on the "h" fluid side and compare the "c" fluid side HTC for the different sets.

As pointed out by Shah (1990), since the original Wilson plot method relies on a number of assumptions, it has some practical limitations:

- (L1) The flow rate and average temperature of fluid "h" (fluid for which the HTC is not determined) must be kept constant so that its thermal resistance ($R_{c,h}$) remains constant during the experiments.
- (L2) The Reynolds exponent for the HTC correlation in fluid "c" needs to be known in advance.
- (L3) The Nusselt correlation for fluid "c" must be expressed by an equation with only one unknown parameter (such as C_c in Eq. (5.5)). This implies that all test data for fluid "c" must be in one flow region with a known exponent for Re.
- (L4) Fluid property variations and the fin thermal resistance are not taken into consideration on fluid "c" side. Most property variations can be considered by representing the Wilson plot data in terms of $1/(kRe^nPr^{1/3})_c$; however, the method cannot account for a $(\mu/\mu_w)_c^{0.14}$ correction term.

- (L5) Fouling on both sides ("h" and "c") of the heat exchanger must remain constant during the experiments, so that the term b in Eq. (5.9) can be assumed constant.

Shah (1990) discusses how to relax all the limitations of the Wilson plot method except the one related to the one flow region for fluid "c" during the complete testing (i.e., L3). Limitation L5 can, in general, be assumed if low-fouling fluids are used.

As pointed out by Rose (2004), the validity of the Wilson plot method will also depend on the accuracy of the data, the number of test points, the number of parameters to be determined, and the range of the data relevant to the HTC correlation. The number of data points should be considerably larger than the number of parameters to be determined. Also, it would be recommended that the data points in the Wilson plot are uniformly distributed. For example, if the number of experimental tests performed for low Reynolds numbers is much higher than for high Reynolds numbers, then the low Reynolds range will have a higher contribution in the least square method and the obtained HTC correlation may not give adequate results for the high Reynolds range.

If experiments are designed such that the dominant thermal resistance is in the "c" fluid side (fluid for which a HTC correlation is being determined), then small variations in the "h" fluid side, wall, or fouling thermal resistances are not expected to affect considerably the HTC correlation parameters for fluid "c." In this case, no confidence can be given to the calculated value for parameter b, which is related to the assumed terms inherent to the constant thermal resistances in Eq. (5.9), including the fluid "h" thermal resistance. In some cases, it might be possible that due to the scatter in the experimental data (related to measurement imprecisions or the model assumptions), the calculated value of b is negative. If this is the case, either a zero value of b might be assumed, or the best estimate of b might be computed from previous existing correlations; accordingly, only the slope of the straight line in Eq. (5.9) needs to be determined. In contrast, if the relative value of the thermal resistance in fluid "h" approaches that in fluid "c," then a more reliable value for the "h" fluid thermal resistance (related to parameter b) can be obtained at the expense of a lower precision in the estimation of the "c" side HTC.

Finally, another limitation (L6) of the traditional Wilson plot method is that it is based on the (unweighted) least square method. As shown in Eq. (5.17), the calculated uncertainty of the fitted parameters only accounts for the standard error of the regression analysis and it does not consider the uncertainties of the measured experimental variables. During the experiments, the values of ΔT_m are expected to have always similar uncertainties, but the values of $\dot{Q}$ would probably be more accurate for those experiments with lower Re_c values (lower mass flow rates) owing to the larger temperature variation of the "c" fluid. As a result, according to Eq. (5.2), experiments with lower Re_c values will provide more accurate values of R_{ov}, and this should be somehow considered when obtaining the parameters of the HTC correlation. As shown later, this limitation can be relaxed by using a weighted least square method to fit the data.

5.3 Modified Wilson Plot Methods

Most of the inherent limitations in the Wilson plot method can be relaxed by using modified versions of the original method. In this section, alternative expressions of the original Wilson plot method are presented in order to overcome some of these limitations.

5.3.1 Reynolds Exponent Limitation

As mentioned in the previous section, the L2 limitation of the original Wilson plot method is that it requires the knowledge of the exponent of the Reynolds number. Although the assumed value of $n = 0.8$ would probably be a good choice for turbulent flow inside a smooth circular pipe, it may not be such a good choice for a shell side HTC or in the case of tubes with augmented surfaces or turbulators. In these cases, the value of this exponent is not known in advance but should be an outcome of the determined HTC correlation. Unfortunately, the appearance of the parameter n in the exponent of the Reynolds number makes the least square minimization problem to be nonlinear. An extensively adopted procedure for solving this numerical hitch consists in setting an *a priori* unknown value for the exponent of the Reynolds number. Then, the parameters m and b are determined from the Wilson plot of Eq. (5.9), as shown in Figure 5.2a. The assumed value of n can be checked *a posteriori* using a modification of the Wilson plot method (Briggs and Young 1969). It consists in taking logarithms on both sides of Eq. (5.9) to obtain Eq. (5.24).

$$\ln\left[\frac{m}{(R_{ov}-b)\left(kPr^{1/3}A/D\right)_c}\right]=n\ln\left(Re_c\right) \tag{5.24}$$

This equation represents the equation of a straight line (in the form $y = mx$) of $\ln[m/((R_{ov}-b)(AkPr^{1/3}/D)_c)]$ (equivalent to y) in terms of $\ln(Re_c)$ (equivalent to x), as shown in Figure 5.2b. The constant n (slope) can be easily obtained by performing a simple regression analysis of the experimental data, using the values of m and b determined from the regression analysis of Eq. (5.9). If the assumed and calculated values of n differ by more than some allowable error, a recalculation of the parameters is required until satisfactory convergence is obtained.

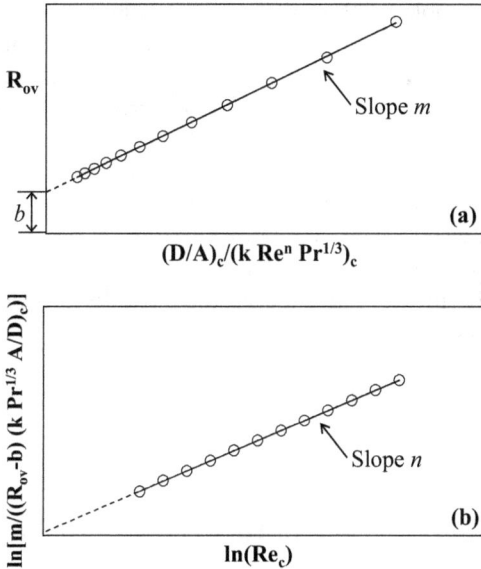

FIGURE 5.2
(a) Wilson plot and (b) modified Wilson plot of Eqs. (5.9) and (5.24).

After completing the iteration process and using the least square (LS) method to fit the data, the values for the parameters b, m (equal to $1/C_c$), and n are obtained. The uncertainties in the estimates of the parameters b and m may be computed using Eqs. (5.19) and (5.20) for the least square fit of Eq. (5.9), with the estimated value of n and considering $v = 3$. The uncertainty in the estimate of parameter n may be calculated with the following analogous formulae:

$$\sigma^2 = \sigma_i^2 = \frac{1}{N-v}\sum_{i=1}^{N}\left(y_i - nx_i\right)^2$$

$$= \frac{1}{N-3}\sum_{i=1}^{N}\left(\ln\left[\frac{m}{\left(R_{ov_i} - b\right)\left(k_i Pr_i^{1/3} A_i / D_i\right)_c}\right] - n\ln\left(Re_{c_i}\right)\right)^2 \quad (5.25)$$

$$u^2(n) = \frac{\sigma^2 N}{N\sum_{i=1}^{N} x_i^2 - \left(\sum_{i=1}^{N} x_i\right)^2} \quad (5.26)$$

Despite that, this widely adopted method is straightforward; it presents the drawback that the original nonlinear least square problem (NLS) of minimizing the merit function given by Eq. (5.10), with $f(b, m, n, x)$, has been replaced by two linear subproblems that are solved sequentially. As pointed out by Khartabil and Christensen (1992), the values obtained for the parameters b, m, and n are not necessarily the same. In some cases, the iterative scheme might converge to values of n that are evidently incorrect. Additionally, the uncertainties of the estimated parameters given by Eqs. (5.19), (5.20), and (5.26) are lower than the actual values, because they are based on independent linear fits.

Nowadays, the NLS can be solved using advanced computer algebra software such as MATLAB®, Mathematica, or Maple, which implements iterative algorithms to solve minimization problems. A good starting guess value for the model parameters is desirable in order to reduce convergence problems, which may be estimated from the results of the two linear fits. Once the model parameters are obtained, Eqs. (5.14)–(5.16) can be used to calculate the uncertainties of the estimated parameters. However, it should be clarified that Eqs. (5.14) and (5.15) are only strictly valid for linear models.

5.3.2 Variable Property Effects or Fin Resistance Limitations

If the fin resistance η_c and the variable viscosity effect $(\mu/\mu_w)_c^{0.14}$ (or temperature ratio in the case of gases) are taken into account in Eqs. (5.3) and (5.6), the following equation is obtained:

$$R_{ov} = \frac{m}{\left(\eta k Re^n Pr^{1/3}(\mu/\mu_w)_c^{0.14} A/D\right)_c} + b \qquad (5.27)$$

Once again, a linear relationship of R_{ov} is obtained as a function of the ratio $(D/A)_c/(\eta k Re^n Pr^{1/3}(\mu/\mu_w)^{0.14})_c$; then, the slope ($m$) and intercept ($b$) can be obtained from a regression analysis of the experimental data. However, the values of η_c and $(\mu/\mu_w)_c^{0.14}$ in Eq. (5.27) depend on the model parameters m and b, so actually a nonlinear regression analysis procedure must be used. In the open literature, the nonlinear regression analysis is usually tackled by successive linear regressions in a trial and error procedure, similar to the following iterative scheme:

1. Assume a value for n, the exponent of the Reynolds number. For a circular tube, this might not be required and a fixed value of $n = 0.8$ might be considered
2. For the first iteration, assume $\eta_c = 1$ and $\mu/\mu_w = 1$
3. Calculate initial estimates for parameters m (which allows to calculate C_c) and b from a Wilson plot of Eq. (5.27)
4. With the value of b, use Eq. (5.3) to calculate the fluid "c" side wall temperature for each (ith) experimental data point based on the "h" and "c" sides' mean temperatures, i.e., $(T_{hm_i} - T_{w_i})/b = (T_{hm_i} - T_{cm_i})/R_{ov_i}$. Then, calculate the viscosity correction term for each data point.

5. Calculate h_{ci} from Eq. (5.6) for each (ith) experimental data point. Then, calculate the fin efficiency and the extended surface efficiency (η_{ci}) using the total (A_c) and finned (A_{fc}) surface areas

6. Obtain new values for parameters m and b from a Wilson plot of Eq. (5.27)

7. Iterate steps 4–6 until the m and b values converge

8. Obtain a new value for n from the modified Wilson plot of Eq. (5.28)

$$\ln\left[\frac{m}{(R_{ov}-b)\left(\eta kPr^{1/3}\left(\mu/\mu_w\right)^{0.14}A/D\right)_c}\right]=n\ln(Re_c) \qquad (5.28)$$

9. Iterate steps 3–8 until the n value converges

As explained previously, this approach does not guarantee that the χ^2 function given by Eq. (5.10) is minimized. It would be preferable to solve directly the NLS in order to find the values of the parameters C_c, b, and n that minimize χ^2.

5.3.3 Constant Thermal Resistance Limitation

The L1 assumption of a constant thermal resistance for one of the fluids might limit the range of data that can be tested in the experimental equipment. A modification of the Wilson plot method can be introduced to avoid this limitation and allow flow rate or average temperature variations on both fluids.

Briggs and Young (1969) applied a modified Wilson plot technique for no phase changes in a shell-and-tube heat exchanger. They considered a Nusselt number correlation such as Eq. (5.5) for the shell and tube sides. The exponent of the Reynolds number was assumed to be equal to 0.8 for the tube side ("c" fluid in the sequel), but unknown for the shell side ("h" fluid). Then, the overall heat transfer resistance was given by:

$$R_{ov}=\left(R_{f,h}+R_w+R_{f,c}\right)+\frac{1}{C_c\left(kRe^{0.8}Pr^{1/3}\left(\mu/\mu_w\right)^{0.14}A/D\right)_c}$$

$$+\frac{1}{C_h\left(kRe^{p}Pr^{1/3}\left(\mu/\mu_w\right)^{0.14}A/D\right)_h} \qquad (5.29)$$

If the constant C_c is known, the previous equation can be manipulated to obtain a Wilson plot function as follows:

$$R_{ov} - \frac{1}{C_c \left(kRe^{0.8} Pr^{1/3} \left(\mu / \mu_w \right)^{0.14} A / D \right)_c} = b + \frac{m}{\left(kRe^p Pr^{1/3} \left(\mu / \mu_w \right)^{0.14} A / D \right)_h}$$

$$(5.30)$$

where $b = R_{f,h} + R_w + R_{f,c}$.

This formulation allows considering test data for varying flow rates and average temperatures on both fluids. A NLS problem needs to be solved in order to determine the parameters b, m (equal to $1/C_h$), and p. As an alternative, the nonlinear analysis may be executed by assuming a value for the exponent p and using successive linear regressions in a trial and error procedure. During the iteration process, the estimated values of b, m, and p are used to evaluate the HTCs and subsequently the wall temperatures required for the viscosity ratio terms.

If R_w is known and $R_{f,h}$ and $R_{f,c}$ are also known or neglected, then the constant C_c can be determined together with C_h and p. In this case, Eq. (5.29) is rewritten as follows:

$$\left(R_{ov} - R_w \right) \left(kRe^{0.8} Pr^{1/3} \left(\mu / \mu_w \right)^{0.14} A / D \right)_c = b + m \frac{\left(kRe^{0.8} Pr^{1/3} \left(\mu / \mu_w \right)^{0.14} A / D \right)_c}{\left(kRe^p Pr^{1/3} \left(\mu / \mu_w \right)^{0.14} A / D \right)_h}$$

$$(5.31)$$

where $b = 1/C_c$ and $m = 1/C_h$.

Eq. (5.31) has three unknowns, b, m, and p, so it should be solved by an iterative scheme. Again, it is not essential to obtain test data for constant flow rate or constant average temperature values on either fluid. Nevertheless, even in case the flow rate and average temperature remain constant on the "c" fluid side, $\left(\mu / \mu_w \right)_c$ variations will still be obtained during the experiments.

5.3.4 Experimental Procedure and Limitations

The experimental procedure is similar to the one described for the Wilson plot method, though limitations L1, L2, and L4 do no longer apply since they can be relaxed by applying the modified Wilson plot methods described previously. An advantage of the Wilson plot method is that it yields a HTC correlation in a straightforward manner, whereas the modified Wilson plot methods require iterative schemes.

Similar comments apply regarding the relative values of the different thermal resistances. For an accurate calculation of a HTC correlation, it would be desirable that the dominant thermal resistance is in the fluid for which the HTC correlation is being determined.

If parameters on both fluids are to be determined, such as in Eq. (5.31), different data sets may be needed for higher precision. One set with dominant "h" fluid thermal resistance data can be used in order to calculate C_c.

A second set with dominant "c" fluid thermal resistance data will be appropriate to determine the parameters C_h and p.

In the more general case, the five constants (C_c, n, C_h, p, and $R_{f,h} + R_w + R_{f,c}$) in Eq. (5.29) could be determined; Khartabil et al. (1988) and Shah (1990) analyzed the required data sets and the solution procedure.

If all the test data are not in the same flow regime (limitation L3), Eq. (5.5) is no longer a proper correlating function since the parameters C_c and n change with the Reynolds number. Briggs and Young (1969) proposed to modify Eq. (5.5) by making the parameters a function of the Reynolds number. An alternative approach consists in considering a linear (or nonlinear) interpolation of Nusselt correlations for the laminar and turbulent flow (Taborek 1990). Churchill and Usagi (1972) proposed the following blending equation for correlating experimental data in the laminar and turbulent flow regimes:

$$Nu = \left(Nu_{\text{lam}}^q + Nu_{\text{tur}}^q \right)^{1/q} \qquad (5.32)$$

The value of the parameter q should be an output of the HTC correlation together with additional parameters for the Nu_{lam} and Nu_{tur} expressions. As it can be seen, the problem becomes nonlinear and the number of parameters to be determined increases.

5.4 Weighted Fits

The L6 limitation of the Wilson plot method was related to the fact that the uncertainties of the experimental data (y_i and x_i) were not taken into account. In order to overthrow this limitation, a weighted least square method (WLS) can be used. For the weighted curve fitting, each data point is weighted inversely proportional to the square of the experimental uncertainty.

Consider the general model $y = f(a, x)$ used for the merit function given in Eq. (5.10). Now, the weighting factors σ_i are estimated considering the uncertainties of the experimental data u_{y_i} and the combined uncertainty $u_{cf(x_{1i},x_{2i},...,x_{ni})}$ on the estimate $f_{x_{1i},x_{2i},...,x_{ni}}$ based on the uncertainties of the input variables ($x_{1i}, x_{2i}, ..., x_{ni}$).

$$\sigma_i^2 = u_{y_i}^2 + u_{cf(x_{1i},x_{2i},...,x_{ni})}^2 \qquad (5.33)$$

For uncorrelated input quantities, the combined uncertainty can be calculated by the following expression (JCGM 2008):

$$u_{cf(x_1,x_2,...,x_n)}^2 = \sum_{j=1}^{n} \left(\frac{\partial f}{\partial x_j} \right)^2 u_{x_j}^2 \qquad (5.34)$$

The uncertainties in the estimates of the fitted parameters $a = (a_1, a_2, ..., a_v)$ can now be determined with Eqs. (5.15) and (5.16), but using the actual σ_i^2 values obtained from Eq. (5.33). For more details regarding the assumptions and limitations of the method, the references cited in this chapter are recommended.

As an example, consider the initial Wilson model given by Eq. (5.9). The number of input variables is $n = 1$ (x), and the number of parameters to be determined is $v = 2$ ($a_1 = m$ and $a_2 = b$). The merit function to be minimized is:

$$\chi^2 = \sum_{i=1}^{N} \frac{(y_i - mx_i - b)^2}{u_{y_i}^2 + m^2 u_{x_i}^2} \tag{5.35}$$

Unfortunately, the appearance of the parameter m in the denominator of Eq. (5.35) makes the minimization problem to be nonlinear, so a nonlinear iterative solution scheme should be implemented. Initial guess values for the parameters (m and b) can be obtained from a Wilson plot by assuming an unweighted least square regression method.

5.5 General Wilson Plot

Rose (2004) reformulated the Wilson plot method as a function of the temperature difference between the hot and cold fluids through the heat exchanger (overall temperature difference), rather than as a function of the overall thermal resistance. If all thermal resistances but $R_{c,c}$ are considered to be constant, then Eq. (5.9) may be entirely reformulated into Eq. (5.36).

$$\Delta T_m = \left(\Delta T_{c,h} + \Delta T_{f,h} + \Delta T_w + \Delta T_{f,c}\right) + \Delta T_{c,c} = b\dot{Q} + \frac{m}{\left(kRe^n Pr^{1/3} A / D\right)_c}\dot{Q} \tag{5.36}$$

Even if the exponent of the Reynolds number is assumed (e.g., $n = 0.8$), one disadvantage of this method is that Eq. (5.36) does not represent the equation of a straight line, so the data can no longer be represented as a straight line on a 2D plot. However, if n is known, Eq. (5.36) becomes linear with respect to b and m; thereby, the parameters b and m can be determined by a simple LS curve fit to a polynomial equation (Bevington and Robinson 2003). Equation (5.36) has the advantage that the uncertainties of ΔT_m (quantity whose residuals are minimized) are generally expected to be approximately equal during the experimental tests. As a result, the standard LS method (without weights) will probably yield similar results than a weighted minimization method, avoiding the need of an iteration process.

If the exponent of the Reynolds number is not known, then Eq. (5.36) becomes nonlinear with respect to b, m, and n. As suggested previously, the nonlinear regression problem can be divided into two simple subproblems that involve linear curve fittings. By setting *a priori* unknown value for n, Eq. (5.36) becomes linear and the parameters b and m can be easily determined. The assumed n value can be checked *a posteriori* by means of a linear regression of Eq. (5.37), using the previous values of b and m. As it happened with the modified Wilson plot methods, an adequate iteration scheme should be implemented. Nevertheless, it should be noticed that the second linear fit of Eq. (5.37) does no longer have the advantage that the quantities whose residuals are minimized have nearly equal uncertainties.

$$\ln\left[\left(\frac{\Delta T_m}{\dot{Q}}-b\right)\frac{\left(kPr^{1/3}A/D\right)_c}{m}\right]=n\ln\left(\frac{1}{Re_c}\right) \qquad (5.37)$$

More complex iterative schemes may be implemented in order to consider the fin resistance η_c for extended surfaces or (μ/μ_w) variations.

5.6 Example Application

5.6.1 Data

Consider that the HTC for internal forced convection inside a tube with turbulators is to be determined applying the Wilson plot method. Water flows inside ("c" fluid) the tube and also on the outer surface ("h" fluid) of the tube, in a counterflow pattern. The HTC on the outer surface of the tube is not known.

Table 5.1 contains reduced data of inlet and exit temperature measurements for both fluids and mass flow measurements for fluid "c," where the subscripts i and e denote the inlet and exit section, respectively. Two sets of experiments are considered, each with the same equipment but different constant values of the "h" fluid flow rate. A copper tube of 10.2 mm inner diameter, 12.7 mm outer diameter, and 1 m length is considered. Each data set is assumed to be controlled in such a way that the "h" fluid mass flow rate and average temperature remained nearly constant, so that the "h" fluid side thermal resistance for each set is expected to be also nearly constant.

The accuracies for the temperature measurements and temperature differences are considered to be 0.1 K. For simplicity, possible inaccuracies associated with heat losses to the surroundings, flow rate measurements, geometric dimensions, or thermodynamic properties are not considered.

TABLE 5.1

Data to Determine the "c" Fluid Side HTC

$M_c \times 10^3$ (kg/s)	T_{hi} (°C)	T_{he} (°C)	T_{ci} (°C)	T_{ce} (°C)	$M_c \times 10^3$ (kg/s)	T_{hi} (°C)	T_{he} (°C)	T_{ci} (°C)	T_{ce} (°C)
Data set n°1									
44.03	33.22	32.85	20.02	27.62	207.2	33.63	32.40	20.02	25.82
56.97	33.27	32.82	20.05	27.38	268.5	33.76	32.28	20.06	25.42
73.72	33.28	32.78	20.00	27.10	347.9	33.91	32.19	20.05	25.03
95.42	33.34	32.72	20.05	26.82	451.0	34.08	32.00	20.00	24.67
123.5	33.45	32.62	20.00	26.50	584.7	34.28	31.79	20.01	24.21
160.0	33.51	32.54	20.01	26.19	758.2	34.47	31.64	20.02	23.76
Data set n°2									
44.12	33.37	32.69	20.01	27.47	208.5	34.16	31.93	20.06	25.27
57.11	33.45	32.59	20.01	27.11	270.3	34.37	31.75	20.00	24.87
73.95	33.52	32.53	20.00	26.80	350.6	34.58	31.50	20.06	24.44
95.78	33.66	32.39	20.05	26.46	454.7	34.82	31.22	20.06	23.96
124.1	33.75	32.31	20.01	26.10	589.8	35.09	30.98	20.03	23.47
160.8	33.95	32.08	20.02	25.70	764.9	35.36	30.75	20.04	23.06

5.6.2 HTC Calculation

From the given data, the heat transfer rate, the logarithmic mean tempera-ture difference, and the overall thermal resistance are determined. All ther-mal and transport properties are calculated with the code REFPROP 9.0 (Lemmon et al. 2010).

A correlation for the "c" fluid side of the type given by Eq. (5.5) is to be determined. However, since small temperature differences are observed in the data, the viscosity correction term $(\mu / \mu_w)_c^{0.14}$ is neglected.

Initially, a value of $n = 0.8$ is assumed for the exponent of the Reynolds number, so Eq. (5.9) is obtained. Thereafter, the Wilson plot method is applied to determine the "c" fluid side HTC.

As turbulators are used inside the tube, the value of $n = 0.8$ for the Reynolds exponent may not be the best choice, so modified Wilson plot methods are also applied in order to determine a better value for the exponent of the Reynolds number. The "c" fluid side HTC is obtained by using different meth-ods to solve the nonlinear problem of determining the constants m, b, and n in Eq. (5.9): splitting the nonlinear problem into the two linear subproblems given by Eqs. (5.9) and (5.24); solving directly the NLS problem in Eq. (5.9); applying a WLS fit to Eq. (5.9); and applying the general Wilson plot method expressed by Eq. (5.36). The results of all these cases are collected in Table 5.2.

5.6.2.1 Wilson Plot Method

Figure 5.3 shows the values of R_{ov} represented as a function of the ratio $(D/A)_c/(kRe^{0.8}Pr^{1/3})_c$ for both sets of data. It can be seen that a linear dependence

TABLE 5.2

Values of Best Fit Parameters and Uncertainties Using Different Methods for the Data in Table 5.1

Method	Set $n°$		Estimate and Uncertainty	σ^2
1: WM-LS, Eq. (5.9)	1	m	22.97(15)	1.49×10^{-9}
		b	$2.76(19) \times 10^{-4}$	
		C_c	0.04353(29)	
	2	m	22.924(98)	5.87×10^{-10}
		b	$5.49(12) \times 10^{-4}$	
		C_c	0.04362(18)	
2: MWM-LS, Eqs. (5.9) and (5.24)	1	m	37.703(82)	1.53×10^{-10}
		b	$4.236(61) \times 10^{-4}$	
		n	0.85803(29)	
		C_c	0.026523(57)	
	2	m	30.708(91)	2.85×10^{-10}
		b	$6.383(84) \times 10^{-4}$	
		n	0.83429(23)	
		C_c	0.032564(97)	
3: MWM-NLS, Eq. (5.9)	1	m	38.7(21)	1.49×10^{-10}
		b	$4.31(16) \times 10^{-4}$	
		n	0.8611(64)	
		C_c	0.0258(14)	
	2	m	29.6(22)	2.78×10^{-10}
		b	$6.28(23) \times 10^{-4}$	
		n	0.8302(87)	
		C_c	0.0337(25)	
4: MWM-WLS, Eq. (5.9)	1	m	37.2(45)	1.59×10^{-10}
		b	$4.21(28) \times 10^{-4}$	
		n	0.856(13)	
		C_c	0.0269(33)	
	2	m	31.2(46)	2.93×10^{-10}
		b	$6.40(39) \times 10^{-4}$	
		n	0.836(17)	
		C_c	0.0321(47)	
5: GWM-NLS, Eq. (5.36)	1	m	36.4(34)	5.79×10^{-3}
		b	$4.17(18) \times 10^{-4}$	
		n	0.854(10)	
		C_c	0.0275(26)	
	2	m	32.1(25)	3.33×10^{-3}
		b	$6.48(16) \times 10^{-4}$	
		n	0.8395(87)	
		C_c	0.0311(24)	

WM, Wilson plot method; MWM, modified Wilson plot method; GWM, general Wilson plot method; LS, linear least square; NLS, nonlinear least square; and WLS, weighted least square.

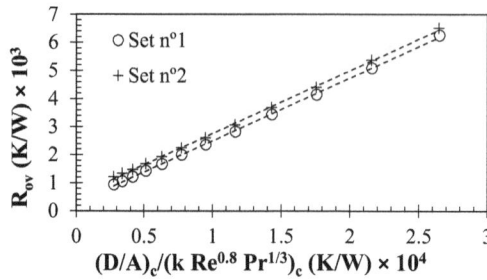

FIGURE 5.3
Wilson plots of Eq. (5.9).

exists between the overall thermal resistance and the ratio $(D/A)_c/(kRe^{0.8}Pr^{1/3})_c$, so the constants ($m$ and b) in Eq. (5.9) were estimated by a simple regression analysis under the Wilson plot method assumptions. The results collected in Table 5.2 (method 1: WM-LS) were obtained for the trio of parameters m, b, and C_c for the two data sets. These values were determined by means of a regression analysis, using the least square method. In the table, the numbers in parentheses indicate the numerical values of the standard uncertainties of the parameters (with a coverage factor $k = 1$) referred to the corresponding last digits of the quoted result (JCGM 2008); for instance, for the results 22.97(15) and 22.924(98) the uncertainties are 0.15 and 0.098, respectively. The uncertainties were determined using Eqs. (5.19) and (5.20). The mean square errors (σ^2) of the fits are also included in Table 5.2.

It is clear that the Wilson plot method predicts different values of the parameter b for the two sets of data ($n°1$ and $n°2$). Parameter b is related to the assumed terms inherent to the constant thermal resistances in Eq. (5.9). The two sets of data considered different flow rate values for fluid "h," so different thermal resistances are expected. On the other hand, the calculated parameters C_c of the HTC correlation are very similar for both sets of data. Focusing on the uncertainties of the parameters calculated from the result of the fits, it is seen that the relative uncertainties for b and C_c are 6.9% and 0.7%, respectively, for the first set of data, and 2.2% and 0.4%, respectively, for the second one.

Results in Figure 5.3 and Table 5.2 show that for most data points, the dominant thermal resistance is in the "c" fluid side. Depending on the experiment, it represents 57%–94% of the total thermal resistance for the data set $n°1$ and 45%–90% for the data set $n°2$. This explains the higher uncertainties obtained for parameter b. As a result, if there is any scatter in the experimental data or if any of the assumptions of the model are not fully satisfied, slight variations in the estimation of parameter C_c might be obtained, but the estimation of parameter b will change considerably. Then, the method gives better results when used for estimating the HTC of the fluid side with the higher thermal resistance and, in general, no confidence can be given to the calculated value of parameter b.

5.6.2.2 Modified Wilson Plot Methods

In the previous section, the exponent of the Reynolds number was assumed to be known and equal to 0.8. A modified Wilson plot method based on the two linear subproblems given by Eqs. (5.9) and (5.24) was solved in order to determine a better value for the exponent of the "c" fluid side Reynolds number.

Shown in Figure 5.4 are the Wilson and modified Wilson plots after completing the iteration process. It is palpable that a linear relationship exists between the experimental data for both figures, so the iteration procedure described in section 5.3.1 was implemented. After completing the iteration process and using the LS method to fit the data, the values shown in Table 5.2 for method 2 were obtained. The given uncertainties in the estimates of the parameters were computed using analogous equations to Eqs. (5.17)–(5.21) but considering $v = 3$. It can be seen that the σ^2 values have been reduced by a twofold factor with respect to method 1, even though the number of degrees of freedom has decreased from 10 (in method 1) to 9 (in method 2). However, the parameters (m, b, and C_c) calculated using method 2 for both sets of data ($n°1$ and $n°2$) are different, and these differences are not explained by their uncertainty values. Then, if an accurate HTC correlation is to be determined, it would be desirable that the same HTC parameters are obtained for different sets of test data or that the differences are within the uncertainties of the estimated parameters. In the present case, the differences between the values of the estimated parameters n and C_c for both sets of data are higher than their standard uncertainties. This is indicative that these results are not fully reliable.

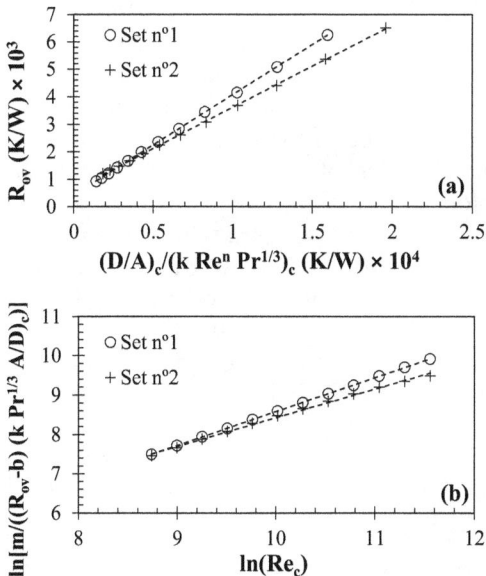

FIGURE 5.4
(a) Wilson plots and (b) modified Wilson plots of Eqs. (5.9) and (5.24).

A recalculation of the parameters C_c, b, and n was performed by solving the nonlinear least square problem (NLS) using the Levenberg–Marquardt algorithm (1963). The new values obtained for these parameters are also listed in Table 5.2 (method 3: MWM-NLS). It can be seen that the mean square errors (σ^2) of the fits have been reduced compared to the values obtained by the two linear fits. Also, the estimated values of the parameters are different than those obtained using the previous method. An important result that emerges is that the standard uncertainties of the parameters n and C_c are much higher than before; in numbers, the uncertainty values have increased by one order of magnitude in the former and by two orders of magnitude in the latter. As a side effect, it is seen that the discrepancies between the results of the fitted parameters n and C_c are closer to their standard uncertainties.

5.6.2.3 Weighted Wilson Plot Method

A recalculation of the parameters m, b, n, and C_c was performed by applying a weighted fit to the modified Wilson plot data given by Eq. (5.9). The nonlinear weighted least square problem (WLS) was solved using the Levenberg–Marquardt algorithm (1963). The new values obtained for the parameters m, b, and C_c are listed in Table 5.2 (method 4: MWM-WLS). In solving the WLS problem, the uncertainties for the experimental data x_i and y_i were determined from the accuracy specifications of the temperature measurements. It was considered that the quoted accuracies provide symmetric bounds ($\pm a$) within which all values of the quantity are assumed to lie and that all values within those bounds are equally probable (uniform distribution), so that the standard uncertainties are equal to $a / \sqrt{3}$. As remarked in JCGM (2008), other distributions may be assumed in other cases.

Results in Table 5.2 show that the estimated parameters for the WLS fit (method 4) are somehow similar to those obtained for the NLS fit (method 3). However, the uncertainties of the estimated parameters have increased by around a twofold factor. Now, the differences of the estimated C_c and n parameters for the two data sets are within the parameters' uncertainty values. In this case, these results will give the experimenter more confidence on the validity of the experimental method and the assumptions considered in the application of the Wilson plot method.

It should be clarified that the σ^2 values obtained for method 4 are slightly higher than for method 3. This is because the WLS method minimizes χ^2 and not σ^2, since different weights are given to the different data points in Eq. (5.10). The $\chi_v^2 = \chi^2/(N-v)$ values for the two sets of data points ($n°1$ and $n°2$) are 0.41 and 0.26, respectively (i.e., on the order of 1).

5.6.2.4 General Wilson Plot Method

The parameters m, b, and C_c were estimated for a modified version of Eq. (5.36) that also includes the exponent of the Reynolds number. The results of this fit are included in Table 5.2 (method 5: GWM-NLS) for the two sets of data points ($n°1$ and $n°2$). Once again, the differences of the estimated C_c and n parameters for the two data sets are within their uncertainty values.

5.6.3 Discussion

The previous results assert the suitability of correlating convection HTC by properly accounting for the quality of the experimental data, rather than using simplified methods based on unweighted linear regression analyses.

In the example considered here, the estimated parameters using method 5 were quite similar to the values obtained using method 4. This is somehow expected because only temperature measurement inaccuracies were considered and they were assumed to be constant for each data point. If other measurement errors are considered (i.e., flow rate measurements), then the results of the general Wilson plot method and a weighted fit method would probably be more different.

Once the parameters of a HTC correlation have been determined, it is possible to calculate the expected values of HTC at any Reynolds number. Figure 5.5a shows the HTC via the Colburn factor (j_H) calculated for the WLS fits of the two data sets. The Colburn factor was selected rather than the Nusselt number (or HTC) because it accounts for the small variations in the Prandtl number of the different data points in Table 5.1.

The $1 - \alpha$ confidence bands for the expected response can be estimated by the following expression (Bates and Watts 1988):

$$f_{j_H}(Re, a) \pm \sqrt{\beta^T \alpha^{-1} \beta} \sqrt{vF(v, N - v; \alpha)} \tag{5.38}$$

where $F(v, N - v, \alpha)$ is the upper α quantile for Fisher's F-distribution with v and $N - v$ degrees of freedom and β is a derivate vector. The components ($i = 1, 2, ..., v$) of β are calculated according to Bates and Watts (1988):

$$\beta_i = \frac{\partial f_{j_H}(Re, a)}{\partial a_i} \tag{5.39}$$

Figure 5.5a also shows the approximate 95% confidence bands of the respective HTC correlations. It can be seen that the j_H predictions are similar for both data sets. The fluid "h" side thermal resistance is higher for the data set $n°2$ than for the data set $n°1$; this explains the wider confidence bands of the HTC correlation for the data set $n°2$.

A reader may notice that the uncertainty values in Table 5.2 and the confidence bands in Figure 5.5a are probably higher than what will be desirable. Both can be reduced by using more accurate instrumentation and a larger number of data points; in fact, only 12 data points were considered for a

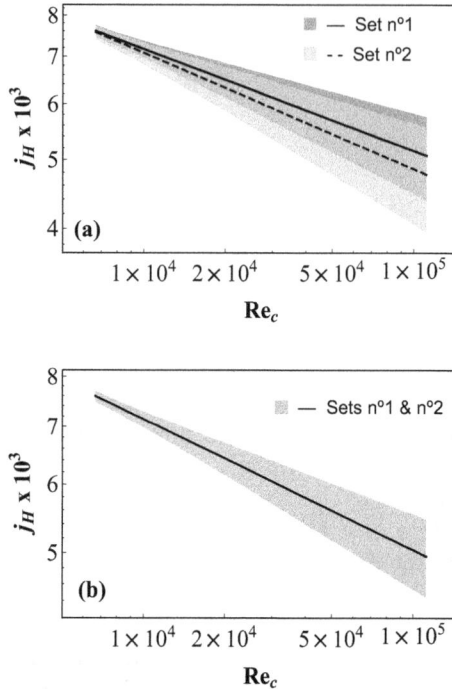

FIGURE 5.5
Fitted Colburn vs. Reynolds number and 95% confidence bands (gray regions) for (a) each data set and (b) both data sets combined.

three-parameter regression model. Figure 5.5b shows that in case all the data points from the two data sets are used for a unique HTC correlation, the confidence bands are reduced.

The confidence bands and the uncertainties calculated previously are only strictly valid for linear models and Gaussian distributions of the errors of the measured data. If this is not the case, sampling techniques such as the Monte Carlo method might be used as an alternative approach (Sieres and Campo 2018).

References

Bates, D.M., and D.G. Watts. 1988. *Nonlinear Regression Analysis and Its Applications.* New York: John Wiley & Sons.

Bevington, P.R., and D.K. Robinson. 2003. *Data Reduction and Error Analysis for the Physical Sciences*, 3rd ed. New York: McGraw-Hill.

Briggs, D.E., and E.H. Young. 1969. Modified Wilson plot techniques for obtaining heat transfer correlations for shell and tube heat exchangers. *Chemical Engineering Progress Symposium Series* 65(92): 35–45.

Churchill, S.W., and R. Usagi. 1972. A general expression for the correlation of rates of transfer and other phenomena. *AIChE Journal* 18(6): 1121–1128.

Fernández-Seara, J., F.J. Uhía, J. Sieres, and A. Campo. 2007. A general review of the Wilson plot method and its modifications to determine convection coefficients in heat exchange devices. *Applied Thermal Engineering* 27(17–18): 2745–2757.

JCGM. 2008. *Evaluation of Measurement Data: Guide to the Expression of Uncertainty in Measurement JCGM 100:2008 (GUM 1995 With Minor Corrections)*. Paris: BIPM Joint Committee for Guides in Metrology.

Khartabil, H.F., and R.N. Christensen. 1992. An improved scheme for determining heat transfer correlations from heat exchanger regression models with three unknowns. *Experimental Thermal and Fluid Science* 5(6): 808–819.

Khartabil, H.F., R.N. Christensen, and D.E. Richards. 1988. A modified Wilson plot technique for determining heat transfer correlations. *Second UK National Conferences on Heat Transfer* 2: 1331–1357.

Lemmon, E.W., M.L. Huber, and M.O. McLinden. 2010. *NIST Standard Reference Database 23: Reference Fluid Thermodynamic and Transport Properties-REFPROP v. 9.0*. Gaithersburg, MD: National Institute of Standards and Technology (NIST), Standard Reference Data Program.

Marquardt, D.W. 1963. An algorithm for least-squares estimation of nonlinear parameters. *SIAM Journal of the Society for Industrial and Applied Mathematics* 11(2): 431–441.

Orear, J. 1982. Least squares when both variables have uncertainties. *American Journal of Physics* 50(10): 912–916.

Rose, J.W. 2004. Heat-transfer coefficients, Wilson plots and accuracy of thermal measurements. *Experimental Thermal and Fluid Science* 28(2–3): 77–86.

Shah, R.K. 1990. Assessment of modified Wilson plot techniques for obtaining heat exchanger design data. *9th International Heat Transfer Conference*, Jerusalem, Israel, 51–56.

Sieres, J., and A. Campo. 2018. Uncertainty analysis for the experimental estimation of heat transfer correlations combining the Wilson plot method and the Monte Carlo technique. *International Journal of Thermal Sciences* 129: 309–319.

Taborek, J. 1990. Design method for tube-side laminar and transition flow regime with effects of natural convection. *9th International Heat Transfer Conference*, Jerusalem, Israel, Paper OPF-11-21.

Uhía, F.J., A. Campo, and J. Fernández-Seara. 2013. Uncertainty analysis for experimental heat transfer data obtained by the Wilson plot method: Application to condensation on horizontal plain tubes. *Thermal Science* 17(2): 471–487.

van Rooyen, E., M. Christians, and J.R. Thome. 2012. Modified Wilson plots for enhanced heat transfer experiments: Current status and future perspectives. *Heat Transfer Engineering* 33(4–5): 342–355.

Wilson, E.E. 1915. A basis for rational design of heat transfer apparatus. *Transactions of the ASME* 37: 47–82.

6

Test Sections for Heat Transfer and
Pressure Drop Measurements: Construction,
Calibration, and Validation

Marilize Everts and Josua P. Meyer

University of Pretoria

CONTENTS

6.1 Introduction

Suitable and accurate experimental data with known uncertainties are of the utmost importance, not only to develop new correlations and to investigate different phenomena, but also to validate and compare the results of numerical and analytical simulations and to prove and explain the fundamentals. Many thorough textbooks on measurement including (but not limited to) those of Dunn (2010), Kutz (2013), Raghavendra and Krishnamurthy (2013), LaNasa and Upp (2014), and Morris and Langari (2016) are available. These books typically cover the principles and theory of measurement, different instruments and sensors, data acquisition systems, and calibration. Furthermore, a large amount of literature on uncertainty analyses, such as the textbook of Dieck (2017) and articles of Kline and McClintock (1953) and Moffat (1988), is also available. When using the keywords "thermocouple calibration procedure" and "pressure transducer calibration procedure" in the field of engineering on Scopus in February 2019, a total of 151 and 148 documents, respectively, were found. However, for the keyword "thermocouple attachment procedure", the number of documents significantly decreased to 4 and none of them specifically focused on attachment techniques. Therefore, many literatures are available on how to calibrate and use measuring instrumentation, but very limited information is available in the literature on how to attach it to the test section in such a way that the accuracy of the results will not be affected.

Although the accuracy of measurements is mainly determined by the accuracy of the measurement equipment, the importance of a well-built, well-calibrated, and well-validated test section should not be overlooked. Unfortunately, almost no information is available in the literature on how to build a test section for accurate heat transfer and pressure drop measurements. Institutional memories and practices are being transferred from one experimenter to another in laboratories or at conferences. However, in many cases, laboratory personnel and students leave without transferring the acquired knowledge. The result is that a new knowledge base then needs to be developed again, which consumes time that could be spent on experiments or data analyses.

The purpose of this chapter is therefore to present the institutional memory that was acquired over 17 years at the University of Pretoria in the form of a guideline that can be used by undergraduate and postgraduate students and professional researchers who need to conduct heat transfer and pressure drop experiments. This chapter specifically focuses on how to build a suitable test section with pressure transducers and thermocouples, as well as how to calibrate pressure transducers, Pt100 probes, and thermocouples. Uncertainties are unavoidable, and any honest experimenter must present the experimental results with an uncertainty analysis. An overview of the uncertainty analysis method, the equations to calculate the uncertainties of the most significant heat transfer and pressure drop variables, and some basic examples are therefore included in this chapter. However, the mathematical content and theory were deliberately kept to the minimum required to understand the basic uncertainty analysis method. Furthermore, because a validation study is essential to have confidence in the accuracy and reliability of the results, suitable correlations that can be used as a guideline have been tabulated. The chapter ends with some interesting challenges that we have encountered over the past 17 years and still need to be explained.

6.2 Our Range and Experience

Over the past 17 years, heat transfer and pressure drop experiments were conducted on a wide range of test sections. The experimental setups were housed in the Clean Energy Research Group laboratory at the University of Pretoria. Table 6.1 contains a summary of the variables and ranges of the parameters in the experiments that were conducted.

TABLE 6.1

Variables and Ranges of the Experiments That Were Conducted

Variable	Range
Tube diameter (mm)	4–19
Tube length (m)	1–9.5
Axial position, x, to diameter, D, ratio	3–1,373
Temperature (°C)	15–65
Pressure drop (kPa)	0.02–579
Heat flux (W/m²)	65–9,500
Reynolds number	300–221,000
Prandtl number	3.1–10

6.3 Construction of Test Section

6.3.1 Test Section Length and Layout

The test section was part of an experimental setup, as described in detail in Meyer and Everts (2018). Several factors played an important role in determining the length of the test section such as the standard/available length of the tubes, the available laboratory space, and equipment and cost. However, the most important factor was definitely whether a developing or fully developed flow had to be investigated. As the Prandtl number of our test fluids was >1 ($3.1 < Pr < 10$), the hydrodynamic entrance length was shorter than the thermal entrance length. Therefore, the thermal entrance length was used as the criterion for fully developed flow. When the focus of the study was on developing flow, the length of the test section was approximately equal to the thermal entrance length or even shorter, owing to space and equipment limitations. When the study focused on a fully developed flow, the test section had to be at least 1.5 m longer than the thermal entrance length to have a sufficient length of tube to conduct the fully developed flow measurements. It was found that a 1.5-m test section was sufficient to conduct pressure drop and temperature measurements.

For a forced convection laminar flow that is hydrodynamically fully developed at the inlet of the test section, the thermal entrance length, Lt, was calculated as a function of the Reynolds number, Re; Prandtl number, Pr; and the inner-tube diameter, D:

$$Lt_{FC} = 0.05RePrD \qquad (6.1)$$

However, when the flow was simultaneously hydrodynamically and thermally developing, a longer thermal entrance length was required (Meyer and Everts 2018):

$$Lt_{FC} = 0.12RePrD \qquad (6.2)$$

For mixed convection conditions, it was found that the thermal entrance length decreased, and therefore, the Grashof number, Gr, had to be incorporated to account for the free convection effects (Everts and Meyer 2020):

$$Lt_{MC} = 0.12 RePrD \left(1 - \frac{Gr^{0.1}}{Pr^{0.5}Re^{0.07}} \right) \tag{6.3}$$

Although mixed convection conditions were expected in many studies, the length of the test section was usually conservatively determined using the thermal entrance length for forced convection conditions.

For a turbulent flow, the thermal entrance length was calculated by the well-known expression:

$$Lt = 10D \tag{6.4}$$

Everts and Meyer (2018b) developed correlations for horizontal tubes with a constant heat flux boundary condition that can be used as a guideline to determine whether the flow is in the laminar, transitional, quasi-turbulent, or turbulent flow regime. To determine whether the flow will be dominated by forced convection or mixed convection, the flow regime maps of Everts and Meyer (2018a) can be used for a constant heat flux boundary condition, while the flow regime maps of Metais and Eckert (1964) can be used for a constant surface temperature boundary condition.

Figure 6.1 contains a schematic of a test section that was used to investigate the heat transfer and pressure drop characteristics of both a developing and fully developed flow. The thermocouples were closely spaced near the inlet of the test section to accurately obtain the temperature profile of the developing flow, while the thermocouple stations were spaced farther apart on the rest of the test section where the flow was expected to be fully developed or close to fully developed. Because this study focused not only on a developing flow, but also on a fully developed flow, the thermocouples were spaced closer to each other again between stations AA and FF, where the flow was expected to be fully developed.

To investigate the pressure drop characteristics of both a developing and fully developed flow, nine pressure taps were fixed to the test section. For the developing flow, the pressure drop was measured across different tube

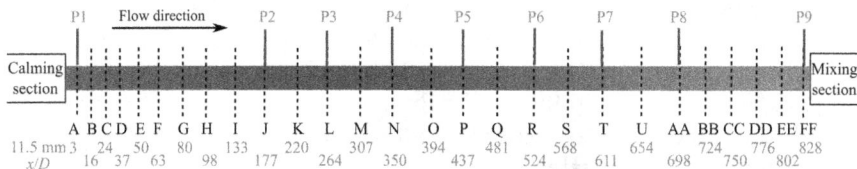

FIGURE 6.1
Schematic of the test section indicating the 27 thermocouple stations, A to FF, as well as the nine pressure taps, P1 to P9, on the 11.5-mm test section.

lengths between pressure tap P1 and the respective pressure tap (P2 to P8). For the fully developed flow, the pressure drop was measured between P8 and P9 only. It was assumed that the pressure drop measurements across the first seven tube lengths (P1 to P8) of the test section could contain a developing and fully developed flow, depending on the Reynolds number, Prandtl number, and heat flux, while the pressure drop measurements between P8 and P9 always contained a fully developed flow only.

6.3.2 Pressure Transducers

6.3.2.1 Differential Pressure Transducers with Interchangeable Diaphragms

Validyne DP15 pressure transducers were usually used owing to the wide range of available diaphragms (0.86 kPa to 22 MPa) and the ability to withstand extreme pressure overloads (200% of the full scale). The inaccuracy of the diaphragms was 0.25% of the full scale. To obtain accurate pressure drop results with low uncertainties, ultra-low-range Validyne DP103 pressure transducers were used to measure pressure drops of <0.86 kPa. Both the DP15 and DP103 pressure transducers had a wet–wet capability and integrated bleed ports. The variable reluctance sensors function by measuring the displacement on interchangeable diaphragms through the resulting change in impedance of two sensing coils. Both the DP15 and DP103 pressure transducers were compatible with the Validyne CD280 carrier demodulator. The CD280 had the option of either four or eight channels to connect the pressure transducers, which improved the versatility of the pressure drop measuring equipment.

6.3.2.2 Pressure Taps

To measure the pressure drop across a specific tube length on the test section, a differential pressure transducer was connected to two pressure taps on the test section. The pressure taps consisted of a 30-mm-long capillary tube with an outer diameter of 3.18 mm that was straightened and silver-soldered onto the test section. A small hole of <10% of the inner diameter of the test section was drilled at 30,000 rpm through the capillary tube and the copper tube. This small diameter was chosen to ensure that the pressure taps did not cause flow obstructions in the test section (Rayle 1959). The high drill speed minimized the formation of burrs on the inside of the test section. Care was also taken to remove any burrs that might have formed by pulling an acetal plug (with a diameter slightly smaller than the inner diameter of the test section) through the tube. To ensure that all the burrs were properly removed, the test section was visually inspected by pulling a borescope through the test section. A bush tap with a quick release coupling was inserted over the capillary tubes, and nylon tubing was used to connect the pressure taps to the differential pressure transducers.

To easily bleed out any air that might have entered the test section, the pressure taps were fixed at the top of the test section. Tam et al. (2013) investigated eight pressure taps that were equally spaced around the periphery of the test section. It was found that for both isothermal and heating conditions, the pressure readings from the peripheral pressure taps were similar and the deviation was <6.6%.

6.3.3 Thermocouples

6.3.3.1 T-Type Thermocouples

T-type thermocouples consist of a copper and a constantan wire. As these thermocouples are mainly used to measure sub-zero temperatures down to –200°C (Morris and Langari 2016), they can only measure temperatures up to 350°C. Furthermore, the temperature range of the thermocouple is also influenced by its insulation. T-type copper–constantan thermocouples with Neoflon PFA insulation were used, and thus, the temperature range of the thermocouples was of –200°C to 150°C (the upper temperature limit was decreased by the insulation material). The copper and constantan wires were separately insulated with blue and red insulation, respectively, before insulating them together with clear PFA insulation. Because the surface temperature measurements usually varied between 15°C and 65°C, a maximum temperature of 150°C was sufficient. The PFA insulation was also preferable because it is durable, and therefore, it not only insulated the thermocouple but also strengthened it. A wide range of thermocouple sizes are available, but we found that thermocouples with a nominal size of 0.6 mm × 1.0 mm were suitable for a wide range of test sections.

6.3.3.2 Thermocouple Preparation

To investigate the influence free convection effects, especially in the laminar and transitional flow regimes, three or four thermocouples were used at each measuring station along the tube length. The thermocouples were very delicate devices because the copper and constantan wires were very thin (~0.6 mm), and therefore, care was taken to prevent any damages. After the three or four thermocouple lengths for a specific measuring station were cut to length, the wires were placed inside a spiral flex. This not only protected the thermocouples and made them easier to handle, but also made them neater and easier to distinguish among the thermocouples of the different measuring stations.

One end of the thermocouple (Figure 6.2a) was prepared by stripping ~1.5 mm of the clear overall insulation, as well as the blue and red thermocouple insulation. The bare copper and constantan wires were then pressed firmly to each other (without twisting them) before soldering them together using tin–lead solder. The thermocouple junctions were then inspected and trimmed to a length of ~1 mm.

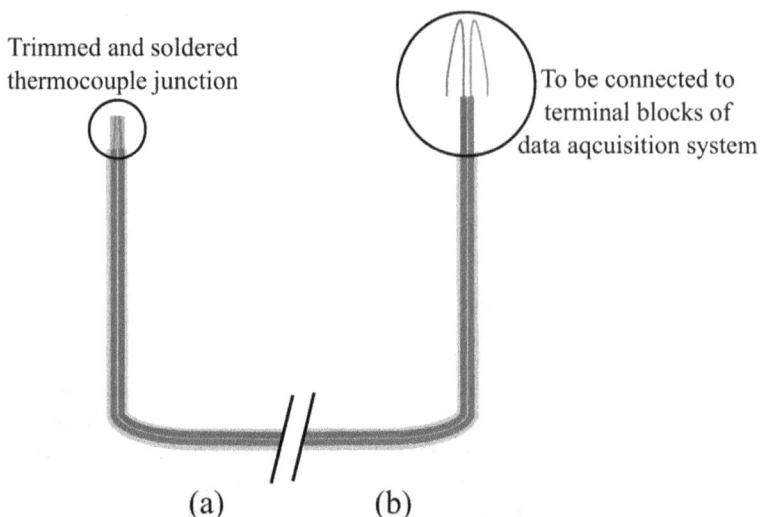

FIGURE 6.2
Schematic of the thermocouple ends that were (a) attached to the test section and (b) inserted into the terminal block of the data acquisition system.

The other end of the thermocouple (Figure 6.2b) was prepared for connection to the terminal blocks of the data acquisition system (Section 6.3.6) by stripping ~30 mm of the clear insulation and 10 mm of the red and blue insulation. A multimeter was used to identify, test, and label the thermocouples by placing one connection on the thermocouple junction and the other onto the bare copper wire. If a current through the wire could be measured, it was known that the two ends corresponded to the same thermocouple. Both ends of the thermocouple were labeled with a waterproof tape to prevent deterioration and to ensure that the labels remained readable throughout the experimental program of approximately 1–3 years.

After the test section had been completely built and insulated, a hairpin bend of ~5 mm was made in the bare copper and constantan wires at the end of the thermocouple (Figure 6.2b). These ends were then fixed into the terminal blocks of the data acquisition system.

6.3.3.3 Copper Test Sections

A two-stage drilling process was used to drill indentations inside the wall of the copper tube. A 0.3 mm indentation (typically for a test section with a wall thickness of ~0.6 mm) was first drilled with a 1-mm-diameter drill and then with a 1.5-mm-diameter drill. The indentation of ~50% of the wall thickness of the tube was an arbitrary choice that we have found worked well. Each indentation was filled with solder by heating the specific location with a blowtorch.

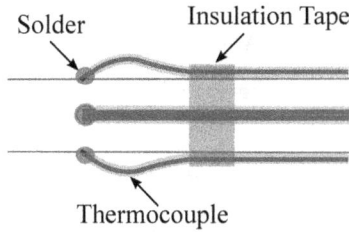

FIGURE 6.3
Schematic of thermocouples soldered to a copper test section.

Two soldering irons were used simultaneously to attach the thermocouple to the test section. The tip of a large soldering iron (300 W) with a tip diameter of ~19 mm was specially modified to heat circular tubes by filing a circular indentation in the tip. This soldering iron was then placed onto the tube wall to heat the tube to ~90°C in order that the solder in the tube indentation started to shine (but not yet to melt).

A second small soldering iron (50 W) with a flat tip that is ~2 mm wide and 0.8 mm thick was used to attach the thermocouple to the test section. Solder was placed on the tip of the small soldering iron. The thermocouple was then pressed onto the indentation (filled with solder) using this soldering iron. This process melted the solder in the indentation and securely fixed the thermocouple in the indentation. The heat was removed, allowing the solder to cool down, and the thermocouples were carefully verified to ensure good contact with the tube. Special care was taken to secure the lead wires of the thermocouples to the test section, to eliminate any strain on the thermocouple junction that might lead to damage. Insulation tape was used to secure the lead wire to the test section ~20 mm from the thermocouple junction. The remaining length of the lead wire was rolled into a bundle and secured to the test section using a small cable tie. Figure 6.3 contains a schematic of the thermocouples fixed to a copper test section.

6.3.3.4 Stainless Steel Test Sections

Because the thermocouples cannot be soldered to stainless steel tubes, Arctic Alumina thermal adhesive (thermal conductivity of 9 W/m K and curing time of 5 min) was used to attach the thermocouples to the stainless steel test sections. In general, the wall thickness of stainless steel tubes was larger than that of the copper test sections, owing to the different method that was used to heat the stainless steel test sections (Section 6.3.7.2). Therefore, 0.5 mm indentations (~50% of the wall thickness of the stainless steel tubes) were drilled into the tube using a 3D printed jig. This ensured that all the indentations were drilled to the same depth. The thermocouples were then glued into the indentations using a thermal adhesive (Figure 6.4). Similar to the copper test sections, the lead wires of the thermocouples were secured with insulation

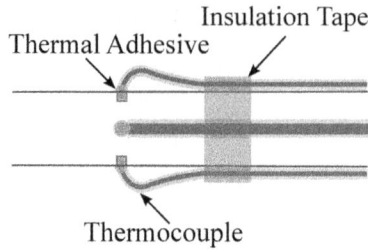

FIGURE 6.4
Schematic of thermocouples glued to a stainless steel test section.

tape ~20 mm from the thermocouple junction to eliminate any strain on the thermocouple junction. The remaining length of the lead wire was rolled into a bundle and secured to the test section using a small cable tie.

6.3.3.5 Tube-in-Tube Heat Exchangers

When temperature measurements had to be taken on the inner tube of a tube-in-tube heat exchanger, it was important to ensure that the thermocouples were embedded into the tube wall (Figure 6.5a) to prevent flow obstructions in the boundary layer and to ensure that the minimum thermocouple length was exposed to the flow (Van der Westhuizen et al. 2014). At each measuring station on the inner tube, four slots (at 90° intervals) were

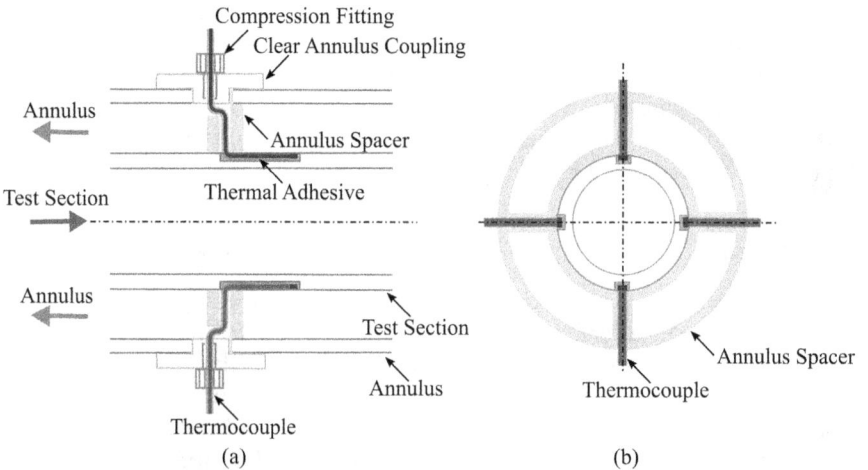

FIGURE 6.5
Schematic of (a) thermocouples attached to the inner tube of a tube-in-tube heat exchanger, annulus spacer, annulus coupling, and compression fitting; and (b) the front view of the annulus spacer with thermocouples.

milled on the outside of the tube. The length and width of the slots were ~20 and 1.2 mm, respectively, while the depth of the slot depended on the tube diameter and wall thickness. As the wall thickness of the copper test sections was in general very small, 60%–75% of the wall thickness was used, as a rule of thumb, for the depth of the slot.

The thermocouple junctions were soldered onto the tube inside these slots (in a similar manner to the process explained in Section 6.3.3.3), and the thermocouple wires were positioned in the slot and covered with a thermal epoxy. The remaining solder and thermal epoxy were carefully filed away. This prevented flow obstructions for ~20 mm downstream of each measuring station (Lawrence and Sherwood 1931).

To minimize the thermocouple length exposed to the flow, 3D printed spacers (Figure 6.5b) were used to route the thermocouple lead wires through the annulus. The spacers had a width of 1.6 mm and were made of polylactide with a low thermal conductivity of 0.13 W/m K to minimize heat conduction between the inner and outer tubes. The purpose of the spacers was thus threefold: (1) to route the thermocouple wires through the annulus, (2) to prevent sagging, and (3) to ensure concentricity in the annulus.

The outer tubes between the spacers were joined using acrylic couplings that also served as a small "window" for visual inspection of the annular flow. To prevent any leaks, the thermocouple wires exited the tube-in-tube heat exchanger though a compression fitting in the acrylic coupling.

The thermocouples on the outer tube of the tube-in-tube heat exchanger were attached as described in Section 6.3.3.3. These thermocouples were used to measure the average fluid temperature at a station. It should be noted that this could only be done when the Reynolds number in the annulus was in the turbulent flow regime.

6.3.4 Pt100 Probes

Ultra-precise immersion Pt100 probes with 1/10 DIN accuracy were used in the flow-calming section (Section 6.3.9) and mixing section (Section 6.3.10) to measure the bulk inlet and outlet temperatures, respectively. It was decided to use Pt100 probes instead of thermocouples to measure these temperatures for three reasons: (1) Only two were usually required (Pt100 probes were much more expensive than thermocouples), (2) the bulk fluid temperature was required and not the temperature at a specific location on the test section, and (3) the inaccuracy of the Pt100 probes was less ($0.06°C$) than that of the thermocouples (inaccuracy of $0.1°C$). The Pt100 probes were made of stainless steel to prevent corrosion and had a 1/8 NPT mounting thread connection. Furthermore, the Pt100 probes had four-wire shielded cables, which enabled us to use sense lines to improve the accuracy of the Pt100 probes with very long cables (owing to the layout of the test section and experimental setup).

6.3.5 Flow Meters

Micro Motion ELITE Coriolis mass flow meters with different capacities were used to measure the mass flow rates through the test sections. The mass flow meters were factory-calibrated to an inaccuracy of 0.05% of the full-scale value. These mass flow meters contained two measuring tubes that were forced to oscillate and produce a sine wave. The tubes vibrated in phase with each other when there was no flow. Once flow passed through the tubes, the Coriolis forces caused the tubes to twist, which resulted in a phase shift. The mass flow rate was directly proportional to the measured time difference between the waves. Therefore, the mass flow meters contained no moving parts and were very robust. Furthermore, they were also suitable for fluids with a wide range of densities and viscosities and could therefore be used not only for water, but also for a wide range of test fluids, such as propylene glycols with different concentrations. These flow meters were also very versatile and had a variety of options for the interfaces and port sizes.

6.3.6 Data Acquisition

A data acquisition system was used to record the data from the Pt100 probes (temperatures), thermocouples (temperatures), pressure transducers (pressure drops), and flow meters (mass flow rates). The data acquisition system consisted of a personal computer using National Instruments LabVIEW software. The data acquisition system also consisted of SCXI (Signal Conditioning eXtensions for Instrumentation) hardware, which included terminal blocks, analogue-to-digital converters, and multiplexers.

A SCXI-1001 chassis with SCXI-1102C thermocouple/voltage input modules (for the pressure transducers, thermocouples, and flow meters), SCXI-1503 RTD modules (for the Pt100 probes), and SCXI-1124 digital-to-analogue converter modules (for the pumps) was used. The pressure transducers and flow meters were connected to SCXI-1308 terminal blocks (recommended for high-precision current measurements) that were connected to the SCXI-1102C modules. The thermocouples were connected to SCXI-1303 terminal blocks (recommended for high-precision thermocouple measurements) that were also connected to the SCXI-1102C modules. A SCXI-1306 terminal block was used to connect the Pt100 probes to the SCXI-1503 module. SCXI-1325 analogue output terminal blocks connected to the SCXI-1124 modules were used to control the pumps via LabVIEW.

The measured raw data were saved as .txt files, and MathWorks MATLAB scripts were in general used for the data processing.

6.3.7 Heating Method

Depending on the experimental setup, we have made use of two different methods of heating: (1) heating wire and (2) in-tube heating.

6.3.7.1 Heating Wire

A constant heat flux boundary condition was applied to the test sections by coiling heating wire around the tube. Single-strand constantan heating wires (which have a constant electrical resistivity of 50 $\mu\Omega$ cm over a wide range of temperatures) were usually used, and the diameter of the wire depended on the required heat flux. The constantan wires were PFA-insulated, and the thickness of the insulation depended on the constantan wire diameter.

To determine the total length of the heating wire, L_{wire}, the overall diameter of the insulated constantan wire, D_{wire}, outer diameter of the test section, D_o, and length of the test section, L, were used:

$$L_{wire} = \frac{\pi L D_o}{D_{wire}} \tag{6.5}$$

The electrical resistance, R_e, of the constantan wire was determined from the heating wire resistivity, ρ_e, length (calculated from Eq. (6.5)), and diameter of the constantan wire (without the insulation), D_c:

$$R_e = \frac{\rho_e L_{wire}}{\frac{\pi}{4} D_c^2} \tag{6.6}$$

For a specific heat flux, $\dot{q}$, the current, I, through the heating wire was determined from:

$$I = \sqrt{\frac{\dot{q}\pi DL}{R_e}} \tag{6.7}$$

and the corresponding required voltage drop, V, was then calculated as follows:

$$V = IR_e \tag{6.8}$$

To prevent burnout, a small current through the heating wire was desired, and therefore, direct current power supplies with high voltage and low current output were used. The power supplies had a maximum power output of 3 kW, maximum voltage of 360 V, and maximum current of 15 A. To obtain the desired heat flux, it was usually required to connect several heating wires in parallel, as this not only reduced the overall resistance of the heating wire but also limited the current flowing through each wire. Furthermore, in some cases, two or more power supplies were connected in parallel when high power inputs were required. In an effort to reduce the effect of electromagnetic interference caused by the currents through the coiled wires, the heating wires were connected in pairs, with opposing

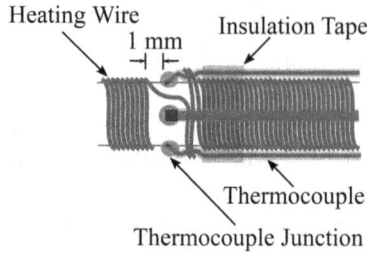

FIGURE 6.6
Schematic of the heating wire coiled at a measuring station.

polarities, to each power supply. According to Lenz's law, the opposing directions of current flow should produce two opposing magnetic fields, which will, in turn, largely cancel each other out (Ulaby et al. 2010).

To ensure uniform heating, the heating wires were tightly coiled along the test section. When coiling the heating wire, it was important to consider how close the heating wire could be to the thermocouple junction before affecting the temperature measurements. Everts (2014) experimentally investigated different coiling techniques and concluded that a gap of ~1 mm between the heating wire and the thermocouple junction was sufficient. Furthermore, when the heating wire was coiled twice over the thermocouple wire, the thermocouple was secured while the temperature measurements remained unaffected. Figure 6.6 shows a schematic of the heating wire coiled around the tube and the thermocouple wire.

The constantan heating wires were coiled around the copper test section only. To prevent losses and to keep the resistance as low as possible, larger shielded cables (1.5 mm^2) were used between the terminals of the power supplies and the constantan heating wires around the test section.

6.3.7.2 In-Tube Heating

The stainless steel tubes had an electrical resistivity, ρ_e, of 74 $\mu\Omega$ cm (at 20°C), and the electrical resistance of the tubes was determined from the inner diameter, D, outer diameter, D_o, and length, L, of the test section:

$$R_e = \frac{\rho_e L}{\frac{\pi}{4}\left(D_o^2 - D^2\right)} \tag{6.9}$$

For a stainless steel tube with a length of 6 m, outer diameter of 6 mm, and inner diameter of 4 mm, the theoretical electrical resistance was calculated to be 0.283 Ω using Eq. (6.9). The actual electrical resistance of the tube was then measured to be 0.282 Ω at the laboratory temperature of 21.8°C, which corresponded well with the theoretical electrical resistance.

As the test sections were heated by passing current through the tube wall, a direct current power supply with a high current output was used. The power supply had a maximum power output of 3 kW, maximum voltage of 40 V, and maximum current of 60 A. Direct current power supplies were chosen because alternating current power supplies would produce a phenomenon called "skin effect." The result is a non-uniform distribution of current across the cross section of the tube, with a maximum current density expected at the outer radius of the tube (Nagsarkar and Sukhija 2014). Thus, the cross section of the tube will not be heated uniformly. For a specific heat flux, the current through the tube wall was determined from Eq. (6.7).

It should be noted that the electrical resistivity of the tube changed during testing owing to the changes in temperature. Therefore, the current inputs were continuously adjusted to ensure that the desired constant heat flux was obtained throughout the experiments. To account for small variations, the power that was supplied to the test section was logged and averaged.

Shielded cables were used between the terminals of the power supply and the stainless steel tubes. The shielded cables were connected to the stainless steel tubes using brass lugs (Figure 6.7) to ensure that a secure electrical connection could be maintained. The tubes and the connectors were covered with Kapton film for electrical insulation.

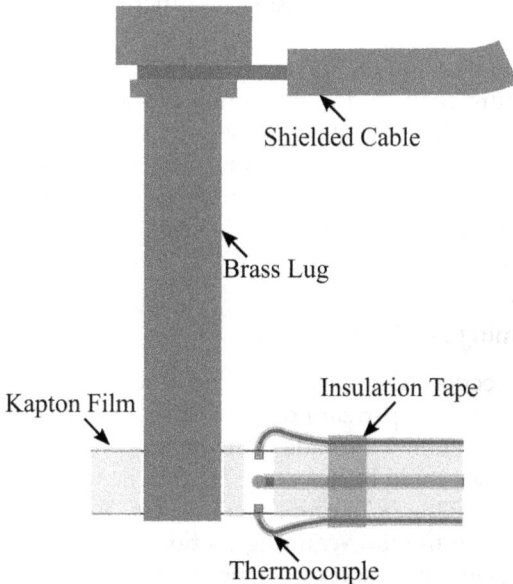

FIGURE 6.7
Schematic of the brass lug that was used to securely connect the power supply to the stainless steel tube.

6.3.8 Insulation

The test sections were thermally insulated using Armaflex Class O insulation. This is a lightweight, flexible, closed-cell insulation with a low thermal conductivity of 0.035 W/m K. One-dimensional heat conduction equations were used to determine the required thickness of the insulation. The ambient temperature of the laboratory (temperature controlled to 22°C) was used as the outer temperature, T_o, while 65°C was used as the inner temperature, T_i. This temperature was the maximum wall temperature on the test section measured or estimated. To account for the natural convection heat transfer inside the laboratory, a convection heat transfer coefficient of 5 W/m^2°C was used. The heat loss through the insulation, $\dot{Q}$ was calculated from:

$$\dot{Q} = \frac{T_i - T_o}{R_{ins} + R_{conv}} = \frac{T_i - T_o}{\dfrac{\ln\left(D_{ins}/D_o\right)}{2\pi L k_{ins}} + \dfrac{1}{h\pi D_{ins} L}} \tag{6.10}$$

where D_{ins} is the outer diameter of the insulation, D_o is the outer diameter of the test section, k_{ins} is the thermal conductivity of the insulation, and L is the length of the test section. The heat loss through the insulation calculated from Eq. (6.10) was compared with the maximum heat input to the test section. The thickness of the insulation was increased until the theoretical heat loss was ~2%.

To prevent axial heat conduction from the test section to the flow-calming section and mixing section, acetal thermal blocks were used. Acetal was used because it has a low thermal conductivity of 0.31 W/m K, and it is also a convenient material for manufacturing. The flange of the flow-calming section (containing a square-edged, re-entrant, or bellmouth inlet) and mixing section was therefore manufactured from acetal.

6.3.9 Flow-Calming Section

A flow-calming section (Figure 6.8) similar to the one used by Ghajar and Madon (1992) was used upstream of the test section to straighten the flow. The flow-calming section was made from clear acrylic plastic to ensure that entrained air bubbles could be detected. The acrylic tube had an inner diameter and length of 172 and 700 mm, respectively. To prevent any temperature gradients inside the flow-calming section, the fluid flowed through a 100-mm cavity filled with a soft nylon mesh before it reached a Pt100 probe, where the average inlet temperature was measured. Possible temperature gradients in the flow-calming section were also reduced by insulating the supply lines to the flow-calming section.

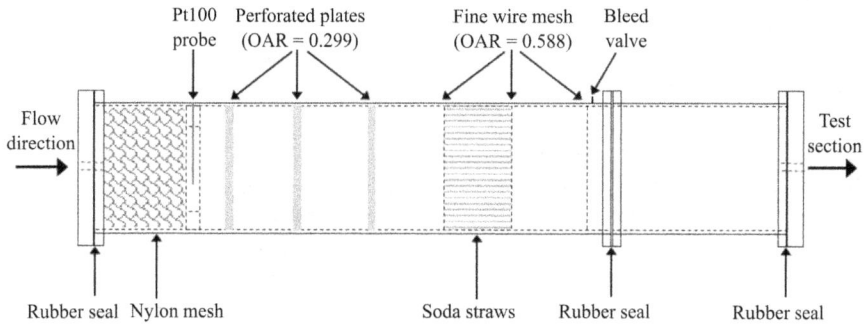

FIGURE 6.8
Schematic of a flow-calming section with a Pt100 probe to measure the inlet water temperature, perforated plates, soda straws, and wire meshes.

Three perforated acrylic plates with an open-area ratio (OAR) of 0.299 (73 holes with a diameter of 11 mm) were inserted after the Pt100 probe and were followed by tightly packed soda straws (inside diameter 5.1 mm, length 102 mm, OAR of 0.855), between two galvanized steel meshes (wire diameter 0.37 mm, OAR of 0.588). The fluid passed through another galvanized steel mesh (wire diameter 0.37 mm, OAR of 0.588) before leaving the flow-calming section. The inlet section was bolted to the flow-calming section and consisted of a clear acrylic tube with inside diameter and length of 172 and 195 mm, respectively. An acetal disc was bolted to the inlet section to obtain a square-edged or re-entrant inlet. Alternatively, the inlet section was replaced with a bellmouth inlet.

A bleed valve was installed prior to the inlet section to bleed the air that entered the flow-calming section. The Pt100 probe connection inside the flow-calming section was used as another bleed valve. The flow-calming section was properly insulated against heat loss using 40-mm-thick insulation with a thermal conductivity of 0.034 W/m K. Peep holes and lids were incorporated into the insulation so that any air bubbles could be detected.

6.3.10 Mixing Section

During laminar flow measurements, significant cross-sectional temperature gradients in the radial and tangential directions developed throughout the test section. Therefore, to obtain a uniform outlet temperature, a mixing section (Figure 6.5a) was inserted after the test section to mix the water exiting the test section. The purpose of the mixing section was twofold: (1) to house the splitter plate mixer and (2) to house a Pt100 probe, which was used to measure the outlet temperature.

The splitter plate mixer design was based on the work done by Bakker et al. (2000), who investigated laminar flow in static mixers with helical splitter plates. The mixer consisted of four copper splitter plates, with a length-to-width ratio of 1.5. The width of the plates was approximately the same as the inner diameter of the test section. The elements were positioned and soldered in order that the leading edge of an element was perpendicular to the trailing edge of the next element. Every splitter plate repeatedly split the thermal boundary layers to ensure a uniform temperature gradient in the radial direction. The splitter plates were placed inside the acetal mixing section, which directed the fluid to flow over and along the entire Pt100 probe after it has been mixed. This ensured that the entire Pt100 probe was exposed to the mixed fluid and also eliminated any stagnant recirculation zones. Diameter D_1 was approximately the same as that of the test section, while diameter D_2 was 8 mm. The diameter of the Pt100 probe was 3 mm; therefore, the hydraulic diameter, D_{h2}, was 5 mm. The significant contraction from 11.5 to 5 mm caused an increase in fluid velocity along the Pt100 probe, which also resulted in enhanced mixing. The mixing section was insulated with 75-mm-thick insulation to prevent any heat loss, and air was bled from the mixer using the Pt100 probe connection to the mixing section.

Figure 6.9b contains a schematic representation of the initial mixing section design that was found to be insufficient. Although the entire cavity was filled with the test fluid and all the air was bled out using the Pt100 connection, the measured outlet temperatures were lower than expected. It was found that this was due to the stagnant recirculation zones in the mixing section, which not only increased the thermal lag but also decreased the fluid temperature. It was therefore decided to change the design to that in Figure 6.9a, which ensured that the entire Pt100 probe was exposed to the fluid and that any stagnant recirculation zones were eliminated.

It was also suggested that the nylon mesh and Pt100 probe at the inlet of the flow-calming section in Figure 6.8 should be replaced by an inlet mixer similar to the one in Figure 6.9a. This is especially recommended when the tube length between the thermostat-controlled bath and the flow-calming section is very long and/or is not sufficiently insulated.

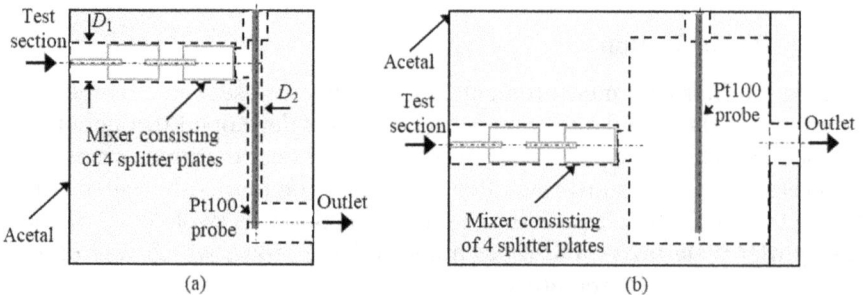

FIGURE 6.9
Schematic of (a) the mixing section with the splitter plates and a Pt100 probe to measure the outlet water temperature and (b) the initial mixing section design.

6.4 Calibration

6.4.1 Pressure Transducers

Differential pressure transducers with interchangeable diaphragms were used to measure the pressure drops inside the test section. Based on the accuracy of the specific diaphragm and pressure transducer that were used, a suitable manometer (with higher accuracy than the pressure transducer) was selected. For example, the diaphragms with a full scale smaller than 2.5 kPa were calibrated using a Betz manometer with an inaccuracy of 2.5 Pa, while diaphragms with a full scale smaller than 10 kPa were calibrated using a low-pressure controlled air manometer with an inaccuracy of 10 Pa. Diaphragms with a full scale larger than 10 kPa were calibrated by connecting a Beta T-140 manometer (full scale of 50 kPa and inaccuracy of 50 Pa) in parallel with the pressure transducer to a water column.

To calibrate the pressure transducer, the zero and span screws of the amplifier were adjusted so that the zero of the manometer corresponded to 4 mA in the LabVIEW program and the full scale corresponded to 20 mA. This usually required several iterations between the zero and full scale of the diaphragm by either setting the required pressure in the manometer or filling the water column to the required height. Once the zero and full scale had been set correctly, the pressure was increased in ~5% intervals from zero to the full scale and then decreased again in 5% intervals back to zero. Owing to hysteresis, slight differences existed between the increasing and decreasing runs; however, they were usually smaller than the uncertainty of the pressure drop measurement (Section 6.6) and were therefore considered to be negligible.

At each pressure drop interval, 200 measuring points were logged once the desired pressure had been obtained. The current signal obtained from the LabVIEW program was converted to a pressure reading in MATLAB via interpolation (between 4 and 20 mA). Figure 6.10 contains sample plots of the pressure drop measurements using an 8.6-kPa diaphragm as a function of (a) the pressure drop measured using the manometer and (b) the current signal. To obtain the calibration equation, a linear curve fit was done in Figure 6.10b to obtain the relationship between the pressure drop measurements, ΔP, and the current signals, ΔP_A.

The gradient, m, with a value of 0.53852 and the y-intercept, c, with a value of –2.1157 resulted in the following calibration equation:

$$\Delta P_{8.6\,kPa} = 0.53852\Delta P_A - 2.1557 \tag{6.11}$$

To confirm that the calibration was successful and that the current signal measured by the pressure transducer was correctly converted to a pressure signal, the pressure drop differences between the manometer and the

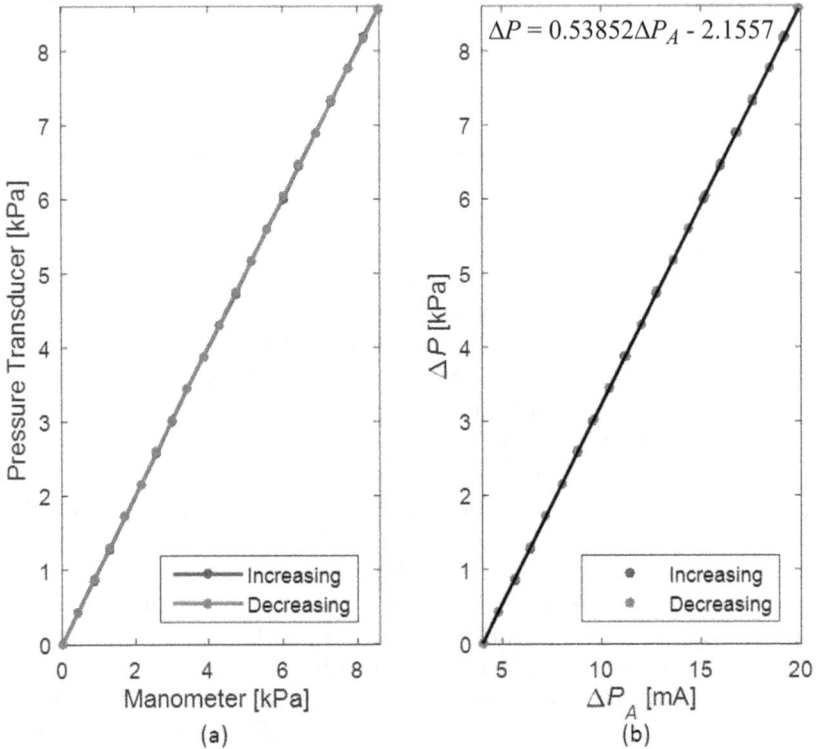

FIGURE 6.10
Pressure drop measurements using an 8.6-kPa diaphragm as a function of (a) the pressure drop measured by the manometer and (b) the current signal.

pressure transducer were also investigated (Figure 6.11). The black dotted lines represent the uncertainty of the pressure transducer diaphragm, which was determined using linear regression analysis (Section 6.6). Depending on the inaccuracy of the manometer that was used, the calculated uncertainty of the pressure transducer was often less than that in the manufacturer's specification. If this was the case, to be conservative, the manufacturer-specified inaccuracy of the pressure transducer was used as the uncertainty. Because all the data points in Figure 6.11 fell approximately within the dotted lines, the calibration was considered to be successful.

The equation of the linear curve fit was used during the data reduction of the actual tests. It should be noted that due to the different densities of air and water, the pressure drop readings were significantly affected when there was air in the system. Because the solubility of gases decreased with increasing temperature, dissolved air bubbles escaped. Therefore, after each start-up of the pump (in the beginning of the day or when the experimental setup had to be switched off and then restarted later in the day), the pressure taps and pressure transducers were bled to ensure that there was no air in the system.

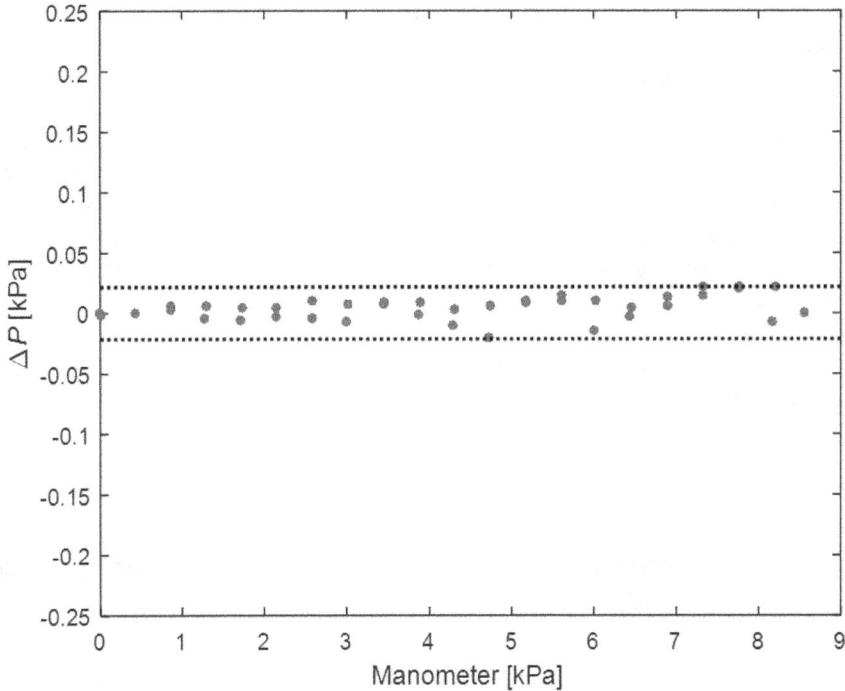

FIGURE 6.11

Pressure drop differences between the pressure measured by the manometer and the pressure transducer. The black dotted lines represent the uncertainty of the pressure transducer diaphragm.

A pressure reading, consisting of 200 measuring points, was then taken during no-flow conditions, and the average was used as the offset. The no-flow condition was obtained by opening the bypass valve and closing the supply valve to ensure that there was water supply for the pump, however, not through the test section. The final pressure transducer calibration equation for the 8.6-kPa diaphragm used in the data reduction program was thus:

$$\Delta P_{8.6\,\text{kPa}} = 0.53852\Delta P_A - 2.1557 - \text{offset} \tag{6.12}$$

6.4.2 Pt100 Probes

The Pt100 probes were calibrated inside a thermostat-controlled bath against a DigiCal DCS2 digital thermometer with an inaccuracy of 0.03°C. As the measuring range of the Pt100 probes during the experiments was usually of 20°C–60°C, the Pt100 probes were calibrated between 15°C and 65°C at 2.5°C intervals. Once the thermostat-controlled bath reached the desired steady-state temperature, a measurement consisting of 200 measuring points was taken and the average of the 200 measuring points was used. The process was then repeated for decreasing temperatures (from 65°C to 15°C) to ensure that a constant curve was obtained and to account for the effect of hysteresis (although it was usually found to be negligible).

Figure 6.12 contains a sample plot of the average temperature measurements of a Pt100 probe against the thermometer measurements. This figure indicates that hysteresis was negligible because there was no significant difference between the temperature measurements of the increasing and decreasing calibration runs and the measured temperatures of the Pt100 probe were very close (within 0.2°C, uncalibrated) to those of the thermometer.

The average calibration factor of the Pt100 probe was obtained by performing a linear curve fit through the data of three sets of measurements (both increasing and decreasing calibration runs). The calibrated temperatures were then determined as follows:

$$T_{cal} = \frac{(T_{uncal} - c)}{m} \tag{6.13}$$

where m and c correspond to the slope (value of 1.0013) and y-intercept (value of 0.10197) of the linear curve fit, respectively (Figure 6.12).

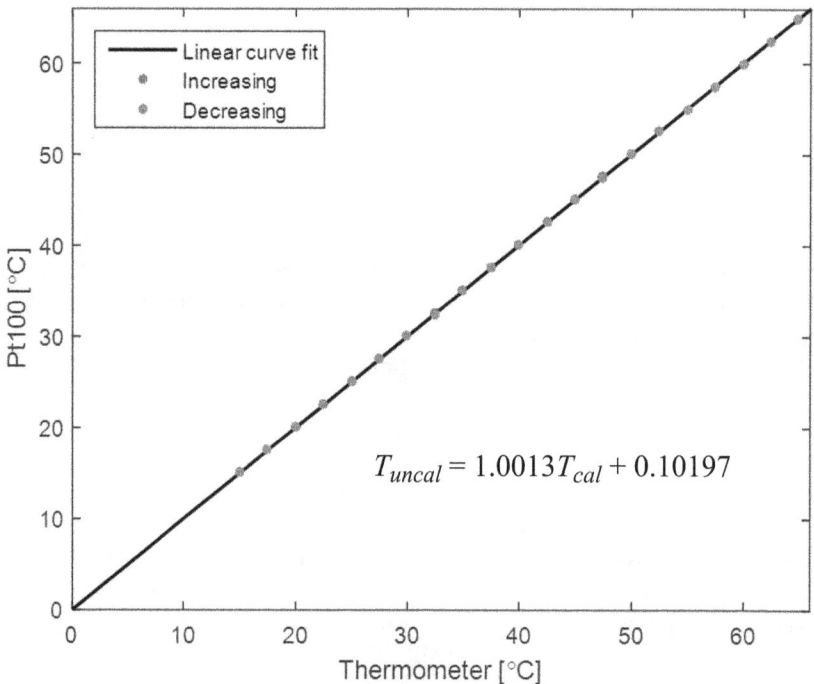

$$T_{uncal} = 1.0013 T_{cal} + 0.10197$$

FIGURE 6.12
Temperature measurements of a Pt100 probe for increasing and decreasing calibration runs.

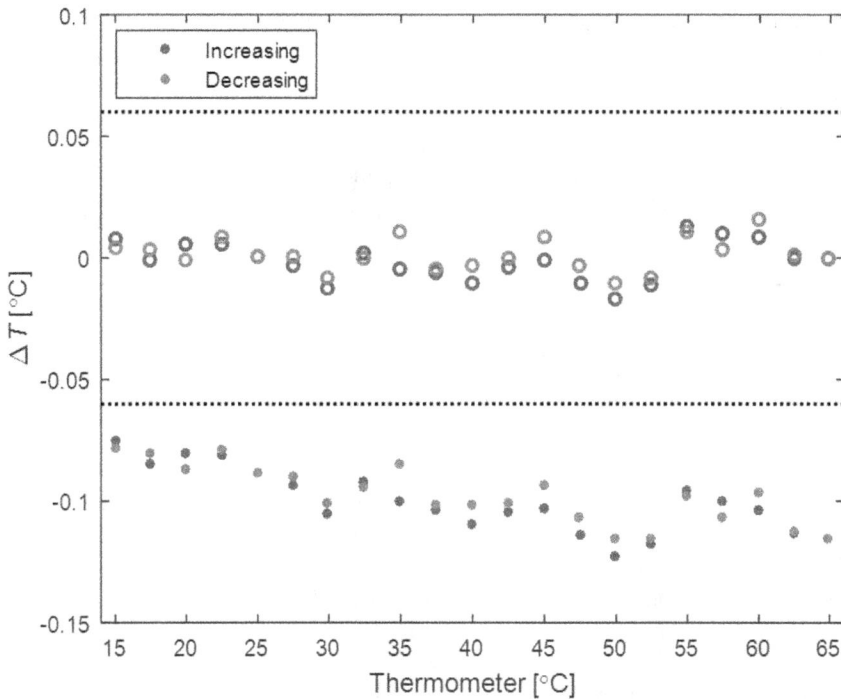

FIGURE 6.13
Temperature difference between the thermometer and the Pt100 probe before (filled markers) and after (empty markers) calibration. The black dotted lines represent the uncertainty of the Pt100 probe.

To confirm that the calibration was successful, the temperature differences between the thermometer and the Pt100 probe before (filled markers) and after (empty markers) the calibration were also investigated (Figure 6.13). The uncertainty of this Pt100 probe was determined to be between 0.0305°C and 0.0311°C, which was less than the manufacturer-specified inaccuracy of 0.06°C. Therefore, a value of 0.06°C was used as the uncertainty of the Pt100 probe and is indicated by the black dotted lines. The maximum temperature difference before calibration was 0.12°C and decreased to 0.016°C after calibration. Because this temperature difference was well within the uncertainty of the Pt100 probe, it was concluded that the calibration was successful.

6.4.3 Thermocouples

Similar to the Pt100 probes, the thermocouples could also be calibrated against a digital thermometer inside a thermostat-controlled bath, before attaching them to the test section. However, the properties of the thermocouple junction were likely to change during the attachment process (either soldered or glued). To improve the accuracy and reliability of the calibration factors, in situ calibration was used. Once the test section was completely built (thermocouples and pressure taps attached and heating wire coiled if necessary) and insulated, the thermocouples were connected to the terminal blocks of the data acquisition system. Mixing sections (Figure 6.9a) were connected at the inlet and outlet of the test section. Insulated supply and return lines from the thermostat-controlled bath were then connected to the mixing sections. Water was therefore circulated through the test section and thermostat-controlled bath only and not through the entire experimental setup, which decreased the thermal lag and calibration time significantly. The flow-calming section was replaced by a mixing section during the calibration process to avoid exposing the acrylic flow-calming section to high temperatures (the maximum operating temperature of extruded acrylic is 70°C). The smaller volume of the mixing section compared with the flow-calming section also assisted in reducing the thermal lag during calibration.

The thermocouples were calibrated using a thermostat-controlled bath and two Pt100 probes (one inside each mixing section), which were calibrated to an accuracy of 0.06°C. The thermocouples were calibrated between 15°C and 60°C at 2.5°C intervals using the water from the thermostat-controlled bath. Therefore, no heating was applied to the test section using the power supplies. Once the thermostat-controlled bath reached the desired temperature and both Pt100 probes measured the same constant temperature (within the uncertainty of the Pt100 probes), a measurement (consisting of 200 measuring points) was taken at a sampling frequency of 10 Hz. The process was also repeated for decreasing temperatures (from 60°C to 15°C) to ensure that a constant curve was obtained and to account for the effect of hysteresis. In a similar manner to those of the Pt100 probes, the calibration factors of the thermocouples were obtained by performing a linear curve fit through the results of three sets of measurements (both increasing and decreasing calibration runs).

We have found that, for long test sections and at very high or very low temperatures (compared with the ambient temperature), it was sometimes challenging for the Pt100 probes to measure the same temperature (within the uncertainty of the Pt100 probes). If the temperature difference was <0.2°C (this was an arbitrary value that we have found to be acceptable), a linear curve fit was performed between the temperature measurement of the Pt100 in the inlet mixing section and the Pt100 inside the outlet mixing section. This linear equation was then used to obtain the reference temperature at each measuring station.

Figure 6.14 contains a sample plot of the uncalibrated and calibrated temperature measurements of the thermocouples along the tube length at temperatures of ~15°C, 30°C, 45°C, and 60°C. The uncertainties of these thermocouples were determined to be between 0.0605°C and 0.0649°C. As this was less than the manufacturer-specified inaccuracy of 0.1°C, a value of 0.1°C was used as the uncertainty of the thermocouples. The solid black line represents the temperatures measured by the Pt100 probes, and the dotted black

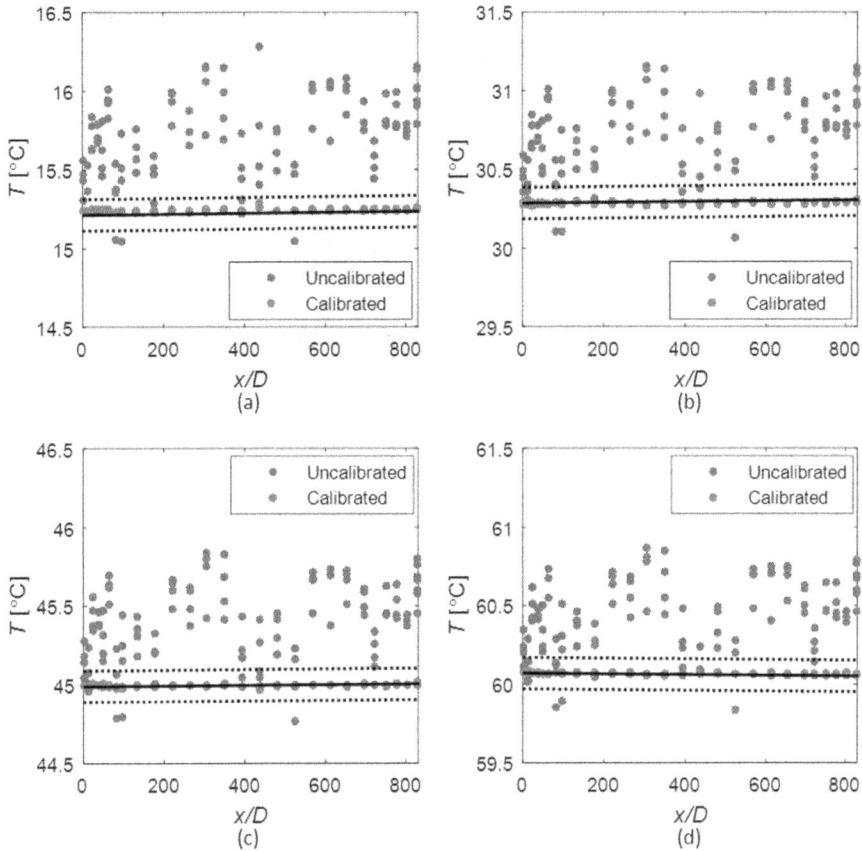

FIGURE 6.14
Local uncalibrated and calibrated temperatures as a function of axial position at temperatures of (a) 15.2°C, (b) 30.3°C, (c) 45.0°C, and (d) 60.1°C. The solid black line represents the temperatures measured by the two Pt100 probes at the inlet and outlet of the test section, and the dotted black lines represent the uncertainty of the thermocouples.

lines indicate the thermocouple uncertainty of 0.1°C. Figure 6.14 confirms that the calibration was successful because all the calibrated temperature measurements were well within the dotted black lines. The average deviation between the thermocouple and Pt100 probe measurements along the test section for temperatures between 20°C and 60°C was 0.43°C before calibration, and it was reduced to 0.009°C after calibration.

It should be noted that a graph similar to that in Figure 6.14 could also be used to identify any faulty thermocouples. If the error of a specific thermocouple was consistently higher (and larger than the uncertainty of the thermocouple) over the entire temperature range, the thermocouple was considered as inaccurate. In many cases, this was due to the thermocouple junction that was damaged during the manufacturing or assembling processes, part of the lead wire that was damaged, or an improper connection in the terminal block. In the cases where a thermocouple was damaged, the data measured by the specific thermocouple were excluded in the data reduction process.

The deviation between the thermocouples and Pt100 probes before and after calibration was also compared for different attachment techniques (Everts 2017). It was found that the deviation of the thermocouples that were soldered to a copper test section was less than that of the thermocouples that were glued to a stainless steel test section using a thermal adhesive. The lower thermal conductivity of the thermal adhesive (9 W/m K) compared with that of the solder (50 W/m K) possibly led to less accurate temperature measurements. Furthermore, the uncertainties of the thermocouples that were glued were also slightly higher (maximum uncertainty of 0.101°C) than those of the thermocouples that were soldered (maximum uncertainty of 0.0649°C). However, as long as the maximum deviation after calibration remained less than the thermocouple uncertainty of 0.1°C, the calibration was considered to be successful.

6.4.4 Drift and Recalibration

The characteristics of instruments such as pressure transducers, thermocouples, and Pt100 probes changed with time, which required the recalibration of these instruments. These changes were caused by temperature changes, mechanical wear, and exposure to dust, dirt, fumes, and chemicals. Although the magnitude of drift was usually associated with the amount of use, some drift may even occur during storage due to the aging effect of the components.

Because several factors such as type of instrument, frequency of use, and environmental conditions have a strong influence on the calibration frequency, it is almost impossible to determine the required calibration frequency theoretically (Morris and Langari 2016). Therefore, practical experimentation is usually required to determine the recalibration frequency. We have found that it was good practice to recalibrate the pressure transducers, thermocouples, and Pt100 probes every 6 months. Although the changes after 6 months appeared to be small and negligible, they did influence the accuracy and reliability of the experimental data over longer periods. When calibrating and especially recalibrating, the use of proper and rigorous calibration procedures and documentation was essential. This made it possible to compare the calibration factors over the time of experimentation. The changes in the calibration factors were generally very small, and large variations usually indicated a faulty instrument.

We have found it very useful to identify approximately three benchmark cases and use them throughout the experiments (approximately every 6 months), especially after calibration or modifications in the experimental setup. The three cases should give a good representation of the experimental results, and typically included one validation case and two other cases (part of the test matrix). The new results of each benchmark case were then added approximately every 6 months and compared with the previous results. If no significant changes occurred, accurate and reliable experimental data could be expected. However, if the changes were significant, the calibration or modifications that were implemented had to be verified.

6.5 Data Reduction

The used data reduction methodology was explained in detail in Meyer and Everts (2018) and is summarized here for convenience.

For a constant heat flux boundary condition, the average axial temperature of the water increased linearly. The mean fluid temperature, T_m, at a specific tube location, x, was obtained using a linear temperature distribution between the measured inlet temperature, T_i, and measured outlet temperature, T_o, of the fluid over the tube length, L:

$$T_m = \left(\frac{T_o - T_i}{L}\right) x + T_i \tag{6.14}$$

The bulk fluid temperature, T_b, along a tube length, $L(x)$, measured from the inlet of the test section, was calculated as follows:

$$T_b = \left(\frac{T_o - T_i}{L}\right) \frac{L(x)}{2} + T_i \tag{6.15}$$

The properties of water (density, ρ; dynamic viscosity, μ; thermal conductivity, k; specific heat, C_p; Prandtl number, Pr; and thermal expansion coefficient, β) were determined using the thermophysical correlations for liquid water (Popiel and Wojtkowiak 1998) at the bulk fluid temperature for the average properties, and at the mean fluid temperature for the local properties at a specific point, x, measured from the inlet of the test section.

The Reynolds number, Re, was calculated as follows:

$$Re = \frac{\dot{m} D}{\mu A_c} \tag{6.16}$$

where $\dot{m}$ is the measured mass flow rate inside the tube, D is the inner-tube diameter, μ is the dynamic viscosity, and A_c is the cross-sectional area of the test section ($A_c = \pi/4 D^2$).

The electrical input rate ($\dot{Q}_e = VI$) remained constant, resulting in a constant heat flux. The heat transfer rate to the water, $\dot{Q}_w$, was determined from the measured mass flow rate, measured inlet and outlet temperatures of the water, and the specific heat, which was calculated at the bulk fluid temperature:

$$\dot{Q}_w = \dot{m}C_p(T_o - T_i) \tag{6.17}$$

The heat transfer rate to the water, $\dot{Q}_w$, was continuously monitored by comparing it with the electrical input rate, $\dot{Q}_e$, of the power supply. The energy balance error, EB, which ideally should be as close as possible to zero for a well-insulated test section, was determined as follows:

$$EB = \left|\frac{\dot{Q}_e - \dot{Q}_w}{\dot{Q}_e}\right| \times 100 = \left|\frac{VI - \dot{m}C_p(T_o - T_i)}{VI}\right| \times 100 \tag{6.18}$$

The heat flux, $\dot{q}$, on the inside of the tube wall was determined from the heat transfer rate to the water, $\dot{Q}_w$, and the inner-surface area, A_s, of the test section along the heated length, L:

$$\dot{q} = \frac{\dot{Q}_w}{A_s} = \frac{\dot{m}C_p(T_o - T_i)}{\pi D L} \tag{6.19}$$

The heat transfer rate to the water was used to determine the heat flux because it was regarded as more accurate than the electrical input rate. Because the energy balance error was not zero and some losses did occur to the ambient air, the electrical input rate was always slightly higher than the heat transfer rate to the water.

The average of the n temperature measurements at a station (with $n=3$ or $n=4$) was used as the average outer surface temperature, $T_{s,o}$, at a specific thermocouple station:

$$T_{s,o} = \frac{T_1 + T_2 + \cdots + T_n}{n} \tag{6.20}$$

The thermal resistance, R_{tube}, across the tube wall was calculated using the following equation:

$$R_{tube} = \frac{\ln\left(\frac{D_o}{D}\right)}{2\pi L k} \tag{6.21}$$

where D_o and D are the measured outside and inside diameters of the tube, respectively.

Because the thermal resistance and heat input were known, the temperature differences across the tube wall, ΔT, could be calculated:

$$\Delta T = \dot{Q}_w R_{tube} \tag{6.22}$$

When the temperature difference across the tube wall was less than the uncertainty of the temperature measurements (usually 0.1°C), it was considered to be negligible. Therefore, the surface temperature determined from Eq. (6.20) was used as the average surface temperature on the inside of the tube at a measuring station. However, when the temperature difference was greater than the uncertainty of the temperature measurements, the inner-surface temperatures were determined as follows:

$$T_s = T_{s,o} - \Delta T = T_{s,o} - \dot{Q}_w R_{tube} \tag{6.23}$$

In the different test sections that were tested, it was found that the thermal resistance of the thin-walled copper test sections was usually negligible. However, it was not negligible for the thick-walled stainless steel test sections.

The average surface temperature, $\bar{T}_s$, along a tube length, $L(x)$, measured from the inlet of the test section, was calculated as follows:

$$\bar{T}_s = \frac{1}{L} \int_0^L T_s(x)\,dx \tag{6.24}$$

The heat transfer coefficients, h, were then determined from the following equation, because the heat flux, $\dot{q}$, surface temperature, T_s, and mean fluid temperature, T_m, were known:

$$h = \frac{\dot{q}}{(T_s - T_m)} \tag{6.25}$$

The Nusselt numbers, Nu, were determined from the heat transfer coefficients as follows:

$$Nu = \frac{hD}{k} \tag{6.26}$$

The heat transfer results were also investigated in terms of the Colburn j-factors to account for the variations in the Prandtl numbers of sequential measurements and to investigate the relationship between heat transfer and pressure drop:

$$j = \frac{Nu}{RePr^{\frac{1}{3}}} \tag{6.27}$$

The Graetz numbers, Gz, were determined as follows:

$$Gz = RePr\frac{D}{x} \tag{6.28}$$

while the Grashof numbers, Gr, were determined using the following equation:

$$Gr = \frac{g\beta(T_s - T_m)D^3}{v^2}$$

(6.29)

where 9.81 m/s^2 (or a more specific measured value) was used for the gravitational acceleration, g, and the kinematic viscosity was obtained from the density and dynamic viscosity ($v = \mu/\rho$).

The modified Grashof numbers, Gr^*, which are a function of heat flux instead of surface-fluid temperature differences, are the product of the Grashof numbers and Nusselt numbers and were determined as follows:

$$Gr^* = GrNu = \frac{g\beta\dot{q}D^4}{v^2k}$$

(6.30)

The Rayleigh numbers, Ra, were determined as the product of the Grashof numbers and Prandtl numbers:

$$Ra = GrPr$$

(6.31)

$$Ra^* = Gr^*Pr$$

(6.32)

The Richardson numbers, Ri, were determined from the Grashof and Reynolds numbers:

$$Ri = \frac{Gr}{Re^2}$$

(6.33)

$$Ri^* = \frac{Gr^*}{Re^2}$$

(6.34)

Equations (6.25)–(6.34) refer to the local values at a specific axial position along the tube length. The average values along a tube length, $L(x)$, measured from the inlet of the test section, were obtained by using the bulk fluid temperature (Eq. 6.15) and average surface temperature (Eq. 6.24) instead of the mean fluid temperature (Eq. 6.14) and local surface temperature (Eq. 6.23).

The friction factors, f, were calculated from the mass flow rate and pressure drop measurements, ΔP, between two pressure taps, which were apart from each other a length $L(x)$, using the bulk fluid properties (Eq. 6.15):

$$f = \frac{2\Delta PD}{L(x)\rho_b V^2} = \frac{\Delta P\rho_b D^5\pi^2}{8\dot{m}^2 L(x)}$$

(6.35)

In general, the percentage error of a measurement or calculated value was determined as %error= $|M_{exp} - M_{cor}|/M_{ref} \times 100$. When the experimental setup and data reduction method were validated, M_{ref} was obtained from existing correlations in the literature, M_{cor}. However, when the accuracies of the correlations were determined, M_{ref} was obtained from the experimental data, M_{exp}. The average percentage error was taken as the average of the absolute errors of the data points.

In a recent study conducted by Everts and Meyer (2018b), other parameters such as the transition gradient and width of the transitional flow regime, which are specifically for the transitional flow regime, were also defined.

6.6 Uncertainty Analysis

Although the aim is to avoid all possible errors in the experimental results, this is not possible in the real world. Therefore, it is important to provide some measure of reliability with the results (Kline and McClintock 1953). Kline and McClintock (1953) made use of the framework of statistical inference to estimate the uncertainty in single-sample experiments, and this formed the basis of uncertainty analyses.

6.6.1 Background

For a variable x_i with a known uncertainty δx_i, the variable and uncertainty were expressed in the following form:

$$x_i = x_i (\text{measured}) \pm \delta x_i \tag{6.36}$$

The value of x_i (measured) was a single measurement, and δx_i represented the standard deviation multiplied by Student's t-variable (Moffat 1988). The result, R, of a measurement was a function of several variables and was calculated from a group of equations:

$$R = R(x_1, x_2, x_3, ..., x_n) \tag{6.37}$$

If only one term was in error, the uncertainty in R was determined as follows:

$$\delta R = \frac{\partial R}{\partial x_i} \delta x_i \tag{6.38}$$

The sensitivity coefficient, δR, was used to determine the effect that x_i had on the overall uncertainty. The root sum square method was used to determine the uncertainty for several independent variables (Moffat 1988):

$$\delta R = \left[\sum_{i=1}^{n} \left(\frac{\partial R}{\partial x_i} \delta x_i \right)^2 \right]^{\frac{1}{2}}$$

(6.39)

The overall uncertainty in a result was usually dominated by only a few terms. Therefore, terms that were three times smaller than the largest term can be considered as negligible (Moffat 1988).

There are two types of errors that arose during measurement processes: bias and precision. The bias error determined the accuracy of the measurement. These errors were normally specified by the manufacturer of the instrument and resulted from calibration and imperfections in the measuring equipment. The precision error relates to the scatter that existed in the data. These errors were caused by variations in the measurement process, electrical noise, etc. A 95% confidence interval was usually used, which means that the magnitudes of the bias and precision errors corresponded to the 95% probability that the actual error will not be more than the estimate. Based on the bias and precision errors, the uncertainty in a single measurement was calculated as follows (Dunn 2010):

$$\delta x_i = \left(b_i^2 + p_i^2 \right)^{\frac{1}{2}}$$

(6.40)

A linear regression analysis was used to obtain the precision of the pressure transducers, Pt100 probes, and thermocouples, and it was used as the bias of the pressure drop and temperature measurements. The regression analysis determines a mathematical relation between two or more variables. The value of x was usually known, and the value of y was obtained from measurements; therefore, the uncertainty resulted from the y variable (Dunn 2010). The uncertainty of the y variable was determined using Eq. (6.41):

$$\delta y = \pm \, tS_{yx} \sqrt{\frac{1}{N} + \frac{1}{M} + \frac{(x_i - \bar{x})^2}{S_{xx}}}$$

(6.41)

where S_{xx} was defined as follows:

$$S_{xx} = \sum_{i=1}^{N} (x_i - \bar{x})^2$$

(6.42)

S_{yx} was obtained by first calculating S_{xy} and the parameters a and b:

$$S_{xy} = \sum_{i=1}^{N} (x_i - \bar{x})(y_i - \bar{y})$$

(6.43)

$$b = \frac{S_{xy}}{S_{xx}} \tag{6.44}$$

$$a = \bar{y} - b\bar{x} \tag{6.45}$$

$$y_{ci} = a + bx_i \tag{6.46}$$

$$S_{yx} = \sqrt{\frac{\sum_{i=1}^{N}(y_i - y_{ci})^2}{N-2}} \tag{6.47}$$

To obtain the uncertainty in the x variable, the uncertainty in y was divided by the slope of the regression line:

$$\delta x = \frac{\delta y}{m} \tag{6.48}$$

For all other instruments, the bias was considered as the accuracy specified by the manufacturer. The precision was obtained from the standard deviation of the measuring points that were logged during each measurement. This was then multiplied by Student's t-variable to fall within the 95% confidence region.

6.6.2 Basic Procedure and Examples

Examples are given to illustrate the basic procedure of the linear regression analysis method (uncertainties of Pt100 probes, thermocouples, and pressure transducers during the calibration process) and the uncertainty analysis method (uncertainties of Reynolds number, Nusselt number, and friction factor during the experiments). Additional details regarding the calculation of the other variables along with examples are available in Everts (2014, 2017). The experimental data of Everts (2017) were used in the examples. It should also be noted that, although Moffat (1988) considered the terms that are three times less than the largest term as negligible, it was decided to include all the terms in the examples.

6.6.2.1 Linear Regression Analysis: Pt100 Probe

Figure 6.15 contains a sample plot of the calibrated Pt100 temperature measurements as a function of the temperatures measured by the thermometer. A total of 21 data points were taken between 15°C and 65°C, and each data point consisted of 200 measuring points. The dotted horizontal and vertical lines represent the average calibrated Pt100 and thermometer measurements,

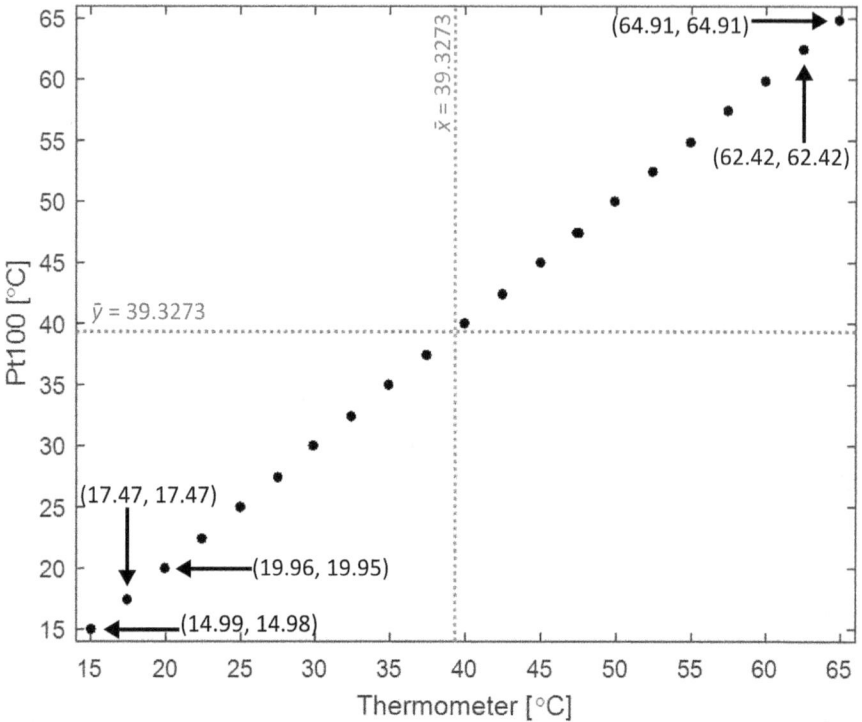

FIGURE 6.15
Sample plot of the calibrated Pt100 temperature measurements as a function of the temperatures measured by the thermometer. The dotted horizontal and vertical lines represent the average calibrated Pt100 and thermometer measurements, respectively.

respectively, across the calibrated temperature range. The slope of the diagonal line fitted through the data points was 1.0007.

S_{xx} was determined using Eq. (6.42) by substituting x_i with each thermometer measurement and $\bar{x}$ with the average of all the thermometer measurements:

$$S_{xx} = \sum_{i=1}^{N}(x_i - \bar{x})^2$$

$$= (14.99 - 39.3273)^2 + (17.47 - 39.3273)^2$$

$$+ (19.96 - 39.3273)^2 + \cdots + (62.42 - 39.3273)^2 + (64.91 - 39.3273)^2$$

$$= 8,966.27$$

S_{xy} was determined using Eq. (6.43) by substituting y_i with each Pt100 probe measurement and $\bar{y}$ with the average of all the Pt100 probe measurements:

$$S_{xy} = \sum_{i=1}^{N}(x_i - \bar{x})(y_i - \bar{y})$$

$$= (14.99 - 39.3273)(14.9825 - 39.3273) + (17.47 - 39.3273)(17.4709 - 39.3273)$$

$$+ (19.96 - 39.3273)(19.95 - 39.3273) + \cdots + (62.42 - 39.3273)(62.4200 - 39.3273)$$

$$+ (64.91 - 39.3273)(64.9105 - 39.3273)$$

$$= 8,966$$

The parameters b and a were then determined using Eqs. (6.44) and (6.45):

$$b = \frac{S_{xy}}{S_{xx}}$$

$$= \frac{8,966}{8,966}$$

$$\approx 1$$

$$a = \bar{y} - b\bar{x}$$

$$= 39.3273 - 39.3273$$

$$\approx 0$$

With both a and b known, y_{ci} of each temperature increment was calculated from Eq. (6.46). Next, S_{yx} was determined as follows, where N was the number of temperature increments:

$$S_{yx} = \sqrt{\frac{\sum_{i=1}^{N}(y_i - y_{ci})^2}{N-2}} = \sqrt{\frac{\text{sum1}}{N-2}}$$

$$\text{sum1} = \sum_{i=1}^{N}(y_i - y_{ci})^2$$

$$= (14.9825 - 14.99)^2 + (17.4709 - 17.47)^2$$

$$+ (19.9546 - 19.96)^2 + \cdots + (62.4200 - 62.42)^2 + (64.9105 - 64.91)^2$$

$$S_{yx} = \sqrt{\frac{\text{sum1}}{19}}$$

$$S_{yx} = 0.0109$$

The precision component, p, of the Pt100 probe uncertainty at 25°C was then calculated from Eq. (6.41):

$$p = \pm \frac{tS_{yx}}{m} \sqrt{\frac{1}{N} + \frac{1}{M} + \frac{(x_i - \bar{x})^2}{S_{xx}}}$$

$$= \pm \frac{2.093 \times 0.0109}{1.0007} \sqrt{\frac{1}{21} + \frac{1}{200} + \frac{(24.95 - 39.3273)^2}{8,966}}$$

$$= \pm 0.0063$$

where t is Student's t-variable and M is the number of measuring points that were logged during each measurement.

The accuracy of the thermometer, which was used during the calibration, was 0.03°C and was used as the bias component, b_i, of the Pt100 probe uncertainty. The overall uncertainty of the Pt100 probe at 25°C was then calculated using Eq. (6.40):

$$\delta x_i = \left(b_i^2 + p_i^2 \right)^{\frac{1}{2}}$$

$$= \left(0.03^2 + 0.0063^2 \right)^{\frac{1}{2}}$$

$$= 0.0307 \,°C$$

The uncertainty of this Pt100 probe decreased from 0.031°C at 15°C to 0.0305°C at 30°C–47.5°C, and then, it increased again to 0.031°C at 65°C. However, as these uncertainties were always less than the manufacturer-specified inaccuracy of 0.06°C, the value of 0.06°C was used as the uncertainty of the Pt100 probe over the entire temperature range.

6.6.2.2 Linear Regression Analysis: Pressure Transducer with a 2.2-kPa Diaphragm

Figure 6.16 contains a sample plot of the pressure transducer measurements as a function of the pressure drops measured by the manometer. A total of 21 data points were taken between 0 and 2.2 kPa, and each data point consisted of 100 measuring points. The dotted horizontal and vertical lines represent the average pressure transducer and manometer measurements, respectively, across the calibrated pressure drop range. The slope of the diagonal line fitted through the data points was 0.9994.

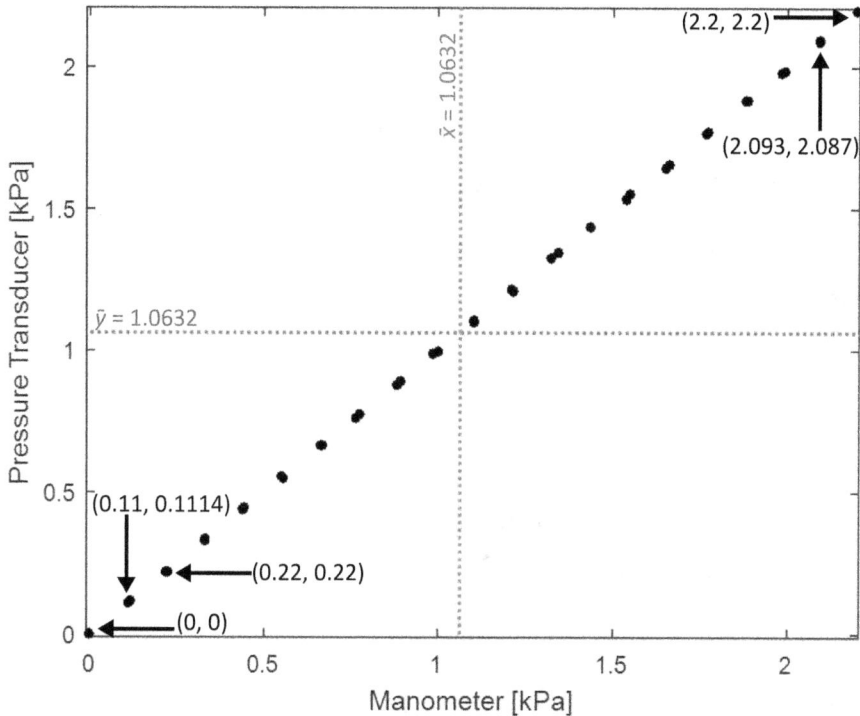

FIGURE 6.16
Sample plot of the pressure transducer measurements as a function of the pressure drops measured by the manometer. The dotted horizontal and vertical lines represent the average pressure transducer and manometer measurements, respectively.

S_{xx} was determined using Eq. (6.42) by substituting x_i with each manometer measurement and $\bar{x}$ with the average of all the manometer measurements:

$$S_{xx} = \sum_{i=1}^{N} (x_i - \bar{x})^2$$

$$= (0 - 1.0632)^2 + (0.1100 - 1.0632)^2$$

$$+ (0.2200 - 1.0632)^2 + \cdots + (2.0930 - 1.0632)^2 + (2.2000 - 1.0632)^2$$

$$= 18.2236$$

S_{xy} was determined using Eq. (6.43) by substituting y_i with each pressure transducer measurement and $\bar{y}$ with the average of all the pressure transducer measurements:

$$S_{xy} = \sum_{i=1}^{N} (x_i - \bar{x})(y_i - \bar{y})$$

$$= (0 - 1.0632)(0 - 1.0632) + (0.1100 - 1.0632)(0.1114 - 1.0632)$$

$$+ (0.2200 - 1.0632)(0.2200 - 1.0632) + \cdots + (2.0930 - 1.0632)(2.0874 - 1.0632)$$

$$+ (2.2000 - 1.0632)(2.2000 - 1.0632)$$

$$= 18.2123$$

The parameters b and a were then determined using Eqs. (6.44) and (6.45):

$$b = \frac{S_{xy}}{S_{xx}}$$

$$= \frac{18.2123}{18.2236}$$

$$= 0.9994$$

$$a = \bar{y} - b\bar{x}$$

$$= 1.0632 - 0.99994 * 1.0632$$

$$= 0.0007$$

With both a and b known, y_{ci} of each pressure drop increment was calculated from Eq. (6.46). Next, S_{yx} was determined as follows, where N was the number of pressure drop increments:

$$S_{yx} = \sqrt{\frac{\sum_{i=1}^{N} (y_i - y_{ci})^2}{N-2}} = \sqrt{\frac{\text{sum1}}{N-2}}$$

$$\text{sum1} = \sum_{i=1}^{N} (y_i - y_{ci})^2$$

$$= (0 - 0.0007)^2 + (0.1114 - 0.1106)^2$$

$$+ (0.2200 - 0.2206)^2 + \cdots + (2.0874 - 2.0924)^2 + (2.2000 - 2.1993)^2$$

$$S_{yx} = \sqrt{\frac{\text{sum1}}{19}}$$

$$= 0.0040$$

The precision component, p, of the pressure transducer uncertainty at 0.67 kPa was then calculated from Eq. (6.41):

$$p = \pm \frac{tS_{yx}}{m} \sqrt{\frac{1}{N} + \frac{1}{M} + \frac{(x_i - \bar{x})^2}{S_{xx}}}$$

$$= \pm \frac{2.08 \times 0.0040}{0.9994} \sqrt{\frac{1}{21} + \frac{1}{100} + \frac{(0.6700 - 1.0632)^2}{18.2236}}$$

$$= \pm 0.0021$$

where t is Student's t-variable and M is the number of measuring points that were logged during each measurement.

The accuracy of the manometer used during the calibration was 0.003 kPa, and it was used as the bias component, b_i, of the pressure transducer uncertainty. The overall uncertainty of the pressure transducer at 0.67 kPa was then calculated using Eq. (6.40):

$$\delta x_i = \left(b_i^2 + p_i^2 \right)^{\frac{1}{2}}$$

$$= \left(0.003^2 + 0.0021^2 \right)^{\frac{1}{2}}$$

$$= 0.0037 \text{ kPa}$$

The uncertainty of this pressure transducer decreased from 0.0041 kPa at 0 to 0.0036 kPa at 0.77–1.21 kPa, and then, it increased again to 0.0043 kPa at 2.2 kPa. However, as these uncertainties were always less than the manufacturer-specified inaccuracy of 0.0055 kPa, the value of 0.0055 kPa was used as the uncertainty of the pressure transducer over the entire pressure range.

6.6.2.3 Uncertainty Analysis: Reynolds Number

Water flowed through a horizontal tube (inner diameter of 11.52 mm and cross-sectional area of 1.0423 m^2) with a mass flow rate of 0.0213 kg/s (the corresponding bulk Reynolds number was 2,683). The viscosity at the bulk fluid temperature was 8.77×10^{-4} kg/ms.

To calculate the Reynolds number uncertainty, the uncertainties of the tube diameter, δD; viscosity, $\delta \mu$; mass flow rate, $\delta \dot{m}$; and cross-sectional area, δA, had to be known. A split-ball instrument and a vernier caliper with an inaccuracy of 20 μm were used to measure the inner diameter of the test section. The viscosity was calculated using the thermophysical equations of Popiel and Wojtkowiak (1998), and the uncertainty was 1%. The mass flow rate was measured using a Coriolis flow meter with an inaccuracy of 0.05% of the full scale; thus, the inaccuracy was 1.3429×10^{-6} kg/s.

The uncertainty of the cross-sectional area was calculated as follows:

$$A_c = \frac{\pi}{4}D^2$$

$$\delta A_c = \left[\left(\frac{\partial A_c}{\partial D}\delta D\right)^2\right]^{\frac{1}{2}}$$

$$= \frac{\pi D}{2}\delta D$$

$$= \frac{\pi \times 0.01152}{2}\times 20\times 10^{-6}$$

$$= 3.62\times 10^{-7}\ \mathrm{m}^2$$

The uncertainty of the Reynolds number was calculated as follows:

$$Re = \frac{\dot{m}D}{\mu A_c}$$

$$\delta Re = \left[\left(\frac{\partial Re}{\partial \dot{m}}\delta \dot{m}\right)^2 + \left(\frac{\partial Re}{\partial D}\delta D\right)^2 + \left(\frac{\partial Re}{\partial \mu}\delta \mu\right)^2 + \left(\frac{\partial Re}{\partial A_c}\delta A_c\right)^2\right]^{\frac{1}{2}}$$

$$= \left[\left(\frac{D}{\mu A_c}\delta \dot{m}\right)^2 + \left(\frac{\dot{m}}{\mu A_c}\delta D\right)^2 + \left(-\frac{\dot{m}D}{\mu^2 A_c}\delta \mu\right)^2 + \left(-\frac{\dot{m}D}{\mu A^2}\delta A_c\right)^2\right]^{\frac{1}{2}}$$

$$= \left[\left(\frac{0.01152}{8.77\times 10^{-4}\times 1.0423\times 10^{-4}}\times 1.3429\times 10^{-6}\right)^2\right.$$

$$+\left(\frac{0.0213}{8.77\times 10^{-4}\times 1.0423\times 10^{-4}}\times 20\times 10^{-6}\right)^2$$

$$+\left(-\frac{0.0213\times 0.01152}{\left(8.77\times 10^{-4}\right)^2\times 1.0423\times 10^{-4}}\times 8.77\times 10^{-6}\right)^2$$

$$+\left.\left(-\frac{0.0213\times 0.01152}{8.77\times 10^{-4}\times\left(1.0423\times 10^{-4}\right)^2}\times 3.62\times 10^{-7}\right)^2\right]^{\frac{1}{2}}$$

$$= 28.8$$

A Reynolds number of 2,683 was used for this experiment; thus, the uncertainty of the Reynolds number was 1.1%.

6.6.2.4 Uncertainty Analysis: Nusselt Number

Water flowed through a horizontal tube (inner diameter of 11.52 mm and length of 9.569 m) with a mass flow rate of 0.0213 kg/s (the corresponding bulk Reynolds number was 2,683). The inlet and outlet temperatures were measured to be 20°C (the standard deviation of the 200 measuring points was 0.03°C) and 31.33°C (the standard deviation of the 200 measuring points was 0.0286°C) using Pt100 probes with a bias of 0.06°C. The thermal conductivity and viscosity at the bulk fluid temperature (25.66°C) were 0.6106 W/m K and 4,180 J/kg K, respectively. The average surface temperature of 30.67°C was obtained from 107 thermocouples (bias of 0.1°C) along the tube length. A heat flux of 3,005.72 W/m² was applied to the tube, and the average heat transfer coefficient and Nusselt number were calculated to be 581.84 W/m²°C and 10.98, respectively.

To calculate the Nusselt number uncertainty, the uncertainties of the tube diameter, δD; specific heat capacity, δCp; thermal conductivity, δk; bulk fluid temperature, δT_b; surface temperature, δT_s; surface area, δA; heat flux, $\delta \dot{q}$; and heat transfer coefficient, δh, were required. The fluid properties were calculated using the thermophysical equations of Popiel and Wojtkowiak (1998), and the uncertainty of the specific heat capacity and thermal conductivity was 0.06% and 2%, respectively. A split-ball instrument and a vernier caliper with an inaccuracy of 20 μm were used to measure the inner diameter of the test section, and a measuring tape with an inaccuracy of 1 mm was used to measure the length of the test section.

The bias, b, of the Pt100 probes that measured the inlet and outlet temperatures was determined to be 0.06°C during the calibration process. The precision, p, was obtained by multiplying the standard deviation of the temperature measurements (obtained from the 200 data points that were logged during each measurement) by Student's t-variable:

$$\delta T_i = \left(b^2 + p^2\right)^{\frac{1}{2}}$$

$$= \left(0.06^2 + (1.981 \times 0.03)^2\right)^{\frac{1}{2}}$$

$$= 0.08445°C$$

Similarly, the uncertainty of the outlet temperature was calculated to be 0.0825°C.

The uncertainty of the bulk fluid temperature was calculated as follows:

$$T_b = \frac{T_i + T_o}{2}$$

$$\delta T_b = \left[\left(\frac{\partial T_b}{\partial T_i} \delta T_i \right)^2 + \left(\frac{\partial T_b}{\partial T_o} \delta T_o \right)^2 \right]^{1/2}$$

$$= \left[\left(\frac{1}{2} \delta T_i \right)^2 + \left(\frac{1}{2} \delta T_o \right)^2 \right]^{1/2}$$

$$= 0.0511°C$$

A total of 107 thermocouples were used to measure the surface temperatures. Similar to that of the Pt100 probes, the bias of the thermocouples was determined to be 0.1°C during the calibration process and the precision was obtained by multiplying the standard deviation of the temperature measurements by Student's *t*-variable:

$$\delta T_s = \left(b^2 + p^2 \right)^{1/2}$$

$$= \left(0.1^2 + (1.981\sigma)^2 \right)^{1/2}$$

The uncertainty of the average surface temperature was then calculated as follows:

$$\overline{T}_s = \frac{T_1 + T_2 + \cdots + T_n}{n}$$

$$\delta \overline{T}_s = \left[\left(\frac{\delta T_1}{n} \right)^2 + \left(\frac{\delta T_2}{n} \right)^2 + \cdots + \left(\frac{\delta T_n}{n} \right)^2 \right]^{1/2}$$

$$= 0.0129°C$$

The uncertainty of the heat flux was calculated as follows:

$$\dot{q} = \frac{\dot{Q}_w}{A} = \frac{\dot{m} C_p (T_o - T_i)}{\pi D L}$$

$$\delta \dot{q} = \left[\left(\frac{\partial \dot{q}}{\partial \dot{m}} \delta \dot{m} \right)^2 + \left(\frac{\partial \dot{q}}{\partial C_p} \delta C_p \right)^2 + \left(\frac{\partial \dot{q}}{\partial T_o} \delta T_o \right)^2 + \left(\frac{\partial \dot{q}}{\partial T_i} \delta T_i \right)^2 + \left(\frac{\partial \dot{q}}{\partial D} \delta D \right)^2 + \left(\frac{\partial \dot{q}}{\partial L} \delta L \right)^2 \right]^{1/2}$$

$$= \left[\left(\frac{C_p (T_o - T_i)}{\pi D L} \delta \dot{m} \right)^2 + \left(\frac{\dot{m} (T_o - T_i)}{\pi D L} \delta C_p \right)^2 + \left(\frac{\dot{m} C_p}{\pi D L} \delta T_o \right)^2 + \left(-\frac{\dot{m} C_p}{\pi D L} \delta T_i \right)^2 \right.$$

$$+\left(-\frac{\dot{m}C_p(T_o-T_i)}{\pi D^2 L}\delta D\right)^2+\left(\frac{-\dot{m}C_p(T_o-T_i)}{\pi D L^2}\delta L\right)^2\Bigg]^{\frac{1}{2}}$$

$$=\Bigg[\left(\frac{4,180(31.3263-20.0004)}{\pi\times 0.01152\times 9.569}\times 1.3429\times 10^{-6}\right)^2$$

$$+\left(\frac{0.0213(31.3263-20.0004)}{\pi\times 0.01152\times 9.569}\times 2.5080\right)^2$$

$$+\left(\frac{0.0213\times 4,180}{\pi\times 0.01152\times 9.569}\times 0.0825\right)^2+\left(-\frac{0.0213\times 4,180}{\pi\times 0.01152\times 9.569}\times 0.0845\right)^2$$

$$+\left(-\frac{0.0213\times 4,180(31.3263-20.0004)}{\pi\times 0.01152^2\times 9.569}\times 20\times 10^{-6}\right)^2$$

$$+\left(\frac{-0.0213\times 4180(31.3263-20.0004)}{\pi\times 0.01152\times 9.569^2}\times 0.001\right)^2\Bigg]^{\frac{1}{2}}$$

$$=30.83\ \text{W/m}^2$$

Once the uncertainties of the bulk and surface temperatures and heat flux were known, the uncertainty of the heat transfer coefficient could be determined:

$$h=\frac{\dot{q}}{(T_s-T_b)}$$

$$\delta h=\left[\left(\frac{\partial h}{\partial \dot{q}}\delta \dot{q}\right)^2+\left(\frac{\partial h}{\partial T_s}\delta T_s\right)^2+\left(\frac{\partial h}{\partial T_b}\delta T_b\right)^2\right]^{\frac{1}{2}}$$

$$=\left[\left(\frac{1}{T_s-T_b}\delta \dot{q}\right)^2+\left(\frac{-\dot{q}}{(T_s-T_b)^2}\delta T_s\right)^2+\left(\frac{\dot{q}}{(T_s-T_b)^2}\delta T_b\right)^2\right]^{\frac{1}{2}}$$

$$=\left[\left(\frac{1}{30.6666-25.6634}\times 30.83\right)^2+\left(\frac{-3005.7195}{(30.6666-25.6634)^2}\times 0.0129\right)^2\right.$$

$$\left.+\left(\frac{3005.7195}{(30.6666-25.6634)^2}\times 0.0511\right)^2\right]^{\frac{1}{2}}$$

$$=9.50\ \text{W/m}^2{}^\circ\text{C}$$

Finally, the Nusselt number uncertainty was determined as follows:

$$Nu = \frac{hD}{k}$$

$$\delta Nu = \left[\left(\frac{\partial Nu}{\partial h} \delta h \right)^2 + \left(\frac{\partial Nu}{\partial D} \delta D \right)^2 + \left(\frac{\partial Nu}{\partial k} \delta k \right)^2 \right]^{\frac{1}{2}}$$

$$= \left[\left(\frac{D}{k} \delta h \right)^2 + \left(\frac{h}{k} \delta D \right)^2 + \left(-\frac{hD}{k^2} \delta k \right)^2 \right]^{\frac{1}{2}}$$

$$= \left[\left(\frac{0.01152}{0.6106} \times 9.50 \right)^2 + \left(\frac{581.8419}{0.6106} \times 20 \times 10^{-6} \right)^2 \right.$$

$$\left. + \left(-\frac{581.8419 \times 0.01152}{0.6106^2} \times 0.0122 \right)^2 \right]^{\frac{1}{2}}$$

$$= 0.284$$

The Nusselt number of this experiment was calculated to be 10.9766; therefore, the Nusselt number uncertainty was 2.6%.

6.6.2.5 Uncertainty Analysis: Friction Factor

Water flowed through a horizontal tube (inner diameter of 11.52 mm and length of 2 m) with a mass flow rate of 0.0213 kg/s (the corresponding bulk Reynolds number was 2,683). The density at the bulk fluid temperature was 996.83 kg/m³. The pressure drop across the tube was measured to be 112.75 Pa (the standard deviation of the 200 measuring points was 1.8 Pa) using a pressure transducer with a bias of 5.5 Pa.

To calculate the friction factor uncertainty, the uncertainties of the pressure drop, $\delta \Delta P$; tube diameter, δD; tube length, δL; density, $\delta \rho$; and mass flow rate, $\delta \dot{m}$, were required. The fluid properties were calculated using the thermophysical equations of Popiel and Wojtkowiak (1998), and the uncertainty of the density was 0.004%. A split-ball instrument and a vernier caliper with an inaccuracy of 20 μm were used to measure the inner diameter of the test section, and a measuring tape with an inaccuracy of 1 mm was used to measure the length of the test section.

The bias of the pressure transducer was determined to be 0.0055 kPa during the calibration process. The precision was obtained by multiplying the standard deviation of the pressure drop measurements (obtained from the 200 data points that were logged during each measurement) by Student's *t*-variable:

$$\delta \Delta P = \left(b^2 + p^2\right)^{\frac{1}{2}}$$

$$= \left(0.0055^2 + \left(1.981\sigma\right)^2\right)^{\frac{1}{2}}$$

The friction factor uncertainty was then calculated as follows:

$$f = \frac{2\Delta P D}{\rho V^2 L} = \frac{\Delta P \rho D^5 \pi^2}{8\dot{m}^2 L}$$

$$\delta f = \left[\left(\frac{\partial f}{\partial \Delta P}\delta \Delta P\right)^2 + \left(\frac{\partial f}{\partial \rho}\delta \rho\right)^2 + \left(\frac{\partial f}{\partial D}\delta D\right)^2 + \left(\frac{\partial f}{\partial \dot{m}}\delta \dot{m}\right)^2 + \left(\frac{\partial f}{\partial L}\delta L\right)^2\right]^{\frac{1}{2}}$$

$$= \left[\left(\frac{\rho D^5 \pi^2}{8\dot{m}^2 L}\delta \Delta P\right)^2 + \left(\frac{\Delta P D^5 \pi^2}{8\dot{m}^2 L}\delta \rho\right)^2 + \left(\frac{5\Delta P \rho D^4 \pi^2}{8\dot{m}^2 L}\delta D\right)^2\right.$$

$$\left. + \left(\frac{-2\Delta P \rho D^5 \pi^2}{8\dot{m}^3 L}\delta \dot{m}\right)^2 + \left(\frac{-\Delta P \rho D^5 \pi^2}{8\dot{m}^2 L^2}\delta L\right)^2\right]^{\frac{1}{2}}$$

$$= \left[\left(\frac{996.8266 \times 0.01152^5 \times \pi^2}{8 \times 0.0213^2 \times 2.027} \times 19.7385\right)^2\right.$$

$$+ \left(\frac{112.7491 \times 0.01152^5 \times \pi^2}{8 \times 0.0213^2 \times 2.027} \times 0.0399\right)^2$$

$$+ \left(\frac{5 \times 112.7491 \times 996.8266 \times 0.01152^4 \times \pi^2}{8 \times 0.0213^2 \times 2.027} \times 20 \times 10^{-6}\right)^2$$

$$+ \left(\frac{-2 \times 112.7491 \times 996.8266 \times 0.01152^5 \times \pi^2}{8 \times 0.0213^3 \times 2.027} \times 1.3429 \times 10^{-6}\right)^2$$

$$\left. + \left(\frac{-112.7491 \times 996.8266 \times 0.01152^5 \times \pi^2}{8 \times 0.0213^2 \times 2.027^2} \times 0.001\right)^2\right]^{\frac{1}{2}}$$

$$= 0.0015$$

The calculated friction factor of this experiment was 0.0310; therefore, the friction factor uncertainty was 4.9%.

6.7 Validation

Proper and rigorous validation is essential to have confidence in the experimental results. When choosing appropriate correlations to validate the experimental data, it is of the utmost importance to take note of the conditions for which the specific correlation was developed (forced or mixed convection, developing or fully developed flow, flow regime, boundary condition, inlet geometry) and the range for which the correlation is valid.

Table 6.2 (laminar flow) and Table 6.3 (turbulent flow) contain a summary of the heat transfer correlations, and Table 6.4 (laminar flow) and Table 6.5 (turbulent flow) contain a summary of the pressure drop correlations that we have found worked well for validating our experimental data. The Nusselt numbers and friction factors in the transitional flow regime were significantly affected by the inlet geometry, axial position, free convection effects, and type of fluid. This explains why the available correlations were suitable for limited ranges only. When the experiments were conducted in the transitional flow regime, the Nusselt numbers and friction factors in the laminar and turbulent flow regimes were validated. When the results in these two flow regimes correlated well with existing correlations, the results in the transitional flow regime were also expected to be accurate and reliable.

6.8 Unexplained Challenges

6.8.1 Flow-Calming Section Contents

To the authors' best knowledge, there is no scientific justification for the specific materials that were used to straighten the flow through the flow-calming section. The flow-calming sections that were used were all based on the flow-calming section design of Ghajar and Madon (1992). In an attempt to determine the significance of the flow-calming section contents, limited heat transfer and pressure drop experiments were conducted in the laminar, transitional, quasi-turbulent, and turbulent flow regimes using a flow-calming section similar to that in Figure 6.8 and an empty flow-calming section (without any contents) (Bashir et al. 2019). It was found that there was no significant difference in the results obtained using the different flow-calming sections in all the flow regimes. Furthermore, Bashir et al. (2019) also investigated the effects of different contraction ratios (from the flow-calming section diameter to the test section diameter). It was found that the transitional flow regime was affected for contraction ratios lower than 33; however, the effect became negligible for higher contraction ratios. The laminar and turbulent flow regimes remained unaffected by the different contraction ratios.

TABLE 6.2

Laminar Nusselt Number Correlations

Morcos and Bergles (1975)

$$Nu = \left\{ 4.36^2 + \left[0.145 \left(\frac{Gr_f^* Pr_f^{1.35}}{Pw_f^{*0.25}} \right)^{0.265} \right]^2 \right\}^{1/2}$$

$$Pw^* = \frac{Pw}{Nu} = \frac{kD}{k_{\text{tube}}t}$$

Fully developed flow
Mixed convection conditions

Shah and London (1978)

$$Nu_{SL} = Nu_1 \times Nu_2 - 1$$

$$Nu_1 = \left[1 + \left(\frac{\pi / (115.2\, z^*)}{\left\{ 1 + (Pr/0.0207)^{2/3} \right\}^{1/2} \left\{ 1 + (220\, z^*/\pi)^{-10/9} \right\}^{3/5}} \right)^{5/3} \right]^{3/10}$$

$$Nu_2 = 5.364 \left[1 + (220\, z^*/\pi)^{-10/9} \right]^{3/10}$$

$$z^* = \frac{\pi}{4Gz}$$

Developing and fully developed flow
Forced convection conditions

Meyer and Everts (2018)

$$Nu = 4.36 + \left(Nu_1^6 + Nu_2^6 \right)^{1/6}$$

$$Nu_1 = \left(0.33 Gz^{0.54} - 0.84 \right) Pr^{-0.2}$$

$$Nu_2 = \left(0.207 Gr^{0.305} - 1.19 \right) Pr^{0.5} Gz^{-0.08}$$

Developing and fully developed flow
Mixed convection conditions

Meyer and Everts (2018)

$$\overline{Nu} = 4.36 + \overline{Nu_1} + \overline{Nu_2}$$

$$\overline{Nu_1} = \frac{1}{L} \int_0^{Lt_{MCD}} Nu_1 dL = \frac{1}{L} \left(-0.84 Pr_b^{-0.2} Lt_{MCD} + 0.72 (Re_b D)^{0.54} Pr_b^{0.34} Lt_{MCD}^{0.46} \right)$$

$$\overline{Nu_2} = \frac{1}{L} \int_{Lt_{MCD}}^{L} Nu_2 dL = \frac{1}{L} \left(0.207 Gr_b^{0.305} - 1.19 \right) Pr_b^{0.42} (Re_b D)^{-0.08} (L - Lt_{MCD})$$

$$Lt_{MCD} = \frac{2.4 Re_b Pr_b^{0.6} D}{Gr_b^{0.57}} \text{ for } L > Lt_{MCD}$$

$Lt_{MCD} = L$ for $L < Lt_{MCD}$
Developing and fully developed flow
Mixed convection conditions

TABLE 6.3

Turbulent Nusselt Number Correlations

Gnielinski (1976)

$$Nu = \frac{(\xi/8)(Re-1000)Pr}{1+12.7\sqrt{(\xi/8)}\left(Pr^{2/3}-1\right)}\left[1+\left(\frac{D}{L}\right)^{2/3}\right]\left(\frac{Pr}{Pr_s}\right)^{0.11}$$

$\xi = (1.8\log_1^0 Re - 1.5)^{-2}$ (Filonenko 1954)

Ghajar and Tam (1994)

$Nu = 0.023\ Re^{0.8}\ Pr^{0.385}\ (\ x/D)^{-0.0054}\ (\mu/\mu_s)^{0.14}$

Meyer et al. (2019)

$$Nu = 0.018Re^{-0.25}\left(Re-500\right)^{1.07}Pr^{0.42}\left(\frac{Pr}{Pr_s}\right)^{0.11}$$

TABLE 6.4

Laminar Friction Factor Correlations

Classical relation by Poiseuille (Cengel and Ghajar 2015)

$$f = \frac{64}{Re}$$

Fully developed flow
Isothermal flow

Tam et al. (2013)

$$f_{app} = \frac{4}{Re}\left(16 + \frac{0.00314}{0.00004836 + 0.0609\left(\dfrac{x/D}{Re}\right)^{1.28}}\right)$$

Developing flow
Isothermal flow

Tam and Ghajar (1998) conducted a study where three different meshes were placed at the outlet of the flow-calming section (25.4 mm before the bellmouth inlet). The purpose of this study was to investigate the effect of different inlet disturbances on the local heat transfer coefficients specifically for a bellmouth inlet. It was found that inlet turbulence levels significantly affected the local heat transfer characteristics and that transition occurred earlier as the inlet turbulence level was increased. Although this study did not focus on the contents of the flow-calming section, it can be concluded that the heat transfer (and most likely the pressure drop) characteristics will be influenced by the flow-calming section contents close to the inlet of the test section.

Although the work of Tam and Ghajar (1998) and Bashir et al. (2019) addressed the inlet turbulence, contents, and contraction ratio of the flow-calming

TABLE 6.5

Turbulent Friction Factor Correlations

Blasius (White 2009)

$$f = 0.316Re^{-0.25}$$

Isothermal flow

Fang et al. (2011)

$$f = 0.25\left[\log\left(\frac{150.39}{Re^{0.98865}} - \frac{152.66}{Re}\right)\right]^{-2}$$

Isothermal flow

Allen and Eckert (1964)

$$f = 0.316Re^{-0.25}\left(\frac{\mu_b}{\mu_s}\right)^{-0.25}$$

Diabatic flow

section, it is not yet sufficient to understand the physics of the flow-calming section. More work is required to develop guidelines regarding the diameter, length, and contents that should be used for the flow-calming section.

6.8.2 Upstream Effects Caused by Exit Mixer

It was found that in almost all our experiments, the temperatures measured by the last thermocouple station on the test section (before the mixing section) were lower than the rest, even when the flow was fully developed. The result was that the Nusselt numbers at the last measuring station were suddenly higher than those at the previous stations. A possible reason for this phenomenon was the upstream effects caused by the exit mixer. Although tube lengths of $x/D = 26$–250 between the last measuring station and the mixing section were allowed, the temperature measurements remained incorrect. Up to now, no information is available on the minimum required tube length to prevent any upstream effects caused by the exit mixer.

6.8.3 Logging Frequency and Sampling Time

During the experiments, the data were logged at frequencies of 10 and 20 Hz for a period of usually 20 s (therefore, ~200 data points). We have found that this was sufficient to obtain data that gave a suitable steady-state measurement in the laminar and turbulent flow regimes. However, in the transitional flow regime, the flow alternated between the laminar and turbulent flow regimes, which led to mass flow rate, temperature, and pressure drop fluctuations. To obtain a good representative steady-state measurement, the number of data points during a measurement was increased to 400.

We briefly tested higher sampling frequencies (up to 500 Hz) but did not find that they produced better results. When sampling at very high frequencies, care should also be taken to prevent the results from being affected by noise caused by the equipment. It is therefore recommended that more work should be dedicated to the sampling frequency and sampling time to determine the optimum sampling frequency and sampling time in each flow regime.

6.8.4 Small Diameter Copper Test Section

As our experiments were conducted using test sections with inner diameters between 4 and 19 mm, our expertise is on macro-tubes and not necessarily on mini- or micro-tubes. When experiments were conducted on copper test sections with an inner diameter of 4.04 mm, outer diameter of 5.04 mm, and lengths of 4.5–8.3 m, challenges existed that were not necessarily a problem in the larger diameter copper test sections or the 4-mm stainless steel test section.

The test sections were heated by coiling heating wire along the test section and passing current through the heating wire. However, once the power supply was switched on, significant scatter existed in the temperature measurements. This problem was solved after the test bench was earthed.

Because the diameter was very small compared with the length, the copper test section was no longer rigid, and care was taken to ensure that the test section was completely straight and level along the entire tube length. It was expected that the local heat transfer coefficients will be affected by bends along the tube length.

Accurate and reliable results were obtained in the laminar and transitional flow regimes, and in the turbulent flow regime for Reynolds numbers <7,000. However, as the Reynolds numbers were increased further, the experimental Nusselt numbers were lower than the Nusselt numbers predicted by the existing correlations. More work is required to explain this phenomenon.

When the pressure taps were attached to the test section, a hole whose diameter was 10% of the inner diameter of the test section, or less, was drilled through the pressure tap and test section. Rayle (1959) concluded that this would ensure that the boundary layer and pressure drop measurements were not affected. Furthermore, care was taken to remove any burrs from the drilling process. However, it was found that temperature measurements of the thermocouples at a measuring station very close to the pressure tap were affected, and the result was that the Nusselt numbers at these measuring stations were too high. A possible reason for this might be that, for small tubes, the hole of the pressure taps should be even <10% of the inner diameter. However, more work is required to develop guidelines regarding the pressure tap holes in small-diameter test sections.

6.8.5 Electric Noise

To prevent noise from influencing the results, it was very important to ensure that good and systematic processes and techniques were implemented throughout the wiring and connection of the equipment and instrumentation. Therefore, shielded cables were used to connect the power supplies to the test sections, and also for the lead wires of the Pt100 probes, pressure transducers, and flow meters. Furthermore, during the planning of the wiring layout, care was taken not to route power cables for instrumentation together with the sense cables of the measuring instruments.

Equipment such as the power supplies, which were known to transmit noise specifically to the thermocouples, was also earthed. Figure 6.17 contains a sample plot of measurements that were taken on a setup that contained a power supply that was not earthed. Although the power of the power supply was switched on, no output power was supplied to the test section yet; therefore, all the thermocouples were expected to measure approximately the same temperature. This figure indicates that significant scatter existed and the difference between the thermocouples was ~1°C, which is one order of magnitude larger than the uncertainty of 0.1°C. Once the power supply was switched off, the scatter decreased to 0.1°C, which was within the uncertainty of the thermocouples. The scatter then increased again to 1°C when the power supply was switched on.

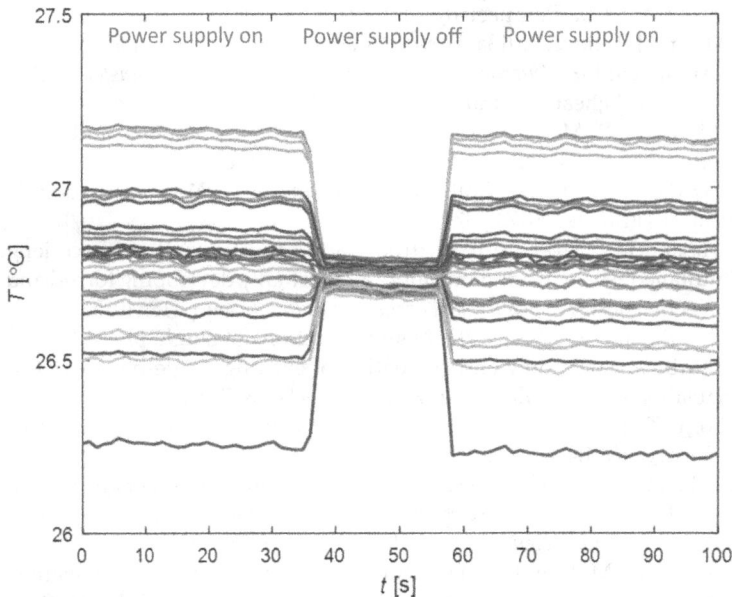

FIGURE 6.17
Temperature as a function of time measured when an unearthed power supply was switched on and off.

References

Allen, R.W., and E.R.G. Eckert. 1964. Friction and heat-transfer measurements to turbulent pipe flow of water (Pr=7 and 8) at uniform wall heat flux. *Journal of Heat Transfer* 86(3):301–310.

Bakker, A., R.D. LaRoche, and E.M. Marshall. 2000. Laminar flow in static mixers with helical elements. Last Modified 15 February 2000 Accessed 24 February. http://www.bakker.org/cfmbook/lamstat.pdf.

Bashir, A.I., J.P. Meyer, and M. Everts. 2019. Influence of inlet contraction ratios on the heat transfer and pressure drop characteristics of single-phase flow in smooth circular tubes in the transitional flow. *Experimental Thermal and Fluid Science* 109:1–16.

Cengel, Y.A., and A.J. Ghajar. 2015. *Heat and Mass Transfer: Fundamentals and Applications*, 5th ed. New York: McGraw-Hill.

Dieck, R.H. 2017. *Measurement Uncertainty: Methods and Applications*. Research Triangle Park: ISA.

Dunn, P.F. 2010. *Measurement and Data Analysis for Engineering and Science*, 2nd ed. Baca Raton, FL: CRC Press.

Everts, M. 2014. Heat transfer and pressure drop of developing flow in smooth tubes in the transitional flow regime. Master of Engineering Master's dissertation, Department of Mechanical and Aeronautical Engineering, University of Pretoria.

Everts, M. 2017. Single-phase mixed convection of developing and fully developed flow in smooth horizontal circular tubes in the laminar, transitional, quasi-turbulent and turbulent flow regimes. PhD thesis, Department of Mechanical and Aeronautical Engineering, University of Pretoria.

Everts, M., and J.P. Meyer. 2018a. Flow regime maps for smooth horizontal tubes at a constant heat flux. *International Journal of Heat and Mass Transfer* 117:1274–1290. doi: 10.1016/j.ijheatmasstransfer.2017.10.073.

Everts, M., and J.P. Meyer. 2018b. Heat transfer of developing and fully developed flow in smooth horizontal tubes in the transitional flow regime. *International Journal of Heat and Mass Transfer* 117:1331–1351. doi: 10.1016/j.ijheatmasstransfer.2017.10.071.

Everts, M., and J.P. Meyer. 2020. Hydrodynamic and thermal entrance lengths for simultaneously hydrodynamically and thermally developing forced and mixed convective flow. *Experimental Thermal and Fluid Science* 118:110153.

Fang, X., Y. Xu, and Z. Zhou. 2011. New correlations of single-phase friction factor for turbulent pipe flow and evaluation of existing single-phase friction factor correlations. *Nuclear Engineering and Design* 241:897–902.

Filonenko, G.K. 1954. Hydraulischer Widerstand von Rohrleitungen (Orig. Russ.). *Teploenergetika* 1(4):40–44.

Ghajar, A.J., and K.F. Madon. 1992. Pressure drop measurements in the transition region for a circular tube with three different inlet configurations. *Experimental Thermal and Fluid Science* 5(1):129–135.

Ghajar, A.J., and L.M. Tam. 1994. Heat transfer measurements and correlations in the transition region for a circular tube with three different inlet configurations. *Experimental Thermal and Fluid Science* 8(1):79–90.

Gnielinski, V. 1976. New equations for heat and mass-transfer in turbulent pipe and channel flow. *International Chemical Engineering* 16(2):359–368.

Kline, S.J., and F.A. McClintock. 1953. Describing uncertainties in single-sample experiments. *Mechanical Engineering* 75(1):3–8.

Kutz, M. 2013. *Handbook of Measurement in Science and Engineering*. Hoboken, NJ: John Wiley & Sons.

LaNasa, P.J., and E. Loy Upp. 2014. *Fluid Flow Measurement: A Practical Guide to Accurate Flow Measurement*. Oxford: Elsevier.

Lawrence, A.E., and T.K. Sherwood. 1931. Heat transmission to water flowing in pipes. *Industrial and Engineering Chemistry* 23(3):301–309.

Metais, B., and E.R.G. Eckert. 1964. Forced, mixed, and free convection regimes. *Journal of Heat Transfer* 86(2):295–296.

Meyer, J.P., and M. Everts. 2018. Single-phase mixed convection of developing and fully developed flow in smooth horizontal circular tubes in the laminar and transitional flow regimes. *International Journal of Heat and Mass Transfer* 117:-1251–1273. doi: 10.1016/j.ijheatmasstransfer.2017.10.070.

Meyer, J.P., M. Everts, N. Coetzee, K. Grote, and M. Steyn. 2019. Heat transfer characteristics of quasi-turbulent and turbulent flow in smooth circular tubes. *International Communications in Heat and Mass Transfer* 105:84–106. doi: 10.1016/j.icheatmasstransfer.2019.03.016.

Moffat, R.J. 1988. Describing uncertainties in experimental results. *Experimental Thermal and Fluid Science* 1:3–17.

Morcos, S.M., and A.E. Bergles. 1975. Experimental investigation of combined forced and free laminar convection in horizontal tubes. *Journal of Heat Transfer* 97(2):212–219.

Morris, A.S., and R. Langari. 2016. *Measurement and Instrumentation: Theory and Application*, 2nd ed. London: Elsevier.

Nagsarkar, T.K., and M.S. Sukhija. 2014. *Power System Analysis*. New Delhi: Oxford University Press.

Popiel, C.O., and J. Wojtkowiak. 1998. Simple formulas for thermophysical properties of liquid water for heat transfer calculations [from 0°C to 150°C]. *Heat Transfer Engineering* 19(3):87–101.

Raghavendra, N.V., and L. Krishnamurthy. 2013. *Engineering Metrology and Measurements*. New Delhi: Oxford University Press.

Rayle, R.E. 1959. *Influence of Orifice Geometry on Static Pressure Measurements*. Bristol: American Society of Mechanical Engineers.

Shah, R.K., and A.L. London. 1978. *Laminar Flow Forced Convection in Ducts*. New York: Academic Press.

Tam, H.K., L.M. Tam, and A.J. Ghajar. 2013. Effect of inlet geometries and heating on the entrance and fully-developed friction factors in the laminar and transition regions of a horizontal tube. *Experimental Thermal and Fluid Science* 44:680–696. doi: 10.1016/j.expthermflusci.2012.09.008.

Tam, L.M., and A.J. Ghajar. 1998. The unusual behavior of local heat transfer coefficient in a circular tube with a bell-mouth inlet. *Experimental Thermal and Fluid Science* 16(3):187–194. doi: 10.1016/S0894-1777(97)10019-X.

Ulaby, F.T., E. Michielssen, and U. Ravaioli. 2010. *Fundamentals of Applied Electromagnetics*, 6th ed. Boston, MA: Pearson.

Van der Westhuizen, J.E., J. Dirker, and J.P. Meyer. 2014. Investigation into using liquid crystal thermography for measuring heat transfer coefficients and wall temperature profiles at inlets and underdeveloped regions. *Heat Transfer, Fluid Mechanics and Thermodynamics (HEFAT)*, Orlando, FL.

White, F.M. 2009. *Fluid Mechanics*, 6th ed. Singapore: McGraw-Hill.

7

Stability Evaluation, Measurements, and Presentations of Convective Heat Transfer Characteristics of Nanofluids

S.M. Sohel Murshed
Instituto Superior Técnico da Universidade de Lisboa

Mohsen Sharifpur, Solomon Giwa, and Josua P. Meyer
University of Pretoria

CONTENTS

7.1 Introduction

Since coining of the term "nanofluids" in 1995 (Choi, 1995) and despite still having scattered data and controversies about the mechanisms of heat transfer in nanofluids (NFs), some progress has been made in some areas of research and development of this novel class of heat transfer fluids (Murshed and Nieto de Castro, 2014). This is due to extensive studies that have been conducted on various thermophysical properties and heat transfer features of nanofluids. Most of the studies in the literature found higher thermophysical properties such as thermal conductivity, viscosity, and heat capacity of nanofluids than those of their base fluids. In addition, these properties also further increase with increasing concentration of nanoparticles (Eastman et al., 1997, 2001; Murshed, 2011, 2012; Murshed et al., 2006, 2008a, 2008b, 2010; Adio et al., 2016a; Tshimanga et al., 2016; Sharifpur et al., 2016). Numerous research works have also been performed on single-phase flow and phase change-based heat transfer features such as the convective heat transfer and flow under natural and forced (laminar, transition, and turbulence) convection of NFs in various geometries (such as cavities, ducts, and channels—macro- to microchannels (MCHs)) as well as their boiling heat transfer evaluation (Lee et al., 2011; Murshed et al., 2008c; Jung et al., 2009; Ghodsinezhad et al., 2016; Garbadeen et al., 2017; Sharifpur et al., 2018). Nanofluids showed even larger enhancement (compared to thermophysical properties) of these single-phase and two-phase heat transfer features (Murshed et al., 2011; Lomascolo et al., 2015). Although research directions of NFs have been expanded rapidly from property characterization to energy harvesting, the primary aim of employing NFs in different engineering applications is to improve the cooling performance of conventional cooling media through increased convection heat transfer.

Due to immense research interest from researchers of multiple disciplines and also due to huge potential in numerous fields and applications, NFs became a very popular research topic in recent years, which can be evidenced from activities and publications related to this topic. For example, Figure 7.1, which provides records of publications on NFs and its various areas extracted and refined from Web of Science (all databases and types of records, searched on December 31, 2018), shows an extensive and exponential increase in research publications on this topic. As mentioned before and illustrated in Figure 7.1, thermal conductivity of NFs received the most research focus and viscosity has also received increasing attention from the researchers. However, despite enormous importance in real applications particularly in the cooling field, convective heat transfer is relatively less studied (Figure 7.1). The main reasons could be comparatively complicated and expensive experimental facilities as well as large volume of NFs required to evaluate their convective heat transfer performance, whereas thermal conductivity and viscosity can easily (may be compromising

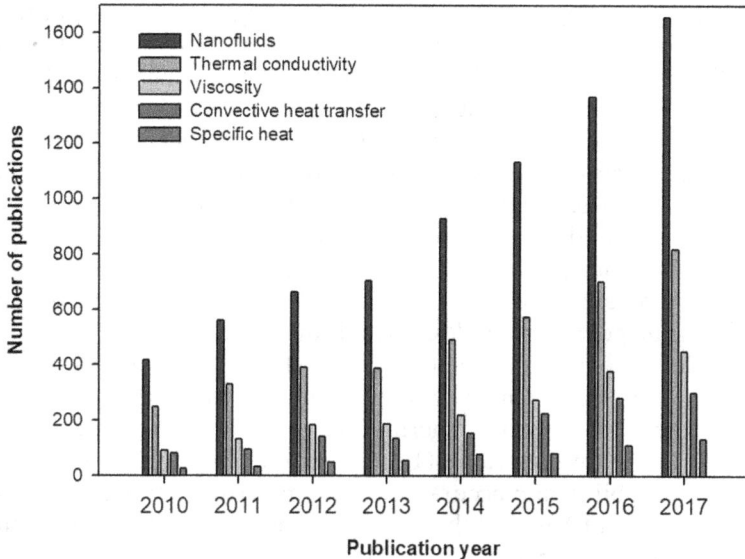

FIGURE 7.1
Publication records of various areas of nanofluids over the past several years (Web of Science).

accuracy) be determined using commercial devices or equipment such as thermal conductivity probe (such as KD2 Pro) and rheometer. It is also to be noted that despite extensive studies on thermal conductivity of NFs, neither the literature results are consistent nor the underlying mechanisms are yet explored. Furthermore, there are some controversies on the reported enhancement and the mechanisms of thermal conductivity of NFs as highlighted by some researchers (e.g., Murshed, 2009).

Despite intensified research on various areas of NFs, the main challenge of this new class of fluids is to achieve long-term stability of nanoparticles (NP) as the stability of nanofluids is critical for their enhanced properties as well as applications. Owing to the Van der Waals attraction forces among the nanosized particles, NP agglomerate, which is a disadvantage that is surmountable (Ghadimi et al., 2011; Babita and Gupta, 2016; Nieto de Castro et al., 2017). Several methods have also been used to enhance and formulate (Adio et al., 2016b) the stability of nanofluids, and monitoring techniques are also available to check the stability of NFs. However, no common and well-accepted protocol or procedure is yet available, and thus, more systematic studies and careful analysis of this crucial factor are of great importance for their sustainable applications.

The pioneering efforts of Tuckerman and Pease (1981) brought forth the use of microchannel heat sinks (MCHS) as cooling devices of which Lee and Choi (1996) were the first to study the convective thermal performance of nanofluids in a MCH. Also, Putra and co-workers (Putra et al., 2003) pioneered the investigation of hydrothermal characteristics of NFs in a cavity.

For both studies using NFs in a MCH and cavity, convective heat transfer enhancement and deterioration, respectively, are reported. Later, Chein and Huang (2005) also analytically determined and analyzed the performance of NFs in a MCH. Subsequent to these foremost investigations, numerous researchers have conducted studies on the convective heat transfer and flow performance of various types of NFs in a different configuration of cavities and shapes of MCHs. Noteworthy is the fact that numerical, theoretical, and experimental methods have been employed in these studies in the literature, of which the numerical method is the most reported. However, this work centers solely on experimental works carried out to examine the use of NFs as thermal transport media in MCHs and cavities.

Nonetheless, the importance of long-term stability of NFs and prominent role of accurate measurement of key parameters in experimental studies, especially those of convective thermal transport and flow performance of NFs in MCHs and cavities, cannot be overlooked. Thus, the achievement and evaluation of stability and accurate measurements of convective heat transfer characteristics of NFs and the way experimental results are presented in the literature are the focus of this study.

7.2 Nanofluid Synthesis and Stability Evaluation

7.2.1 Nanofluid Synthesis and Preparation

Nanofluids are commonly prepared/produced through two basic routes: single-step and two-step routes. The single-step method consists of both the synthesis of the nanoparticles (physical or chemical path) and their dispersion into the base fluid in a single step. One of the main advantages of this route is that better dispersion and stability of NFs can be achieved, whereas the main limitation is the production of small quantity of NFs. Commonly used single-step route includes chemical vapor deposition (CVD) and direct evaporation and condensation (DEC) techniques. On the other hand, the two-step technique involves synthesizing or acquiring (from company) the nanoparticles in the first step and then dispersing those acquired nanoparticles of desired concentration in base heat transfer fluids such as water and ethylene glycol. The two-step techniques are the most widely used route as it is cost-effective and can produce NFs of large quantity for industrial application. There are many companies in the market selling wide varieties of nanoparticles, and there are also some comparatively cheaper nanoparticles for the preparation of NFs. However, the main drawbacks of NF preparation through this route are as follows: Nanoparticles are usually not well dispersed and get agglomerated, and thus, the sample NFs need to undergo

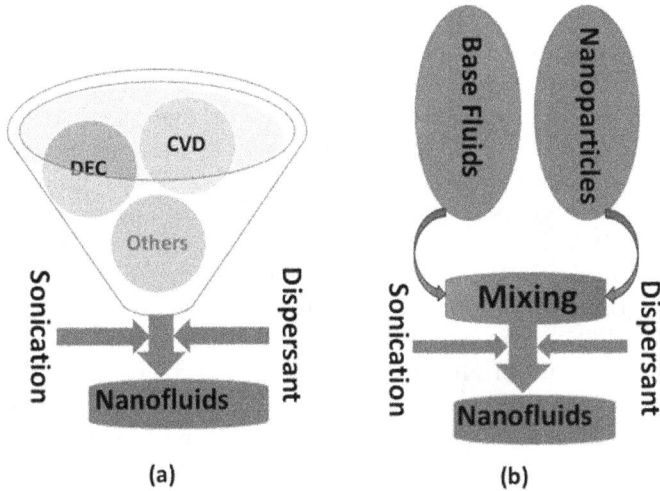

FIGURE 7.2
Schematic of (a) single-step and (b) two-step nanofluid synthesis methods.

some stability testing procedures. In both routes, besides employing (ultra) sonication, researchers also commonly add small concentration of various types of dispersants (also known as surfactants) to nanofluids in order to minimize the agglomeration of nanoparticles. For better understanding, a schematic of both single-step and two-step NFs preparation processes is illustrated in Figure 7.2. In addition to these two routes, there are some other techniques that are also employed by researchers to prepare NFs in small quantity, and most of them are chemical solution- or reaction-based routes. More detailed information on NF synthesis, preparation, and nanoparticle dispersion can be found in the literature (e.g., Murshed and Nieto de Castro, 2014; Babita and Gupta, 2016), and it will not be elaborated further.

7.2.2 Stability Evaluation of Nanofluids

As mentioned before, one of the main challenges toward the development and real applications of NFs is their long-term stability. Stability of NFs is not only necessary for their optimum properties but also an important requirement for their heat transfer and other static or flow-based applications. Stability of nanofluids is the degree of dispersion (e.g., homogenous, non-homogeneous, original or changed structures, and agglomeration) of the nanoparticles into the base fluids. Suspension of nanoparticles into base fluids introduces electric charges into the base fluid as an electrical double layer (EDL) is formed around the surface of the particle. Thus, NFs are mostly electrically conducting media. The formation of EDL is influenced by the size and surface charge of the nanoparticles, volume fraction, and concentration of ions present in the base fluid. Stability due to the homogenization of nanoparticles in the base fluids is paramount to the application of nanofluids in convective heat transfer studies. The stability of nanofluids is strongly connected to the thermal (mostly thermal conductivity, density, specific heat, and viscosity) and flow properties as well as thermal efficiency in addition to micro-/macrostructures of NPs in base fluids (Ghadimi et al., 2011; Murshed and Nieto de Castro, 2014; Nieto de Castro et al., 2017; Sharifpur et al., 2017).

To avoid agglomeration and deposition when the two-step technique of nanofluid formulation is engaged, researchers have used the techniques of surfactants addition, modification of surface, sonication, and pH control to enhance the stability of NFs (Ghadimi et al., 2011; Murshed and Nieto de Castro, 2014; Babita and Gupta, 2016; Adio et al., 2016b). Different sonication times ranging from several minutes to hours have been reported for the dispersion of various nanoparticles during the preparation of sample NFs. The surfactants are also widely added to reduce the interfacial tension between nanoparticles and base fluid by increasing the EDL. Depending on both the type of nanoparticles and the base fluid, various types of surfactants such as cetyl trimethyl ammonium bromide, sodium dodecyl sulfate, gum arabic, sodium dodecyl benzene sulfonate, and oleic acid have been used for NFs' stability. When the pH of NFs is far from the isoelectric point, the stability is observed to be better; thus, in the nanofluid formulation, the pH technique is also used. Lastly, the nanoparticles' surface can be modified or functionalized to improve the stability of the resultant NFs. Stability can also be achieved without using surface modification, pH conditioning, and surfactants, but it usually involves longer duration of sonication or adding vanishing concentration of very small nanoparticles (e.g., 1–10 nm).

Besides achieving stable NFs, there is the need to monitor the level and duration of stability as this is very important in terms of the application as heat transport media. The stability can be inspected using various ways including UV spectrophotometry, sedimentation experiment, centrifugation, visualization, viscosity, thermal conductivity, and zeta potential techniques. With regard to convective heat transfer studies (MCHs and cavities) considered in this work, only UV spectrophotometry, visualization, viscosity, and zeta potential methods have been used. The most used is the visual inspection method, followed by zeta potential, UV spectrophotometry, and viscosity. Also, the morphology of the NFs can be checked using light scattering, scanning electron microscopy, transition electron microscopy, and optical microscopy methods. The visual monitoring of the stability of NFs spans a period of hours to few months of which if no sedimentation is noticed, and the fluids are considered to be stable (Ghadimi et al., 2011; Babita and Gupta, 2016). As a rule of thumb, it is considered that nanofluids with an absolute zeta potential of >30 mV are physically stable, and nanofluids having a zeta potential (absolute) of >30 mV exhibit excellent stability (Nieto de Castro et al., 2017).

7.3 Thermophysical Properties of Nanofluids

In this section, research findings and progress (experimental only) on some thermophysical properties such as thermal conductivity, viscosity, and specific heat of nanofluids, which are very crucial for convective heat transfer performance evaluation, are briefly overviewed.

7.3.1 Thermal Conductivity

As mentioned and shown before (Figure 7.1), among the properties of nanofluids, thermal conductivity has been the main focus of research among researchers. Apart from the controversial and inconsistent findings, most of the studies found anomalously enhanced thermal conductivity of nanofluids (Eastman et al., 1997, 2001; Murshed et al., 2005, 2008a,b, 2010; Angayarkanni and Philip, 2015; Tshimanga et al., 2016). Some representative literature results of the enhanced thermal conductivity (ratio of thermal conductivity of nanofluids over its base fluid, k_{nf}/k_f) of various types of nanofluids as a function of nanoparticles concentration (vol.%) at room temperature are demonstrated in Figure 7.3. It can be confirmed that although results from various groups are not consistent, nanofluids exhibit considerably higher thermal conductivity compared to their base fluids even when the concentration of nanoparticles is very low. Figure 7.3 also shows that the enhanced thermal conductivity further increases with volumetric loading of nanoparticles (until some critical concentration). Not only the concentration but also the size, shape, and type (material) of nanoparticles as well as base fluid properties also influence the thermal conductivity of nanofluids. Nonetheless, although there

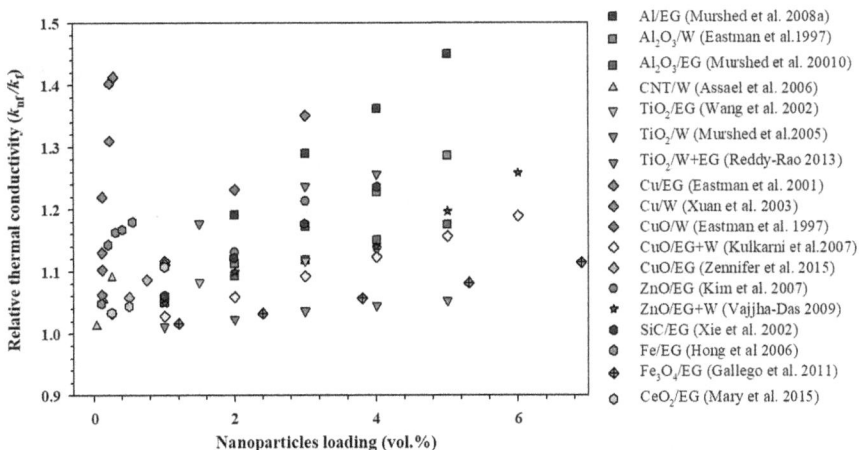

Legend:
- Al/EG (Murshed et al. 2008a)
- Al₂O₃/W (Eastman et al.1997)
- Al₂O₃/EG (Murshed et al. 2010l)
- CNT/W (Assael et al. 2006)
- TiO₂/EG (Wang et al. 2002)
- TiO₂/W (Murshed et al.2005)
- TiO₂/W+EG (Reddy-Rao 2013)
- Cu/EG (Eastman et al. 2001)
- Cu/W (Xuan et al. 2003)
- CuO/W (Eastman et al. 1997)
- CuO/EG+W (Kulkarni et al.2007)
- CuO/EG (Zennifer et al. 2015)
- ZnO/W (Kim et al. 2007)
- ZnO/EG+W (Vajjha-Das 2009)
- SiC/EG (Xie et al. 2002)
- Fe/EG (Hong et al 2006)
- Fe₃O₄/EG (Gallego et al. 2011)
- CeO₂/EG (Mary et al. 2015)

FIGURE 7.3
Representative thermal conductivity results of wide variety of nanofluids from literature.

are some factors that are mostly presumptive, the actual underlying mechanisms for such increase in the measured thermal conductivity are not yet fully understood. In addition, there are some controversies on the feasibility or significance of some mechanisms (Murshed, 2009). Although a large number of predictive models were developed, so far no model is either widely accepted by the researchers or capable of predicting thermal conductivity of a wide range of nanofluids (Aybar et al., 2015).

7.3.2 Viscosity

When it is about fluid flow and convective heat transfer, viscosity is a key property as it is related to the power (i.e., cost also) besides other issues necessary for their numerous practical applications. For instance, the understanding of convective heat transfer and pumping power in flow systems is directly related to the viscosity of the fluids. However, compared to the extensive studies on thermal conductivity, researchers have not paid much attention on viscosity of these new heat transfer fluids that contain nanoparticles.

In order to have an overview of viscosity findings of NFs, the results from the representative studies on various types of NFs as well as comparison of predictions of the above classical models with respect to loading of nanoparticles are presented in Figure 7.4. As can be seen from these results (Figure 7.4), nanofluids obviously exhibit significantly higher viscosity compared to their base fluids and it further increases with increasing concentration of nanoparticles. It can also be confirmed that the existing classical models such as those developed by Einstein (1906), Krieger and

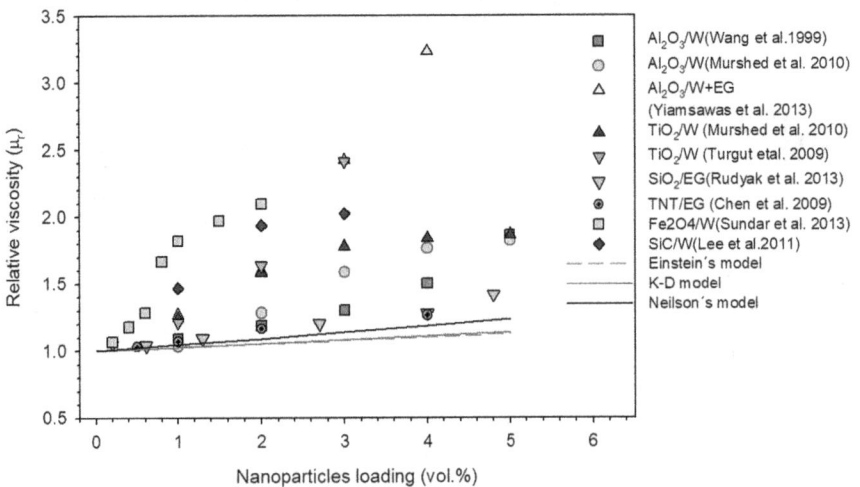

FIGURE 7.4
Relative viscosity of various nanofluids and predictions of classical models as a function of loading of nanoparticles.

Dougherty (K-D) (1959), and Nielsen (1970) are unable to predict these viscosity results of NFs. It is noted that all these classical models are mainly valid for very dilute suspension and only take into account the concentration of solid particles. In addition, all recently proposed empirical correlations are obtained by fitting their own experimental results, and in most of the cases, they are neither widely accepted nor applicable for other types of NFs (Sundar et al., 2013; Murshed et al., 2008c; Meyer et al., 2016). Therefore, no comparisons with those recent empirical models are made here.

It is also evidenced that even for the same type of nanofluid, the viscosity results vary considerably among different research groups. There are a number of factors such as use of nanoparticles of different size and purity, use of different measurement equipment or geometry, shear rate range considered, scale of agglomeration, and different dispersion and sample preparation methods behind such scattered results. Furthermore, among available literature studies, the influence of nanoparticle loading on the viscosity of NFs has been investigated most. In addition to nanoparticle loading, it is also important to identify the effects of other factors such as temperature, base fluids, dispersion, and particle size, type, and shape on the viscosity of NFs. A comprehensive analysis and evaluation of these factors and various other aspects, and an extensive review of models as well as research challenges on viscosity of NFs can be found in our previous articles (Meyer et al. 2016; Murshed and Estellé, 2017).

7.3.3 Specific Heat

Specific heat is particularly important in evaluating their thermal performance such as heat transfer rates under flow conditions or any cases where convection is important. Determination of the effective specific heat of NFs is also crucial for their practical application in thermal system management as well as enthalpy calculation and energy storage. However, a comparatively very small number of research studies (as shown in Figure 7.1) have been performed on this property of NFs. Table 7.1 summarizes the key literature results of the specific heat of various types of NFs. It can be evidenced from the findings presented in Table 7.1 that NFs exhibit both an increase and a decrease in specific heat. Such contradictory (opposite) results not only impede making conclusions but also cannot help exploring the underlying mechanisms. Despite the fact that solids have a small value of specific heat (compare to liquid), it is interesting that many studies found an increase in specific heat of base fluids due to the addition of nanoparticles (i.e., solid) to it and no physics-based explanation was provided for such results. Nonetheless, an increase in this property of NFs is very good for their applications. More details on literature findings on the specific heat of NFs can be found elsewhere (Shahrul et al., 2014; Riazi et al. 2016), and thus, they will not be elaborated further.

TABLE 7.1

Summary of Selected Literature Data of Specific Heat (SH) of Nanofluids

Researchers	Nanofluids	Volume/ Weight (%)	Results: SH Increase (+)/ Decrease (–) (%)
Zhou and Ni (2008)	Al_2O_3/W	1.43–21.7 (v.%)	4.8–55 (–)
Kulkarni et al. (2008)	$Al_2O_3/(EG/W: 50/50)$	2–6	4.84–16.14 (–)
Starace et al. (2011)	Al_2O_3/MO	0.001–0.005	2–3(+)
Shin and Banerjee (2014)	$Al_2O_3/ACaSEu$	1 (w.%)	32 (+)
Kumaresan and Velraj (2012)	MWCNT/(EG/W: 30/70)	0.15–0.45	2.31–9.35(+)
Pakdaman et al. (2012)	MWCNT/HTO	0.1–0.4	21.2–42 (–)
Saeedinia et al. (2012)	CuO/EO	0.2–3	16–29 (–)
Pak and Cho (1998)	TiO_2/W	0.99–3.16	0.8–2.6 (–)
Vajjha and Das (2009a)	ZnO/(EG/W: 60/40)	1–7	4.2–18 (–)
Nelson et al. (2009)	Graphene/PAO	0.3 and 0.6	50 (+, max)
Ghazvini et al. (2011)	Nanodiamond/EO	0.2–2	7–30 (+)

EG, ethylene glycol; EO, engine oil; MO, mineral oil; HTO, heat transfer oil; PAO, polyalphaolefin.

7.4 Convective Heat Transfer of Nanofluids— Experimental Studies

7.4.1 Natural Convection and Channel Flows

7.4.1.1 Measurements and Accuracy

Two sets of parameters are measured for the natural convection performance of NFs in cavities as provided in Table 7.2. These are the thermal properties of the NFs (such as ρ, μ, κ, and shear rate) and the physical properties (temperature and mass flow rate) of the fluids used to supply heat to the cavities. The accuracy (prescribed by the manufacturer) of the instruments used to measure these parameters (thermal and physical properties) is also presented in Table 7.2. For the studies on the convective heat and flow of NFs in MCHs, the measured parameters are volume flow rates, temperatures, and pressures in addition to the thermal properties of the NFs (Table 7.3). The accuracy associated with these parameters as specified by the manufacturers of the instruments is also given in Table 7.3. It is pertinent to mention that various types of thermocouples (J type, K type, and T type) are calibrated and used for temperature measurements in the convective heat transfer studies of NFs in cavities and MCHs.

7.4.1.2 Data Reduction and Presentations

The experimentally obtained thermal properties of NFs, temperatures, volume and volume/mass flow rates, and pressures as they apply to the studies conducted on either forced convective flow in MCHs or natural convection

TABLE 7.2

Accuracy, Uncertainty, Properties, and Stability Related to Convective Heat Transfer in Different Cavities

Researchers	Nanofluids	Accuracy (Provided by the Manufacturer)	Uncertainty	Measured Property	Estimated Properties	Cavity Type	Stability Inspection
Li et al. (2015)	ZnO/DIW/EG+PVP (75:25; 85:15; 95:5) at 5.25 wt.%	κ= ±2%–3%	U_m =±4.4%	T, $\dot{m}$, κ, and μ	Ra, Nu, and h.	Square	V
Putra et al. (2003)	Al₂O₃/DW and CuO/DW (1 and 4 vol.%)	T = 0.2°C and 0.5°C	T = 0.5%–4.0%; P_{in} = 4.0%	T, $\dot{m}$, ρ, μ, κ, and γ	Ra, Nu, and h.	Horizontal cylinder	V
Mahian, et al. (2016)	SiO₂/W (0.5, 1.0, and 2.0 vol.%)	κ = ±3%; T = ±1.5°C; A = ±0.9%; P_{in} = ±2.0%; $\dot{m}$ = ±1.0%	U_m =±4.4%	T, $\dot{m}$, ρ, μ, and κ	Gr, Nu, h, and Ra	Inclined square and triangular	V
Mahrood et al. (2011)	Al₂O₃ and TiO₂/aqueous CMC (0.1 ≤ φ ≤ 1.5 vol.%)	μ = ±3%; pH = ±0.003;	h = 5.3% and Nu = 4.62	N/A	q', Nu, h, and Ra	Vertical cylinder	N/A
Ilyas et al. (2017)	f-Al₂O₃/THO (0.05 ≤ φ ≤ 3 wt.%)	T = 0.1°C	q' =2% and h ≤ 2.1%	T, $\dot{m}$, ρ, μ, C_p, and κ	Q; Gr; Pr; Nu; h; and Ra	Rectangular	V (functionalized using oleic oil)
Joubert et al. (2017)	Fe₂O₃/DIW (0.05 ≤ φ ≤ 0.3 vol.%)	T = ±0.1°C; ρ = ±0.01%; μ = ±1%; κ = ±0.5%; C_p = ±1%.	h = ±3.88%; Nu = ±4.22%; Ra = ±5.08%	T, $\dot{m}$, and μ	Q, Nu, h, and Ra	Square under magnetic field intensity	V and VIS (with surfactant)
Moradi et al. (2015)	Al₂O₃/DIW and TiO₂/DIW (0.1 ≤ φ ≤ 1.5 vol.%)	μ = ±3%	h = ±14.1%; Nu =±14.62%; Q =±7.69%	T and $\dot{m}$	q', Nu, h, and Ra	Inclined vertical cylindrical	V
Solomon et al. (2017b)	Mango bark/DIW (0.01-0.5 vol.%)	T = ±0.2°C	h = 2%; Nu = 5%	T and $\dot{m}$	Q, Nu, Ra, and h	Square with porous media	UV, VIS, and V
Giwa et al. (2018)	Al₂O₃-MWCNT/DIW (0.1 vol.% at 95:5 and 90:10 wt.% ratios)	N/A	Nu = 2%–5%	T, $\dot{m}$, κ, and μ	Q, Nu, Ra, and h	Square	UV, VIS, and V (with surfactant)

V, visual inspection; UV, UV–Vis spectrophotometry; VIS, viscosity; U_m, uncertainty of experiment.

TABLE 7.3

Accuracy, Uncertainty, Properties, and Stability Related to Convective Heat Transfer in Various Microchannels

Researchers	Nanofluids	Accuracy	Uncertainty	Parameter Measured	Parameter Estimated	Flow Regime and Shape/Type	Stability Inspection
Sivakumar et al. (2017)	EG-based CuO, water-based CuO, and Al_2O_3 (0.01–0.3 vol.%)	$P = \pm1.6\%$, $\dot{v} = \pm1\%$, $T = \pm0.4\%$, $P_w = 1\%$	$Q = 6.7$, $h = 6.0$, $Nu = 5.99$, $f = 7.3$	$\dot{v}, T, P$	Re, Q, h, Nu, Re, f, Pr	[L] Serpentine	N/A
Yu and Liu (2013)	Al_2O_3/DIW (1–5 vol.%) and Al_2O_3/PAO (0.65 and 1.3 vol.%)	$T = \pm0.3°C$, $\dot{v} = 1\%$, $P = 2\%$	$h = 2.2\%$–9.6% and $f = 5.8\%$–28.9% (Al_2O_3/DIW); $h = 1.6\%$–3.5% and $f = 5.2\%$–13.8% (Al_2O_3/PAO)	$\dot{v}, T, P$	h, Nu, f, Q, Re	[L] Round	Zeta potential
Duangthongsuk and Wongwises (2017)	SiO_2/DIW (0.3, 0.6, 0.8 vol.%)	N/A	$P = \pm0.05\%$, $T = \pm0.1\%$, $\dot{m} = \pm0.6\%$, $h = \pm8\%$	$\dot{v}, T, P$	$Re, \Delta P, PP, Nu, Pr$	[L] Multiple zigzag flow channel structures	N/A
Gui et al. (2018)	Fe_3O_4/DIW (0.2–0.4 vol.%)	N/A	$Nu = 39\%$ (bias and precision)	$\dot{v}, H, P, T, \rho, \mu, \kappa$, contact angle	Re, h, Nu, Q	[L] Rectangle under magnetic field	N/A
Ahmed et al. (2016)	Al_2O_3/DW (0.3–0.9 vol.%) and SiO_2/DW (0.3–0.9 vol.%)	N/A	$Re = 3.77\%$, $f = 4.25\%$, $Nu = 2.83\%$	$\dot{v}, T, P$	$Re, Q, P_w, h, Nu, f, R_{th}, \Delta P$	[L] Rectangular- and triangular double-layered MCH	Visual inspection
Peyghambarzadeh et al. (2014)	CuO/W (0.1 and 0.2 vol.%) and Al_2O_3/W (0.5 and 1 vol.%)	$T = \pm0.2°C$	$Q = 0.8\%$, $T = \pm0.2$, $h = 15.2\%$, $Nu = 15.5\%$, $Re = 6.1\%$, $f = 6.7\%$	$Q, \dot{v}, T, P$	h, Nu, f, x, Re	[L] Rectangular	Visual and pH
Karami et al. (2018)	Fe_3O_4/DIW (0.1–2 vol.%)	$\dot{v} = \pm<1\%$, $T = \pm0.4\%$, $H = \pm0.05$	$T = \pm0.15°C$, $\dot{m} = 2.1\%$, $H = \pm0.05\%$, $Re = 3.3\%$–7%, $Nu = 3.4\%$, $h = 3.4\%$	$\dot{v}, T, P$	$Re, h, T_{outr}, Pr, f, \Delta P$	[L] Cylindrical pit	N/A

[L], laminar flow; H, magnetic field strength; PP, pumping power; P_w, wattage.

in cavities are engaged in the data reduction. In the absence of experimental thermal properties' data, empirical models or existing experiment-derived correlations are used. This is mainly the case for most studies. Various important parameters pertaining to the convective thermal transport of NFs such as h, Ra, Re, f, Nu, Gr, Pr, and $\bar{Q}$, are estimated using the appropriate equations. These estimated parameters are presented in Tables 7.2 and 7.3, and the equations are expressed below:

$$Q = \dot{m}C_p\left(T_{in} - T_{out}\right) \text{ or } Q = \frac{\dot{v}\rho C_p\left(T_{out} - T_{in}\right)}{A} \tag{7.1}$$

$$h_{nf,bf} = \frac{Q}{A\left(T_H - T_C\right)} = \frac{Q}{A\left(T_w - T_f\right)} \tag{7.2}$$

$$Nu_{nf,bf} = \frac{hL_c}{\kappa_{nf,bf}} \text{ or } Nu_{nf,bf} = \frac{hD_h}{\kappa_{nf,bf}} \tag{7.3}$$

$$Ra = \frac{g\beta_{nf,bf}\left(T_H - T_C\right)\left(\rho_{nf,bf}\right)^2\left(C_p\right)_{nf,bf} L_c^3}{\mu_{nf,bf}\left(\kappa\right)_{nf,bf}} \text{ or}$$

$$Ra = \frac{g\beta_{nf,bf}\left(T_H - T_C\right)\left(\rho_{nf,bf}\right)^2\left(C_p\right)_{nf,bf} L_c^3}{\vartheta_{nf,bf}\left(\alpha\right)_{nf,bf}} \tag{7.4}$$

$$Re_{nf,bf} = \frac{\rho_{nf,bf}u_m D_h}{\mu_{nf,bf}} \tag{7.5}$$

$$Pr = \frac{\mu_{nf,bf}\left(C_p\right)_{nf,bf}}{\kappa_{nf,bf}} \text{ or } Pr = \frac{\vartheta_{nf,bf}}{\alpha_{nf,bf}}$$

$$\text{Where; } \alpha_{nf,bf} = \frac{\kappa_{nf,bf}}{\left(\rho C_p\right)_{nf,bf}} \tag{7.6}$$

$$Gr = RaPr \tag{7.7}$$

$$f = \frac{2D_h\Delta P}{\left(\rho_{nf,bf}\right)(L)\left(u_m^2\right)} \tag{7.8}$$

$$R_{th} = \frac{T_w - T_f}{\dot{Q}} \text{ or } R_{th} = \frac{1}{hA_{conv}} \tag{7.9}$$

$$\varepsilon = 1 - \frac{\left(T_w - T_f\right)_{nf}}{\left(T_w - T_f\right)_{bf}} \tag{7.10}$$

where T_{in} and T_{out} are temperature (°C) at the inlet and exit of MCH/heat exchanger, respectively; T_H, T_C, T_w, and T_f are temperature (°C) at the hot side, the cold side (of cavity), wall surface, and fluid (of MCH), respectively; $\dot{m}$, $\bar{v}$, and u_m are mass flow rate (kg/s), volume flow rate (m³/s), and mean velocity of flow (m/s), respectively; C_p is specific heat (J/kg·°C); Q and $\bar{Q}$ are heat transferred (W) and heat flux (W/m²), respectively; ρ, μ, κ, β, $\dot{v}$, and α are density (m³/kg), dynamic viscosity (Pa·s), thermal conductivity (W/m·°C), thermal expansion coefficient, kinematic viscosity (m²/s), and thermal diffusivity (m²/s), respectively; A and A_conv are area of cavity or MCH and area of convection (m²), respectively; L and L_c are length and hydraulic length (m), respectively; D_h is the hydraulic diameter of MCH (m); h is the coefficient of convective heat transfer (W/m²·°C); f and ΔP are the friction factor and pressure drop (Pa), respectively; Ra, Re, Nu, Gr, and Pr are the Rayleigh, Reynolds, Nusselt, Grashof, and Prandtl number, respectively; R_th and ε are thermal resistance (°C/W) and effectiveness, respectively; g is gravitational acceleration (m/s²); and subscripts *nf* and *bf* indicate nanofluid and base fluid, respectively.

The manufacturers' specified accuracies for the flow meters, thermocouples, and pressure transducers used in the experiments coupled with the estimated parameters are used in calculating the uncertainties associated with an individual convective heat transfer parameter as it relates to the measurements made in the study conducted. These uncertainties are calculated using mainly the well-known Moffat or Kline and McClintock method with most studies calculating the uncertainty subject to random error (precision) (Ali et al., 2013a; Ahmed et al., 2016). Few works calculated uncertainty based on the systematic error (bias), and very few estimated uncertainties using both bias and precision errors (e.g., Gui et al., 2018).

Data (measured and estimated) are principally presented in three categories. The first presentation of data involves the stability test, which mainly comprises the results of nanoparticles' distribution, zeta potential and pH values, viscosity and absorbance (against time), and transmission and scanning electron microscopy. The second is the presentation of thermal properties of NFs using experimentally obtained values or from empirical/experiment-derived correlations as a function of temperature and volume concentration in most cases. The last is the presentation of both estimated (reduced) and experimental data relating to the convection heat and flow properties for studies on convective flow of NFs in MCHs or hydrothermal characteristics of NFs in cavities. The Nu and h are often presented as a function of Ra or Re (depending on the study in question) and volume concentration.

7.4.1.3 Empirical Correlations

The development of empirical correlations from the experimental data acquired is fairly reported for convective heat transfer and flow performance of NFs in cavities and MCHs. In the case of natural convection heat transfer of NFs in enclosures, obtained Nu data are correlated with Ra, volume/weight concentration (φ), aspect ratio (AR), Pr, κ, β, and other parameters as presented in Table 7.4. However, these proposed correlations are developed

TABLE 7.4

Correlations for Convective Heat Transfer in Cavities and Microchannels

Researchers	Nanofluids	Geometry	Correlation
Convective Heat Transfer in Cavities			
Ali et al. (2013a)	Al_2O_3/W	Vertical cylinders	$Nu = \left(7.899 - 8.571 \times 10^{-9} Ra^*\right) \times \left(1.0 - 15.283\varphi + 387.681\varphi^2\right) AR^{0.5}$
Ho et al. (2010)	Al_2O_3/W	Vertical squares	$Nu_{nf} = CRa_{nf}^n \left(Pr_{nf,h}/Pr_{nf}\right)^m \left(\beta_{nf,h}/\beta_{nf}\right)^p$
De la Peña et al. (2017)	AlN and TiO_2/mineral oil	Opened vertical annular	$Nu = 0.496 Ra^{*0.17} K^{\left(\frac{1.582}{k} + 2.463\right)} AR^{-0.54l}$; maximum deviation = 1.6%
Ali et al. (2013b)	Al_2O_3/W	Vertical cylinders	$Nu = 1.426\left(Ra^*\right)^{0.119}\left(1 + 44.097\varphi - 6943.36\varphi^2\right) AR^{0.137}$
Ghodsinezhad et al. (2016)	Al_2O_3/W	Square	$Nu = 0.6091\left(Ra\right)^{0.235}\left(\varphi\right)^{0.00584}$ (for $\varphi \le 0.1$)
			$Nu = 0.482\left(Ra\right)^{0.2356}\left(\varphi\right)^{-0.026}$ (for $\varphi \ge 0.1$); $R^2 = 0.94$
Ilyas et al. (2017)	f-Al_2O_3/THO	Rectangular	$Nu = C\left(Ra\right)^{0.04}\left(1 - \varphi\right)^{-0.015}$; $228 \le Pr \le 592$; $0.97 \le 1 - \varphi$ (wt. frac.) ≤ 1; $C = 4.17\left(\frac{Pr_r}{Pr_r - 0.343}\right)^{2.51}\left(\frac{Pr_r}{K_r^{3.76}\beta_r^{3.483}}\right)$
Nnanna (2007)	Al_2O_3/DIW	Rectangular	$Nu = 16.4 e^{-Ra_c\varphi\left(c^{-m\varphi}\right)} \varepsilon = 4 \times 10^{-7}$; $m = 11$; $10^5 \le \varphi Ra e^{-m\varphi} \le 10^6$

(Continued)

TABLE 7.4 (Continued)

Correlations for Convective Heat Transfer in Cavities and Microchannels

Researchers	Nanofluids	Geometry	Correlation
Convective Heat Transfer in Microchannels			
Manay and Sahin (2016)	TiO_2/DW	Rectangular	$Nu = 3.65\left(RePr\ D_h/L\right)^{0.2}\left(1+\varphi\right)^{13.47}$; $f = 45.79\left(Re\right)^{-0.99}\left(\dfrac{D_h}{L}\right)^{-0.012}\left(1+\varphi\right)^{11.41}$
Jung et al. (2009)	Al_2O_3/DW and EG/DW(50/50)	Rectangular	$Nu = 0.014\varphi^{0.095}Re^{0.4}Pr^{0.6}$
Azizi et al. (2016)	Cu/W	Cylinder	$Nu = 0.293Re^{0.485}Pr^{0.2}\left(1+13.2\varphi\right)^{8.1}$, $f = 16.5\left(\varphi Re\right)^{0.11}Re^{-0.928}$
Duangthongsuk and Wongwises (2017)	SiO_2/DIW	Multiple zigzag flow channel structures	$Nu = 1.07 \times 10^{-7}Re_{Dh}^{0.336}Pr^{7.603}$, $P = 340,350\dot{m}^{1.666}$ kPa
Zhang et al. (2013)	Al_2O_3/DIW	Circular	$Nu = 0.2521Re^{0.397}Pr^{0.432}\left(1+\varphi\right)^{9.836}$; $f = 6.3265Re^{-0.656}\left(1+\varphi\right)^{4.835}$
Selvakumar and Suresh (2012)	CuO/DIW	Rectangular	$Nu = 0.346Re^{0.3143}Pr^{0.638}\left(1+\varphi\right)^{115.341}$

for different configurations of cavities under dissimilar thermal conditions and various types of NFs at dissimilar volume/weight concentrations as well as at different AR.

Similarly, estimated data obtained from the experiments on the convection cooling of MCHs using NFs are employed to propose correlations for the prediction of Nu and f. As can be seen in Table 7.4, Nu is correlated as a function of Re, Pr, D_h/L, and φ, whereas f is related to Re, D_h/L, and φ. These obtained correlations are for different configurations, sizes, and characteristics of MCHs using various NFs. In addition, the thermal conditions under which the studies are conducted differ from one another, and all the correlations are developed at laminar flow regimes.

7.4.1.4 Summary of Findings of Natural Convection in Cavities

Publications available in the public domain have shown that various kinds of NFs in different shapes of enclosures have been experimentally investigated on their convective thermal transport behaviors. The most studied enclosure in this regard is the square cavity, while Al_2O_3/deionized water (DIW) is the most investigated NF. Experimental works on the convective heat transport characteristics of NFs in cavities give contradicting results. A few studies, especially the early ones in which higher volume/weight concentrations are used in the formulation of the tested NFs, reported deterioration of convective heat transfer coefficients of NFs compared to those of the base fluid. Using square cavity filled with ZnO/DIW-EG (5.25 wt.%) NF and a horizontal cylinder with AR of 0.5 and 1 containing Al_2O_3/DW (1–4 vol.%) nanofluid, Li et al. (2015) and Putra et al. (2003), respectively, showed attenuation in heat transfer when NFs are used as thermal transport media in the cavities compared with the base fluids. However, several studies have demonstrated that the use of NFs in various configurations of cavities enhanced their convective heat and flow performances, even under variations in thermal conditions.

In the work of Ghodsinezhad et al. (2016), a differential square cavity containing Al_2O_3/DIW nanofluid was reported to maximally augment h by 15% compared with DIW at 0.1% volume concentration of the nanofluid. Garbadeen et al. (2017) and Ho et al. (2010) employed MWCNT/DIW (0–1 vol.%) and Al_2O_3/W (0.1–4 vol.%) NFs in square cavities, respectively, and found maximum heat transfer rate at 0.1 vol.%, which agrees with the work of Ghodsinezhad et al. (2016). The effect of AR (1, 2, and 4) on the convective heat transfer behavior of Al_2O_3/DIW NFs contained in cavities was investigated by Solomon et al. (2017a). The results showed that the h and Nu are strongly dependent on AR. The highest heat transfer for each AR of the cavity is noticed to be related to the volume concentration. With the cavity having $AR = 1$ (square cavity), the highest heat transfer and h are achieved at 0.1 vol.%. Furthermore, Solomon et al. (2017c) studied the influence of porous media contained in a square cavity saturated with Al_2O_3/(EG/DIW: 60/40%) nanofluid at $0.05 \leq \varphi \leq 0.4$ vol.% on the natural convection heat transport behavior. They reported that heat transfer in the porous cavity containing the nanofluid is a function of concentration, porous media, and NFs. The highest enhancement of 10% compared to the base fluid is observed at 0.05 vol.% and $\Delta T = 50°C$.

Outside the square cavities, Mahrood et al. (2011) employed a vertical cylinder cavity with $AR = 0.5$, 1.0, and 1.5 to investigate the natural convection thermal transport of nanofluids with Al_2O_3 and TiO_2 nanoparticles dispersed in 0.5 wt.% aqueous solution of carboxymethyl cellulose (CMC). Their result showed that heat transfer is enhanced below 0.5 and 1 vol.% with optimum values at 0.1 and 0.2 vol.%, for aqueous CMC-based TiO_2 and Al_2O_3 NFs, respectively. TiO_2/aqueous CMC is noticed to be a better heat transfer medium than Al_2O_3/aqueous CMC NF. Also, increasing AR is observed to enhance heat transfer for both NFs. The hydrothermal behavior of Al_2O_3/DIW (0.2–8 vol.%) NF contained in a rectangular cavity was studied by Nnanna (2007). He found that heat transfer is augmented for nanofluid concentration range of $0.2 \leq \varphi \leq 2$ vol.%, but it deteriorates when the concentration is >2%. Optimum improvement of heat transfer is observed at 0.2 vol.%. With an inclined (30°, 60°, and 90°) cylindrical cavity containing Al_2O_3/DIW and TiO_2/DIW ($0.1 \leq \varphi \leq 1.5$ vol.%) NFs, the natural convection characteristics of the fluids were examined by Moradi et al. (2015). Maximum enhancements of Nu (6.76% and 2.33% relative to DIW) occurred at 0.2 and 0.1 vol.% for Al_2O_3/DIW and TiO_2/DIW NFs, respectively. Nu is noticed to increase with an increase in AR.

Special cases of using a magnetic field to investigate the possible enhancement of convective heat transfer performance of hybrid NFs and bio-nanofluid in cavities have also been carried out. It is worthy to note that an external magnetic field can be utilized in controlling and enhancing thermal and convective properties of ferrofluids (magnetic NFs) and hybrid NFs are recognized for their improved properties. Joubert et al. (2017) investigated the effects of φ, external magnetic field intensity, and configurations on the natural convection of Fe_2O_3/DIW (0.05–0.3 vol.%) nanofluid inside a square cavity. The result showed that both the magnetic field intensity and configuration considerably influence the nanofluid convective heat transfer behavior. Highest augmentation of heat transfer performance is achieved at a magnetic field configuration (b) with 700 G magnets at $Ra = 3.18 \times 10^8$. Giwa et al. (2018) engaged hybrid NFs (Al_2O_3-MWCNT (95:5 and 90:10 ratios)/DIW) at 0.1 vol.% filled into a square cavity to study the natural convection heat transfer performance. They noticed that the use of the hybrid NFs improved heat transfer performance in the cavity compared with both DIW and single-particle NFs of Al_2O_3/DIW. The Al_2O_3-MWCNT (90:10 ratio)/DIW NF is found to have the highest enhancement of 9.8% and 19.4% for h and Nu, respectively. Bio-nanofluids ($\varphi = 0.01$–0.5 vol.%) synthesized from mango bark nanoparticles suspended in DIW were investigated for their natural convection thermal transport behavior in a square cavity (Solomon et al., 2017b). The results showed the deterioration of heat transfer capability of the bio-nanofluid in relation to DIW, though volume concentration of 0.2% gave the highest heat transfer performance. In all these studies, both the buoyancy and Nu are found to be strongly connected to Ra.

7.4.1.5 Convective Heat Transfer in Microchannels

The experimental investigation of NFs as thermal transport media in MCHs has been studied to an extent. These studies entail the effect of NF types, sizes and volume/weight concentrations, variation in heat flux and flow rate, surface roughness, configurations, and sizes and heights of MCHs on the convective cooling of these devices using NFs. Both the laminar and turbulent flows have been reported in the literature with the latter scarcely studied compared to the former. The convective heat transfer of Al_2O_3–water NF in the rectangular MCH is well studied in relation to other NFs and MCH types. For instance, Lee and Mudawar (2007) investigated the thermal cooling performance of Al_2O_3/DIW and Al_2O_3/HFE 7100 NFs in a rectangular MCH under laminar and turbulent flows for single- and two-phase cooling. They reported that h is enhanced for single-phase laminar flow, but it degrades for turbulent flow. It is also observed that at the same Re, the ΔP of the NFs is increased as φ rises, in comparison with the base fluids. The two-phase cooling of MCH using NFs is found not to be feasible as evaporation caused clustering of nanoparticles. Al_2O_3/DW (distilled water) and Al_2O_3/(EG/DIW: 50/50) NFs at 0.6–1.8 vol.% in a rectangular MCH to examine the convective cooling behavior of the NFs was used by Jung et al. (2009). The results revealed that the highest enhancement of h up to 32% at 1.8 vol.% for Al_2O_3/DW NF is achieved in relation to DW. However, Al_2O_3/EG-DW NFs are observed to be unstable. Nu is found to increase with increasing Re. The friction factors of Al_2O_3/DW NFs are identical to those of DW NFs.

The work of Manay and Sahin (2016) evaluated the effect of φ (0.25–2 vol.%) and MCH heights (200–500 μm) on the convective heat transfer and ΔP behavior of TiO_2/DW NF in a rectangle-shaped MCH. Their results showed that increasing channel height reduces the rate of heat transfer and increases ΔP. An increase in φ leads to enhancement of h, with the maximum h occurring at 1.5 vol.%, channel height of 200 μm, and peak mass flow rate. Reduction in channel height from 500 to 200 μm and increasing concentration from 0.25 to 2 vol.% yield a 10%–50% and 18% increase in f, respectively. At a channel height of 300 μm, 0.25 vol.%, and $Re = 200$, optimum heat transfer is obtained.

Peyghambarzadeh et al. (2014) examined the cooling effect of CuO/water (W) (0.1 and 0.2 vol.%) and Al_2O_3/W (0.5 and 1 vol.%) NFs in a rectangular MCH. They reported that the heat transfer characteristics and ΔP of both NFs are improved compared to W. It is noticed that h is augmented with nanoparticle concentration with 27% and 49% for CuO (0.2 vol.%) and Al_2O_3 (1 vol.%) NFs, respectively. For both NFs, f increases compared with W. Azizi et al. (2016) examined the thermal performance of Cu/W (0.05–0.3 wt.%) NF in a cylindrical MCH. In comparison with W, R_{th} is reduced by 21% whereas

Nu (at the entrance) and *f* are increased by 43% and 45.5%, respectively, for the NFs as the concentration rises from 0.05 to 0.3 wt.%. At 0.3 wt.%, *h* is enhanced by 80% as the flow rate increases.

With a single-layer MCH having a high ΔP and not providing a uniform temperature across the channel length, these shortcomings are addressed using double-layered MCHS. Ahmed et al. (2016) employed rectangular and triangular double-layered MCHS (TDLMCH and RDLMCH) to investigate the convective cooling performance of Al_2O_3/DW and SiO_2/DW (0.3–0.9 vol.%) NFs. The use of zigzag TDLMCHS reduced the wall temperature and R_{th} by 27.4% and 16.6%, respectively, at low pumping power and the considered geometry parameters, compared to the RDLMCHS. Across the channel length, zigzag TDLMCHS demonstrate better temperature uniformity. No considerable difference in ΔP is noticed between the MCHS. Al_2O_3-DW NF for 0.9 vol.% has better heat transfer ability than SiO_2/DW NF, whereas the SiO_2-DW nanofluid has a lower pumping power. Sivakumar et al. (2017) experimented the convective heat transfer behavior of three types (CuO/EG, CuO/W, and Al_2O_3/W at 0.01–0.3 vol.%) of NFs in a serpentine-shaped MCH under laminar flow. The results showed that heat transfer rates of the NFs are better than those of the base fluids with CuO/EG NF offering the highest *h*. It is observed that as the h_d diminishes, ΔP increases, and thus, *h* is enhanced for the NFs and base fluids. The serpentine shape is noticed to augment heat transfer in MCH.

Duangthongsuk and Wongwises (2017) investigated the influence of NF (SiO_2/DIW for 0.3–0.8 vol.%), MCHS with multiple zigzag configurations (cross-cutting zigzag heat sink (CCZ-HS) and continuous zigzag heat sink (CZ-HS)), as well as *Re* on heat transfer and flow characteristics. They noticed that using the NF in the MCH improved heat transfer by 3%–15%, relative to DIW. Also, using CCZ-HS improved thermal performance by 2%–6% compared to CZ-HS, even at a given pumping power. Increasing nanofluid concentration is observed to lead to a small increase in ΔP for CCZ-HS. The effect of an external magnetic field on the thermal and flow behavior of NFs in MCHS has only been experimented recently. Gui et al. (2018) characterized a ferrofluid (Fe_3O_4/DIW NF at 0.2–0.4 vol.%) and investigated its heat transfer behavior in a rectangular MCH under magnetic field excitation. The results demonstrated that *Nu* diminishes as the magnetic field intensity increases for all NFs. At 10 mT, heat transfer enhancement is observed for some samples of the NFs in comparison with DIW. In addition, Karami et al. (2018) studied the influence of φ (0.1–2 vol.%), magnetic field types (static and rotational magnetic field (8–31 rad/s)), magnetic field strength (0–1,100 mT), and flow rates (2–8) on the heat transfer and flow characteristics of a ferrofluid (Fe_3O_4/DIW NF) under laminar condition. They reported that heat transfer is diminished using the static magnetic field (SMF) but enhanced with a rotational magnetic field (RMF). An increase in ϕ is noticed to enhance *h*. Maximum *h* is achieved at 1,100 mT, 18 rad/s, and 2 vol.%. When the rotation of the RMF is above 18 rad/s, heat transfer is found to be insignificant. At the same *Re*, ΔP and f are reduced when RMF is used to excite the MCH instead of SMF. In relation to DIW, ΔP and f are slightly increased with φ for the NFs and further increased in the presence of a magnetic field.

7.5 Conclusions and Remarks

Although nanofluids are a popular research field and have attracted enormous research attention worldwide, the progress toward their real applications is not very satisfactory. This is mainly due to a number of issues and challenges that researchers have yet to overcome. One of the main challenges is the long-term stability of NFs. Another critical issue of nanofluids research is the lack of accurate measurements of its properties and characteristics, namely thermal conductivity and convective heat transfer coefficients. Thus, in order for the real development of this field, it is of great importance to address and resolve these challenges, and this study broadly covers them besides reviewing and analyzing other areas such as thermophysical properties and convective heat transfer of NFs in MCHs and cavities.

It is demonstrated that despite the significant increase in thermophysical properties, the literature data are quite scattered and no well-accepted factors or mechanisms behind such anomalous results are yet elucidated. Although nanofluids showed considerably higher thermal conductivity compared to their base heat transfer fluids, their applications in flow conditions such as convective channel cooling or other areas are of serious concern due to the observed even higher viscosity. Thus, there is a need to have optimum NFs in terms of nanoparticle loading and properties by trading off high viscosity with enhanced thermal conductivity or other desired properties.

The stability of NFs can be viewed to be significant to the measurements undertaken for studies on the natural heat transfer behaviors in cavities and convective flow in MCHs. The degree of correctness of the measured thermal properties of the NFs while experimenting and the subsequent data reduction are strongly related to the NFs' stability. The wide use and reportage of visual inspection in the literature to justify the stability of NFs should be downplayed. This method alone is not enough to ascertain the stability of the NFs. Other methods such as zeta potential, absorbance, and monitoring of the viscosity and thermal properties of the NFs should also be conducted to substantiate this claim. In addition, it can suggest that the zeta potential method is to be conducted before and after the experiment for the NFs used in the experiment. This is to properly monitor the stability of the NFs employed in the experiment. Again, the absorbance test should be carried out throughout the run of the experiment for adequate (timeous) checking of the nanofluids' stability while running the experiment.

Another aspect of great concern in the experimental studies is the inconsistency and incomplete report of the sonication of NFs, which is peculiar to the two-step technique of NF formulation. The majority of literature studies, if not all, do not give complete information about how the sonication of NFs was carried out. In most cases, the sonication duration is the only parameter given, and this does not allow repeatability of the experiment. Data on the amplitude, frequency, or pulse-on time and pulse-off time are missing, without which it is not possible to verify or attain the same level of stability as reported in those literature studies. With inconsistency in the stability of NFs, experimentally obtaining the same thermophysical properties of NFs also becomes impossible even for the same type of NF.

The deployment of varieties of NFs as cooling media in cavities and MCHs has shown that natural convection heat transfer is augmented at certain nanoparticle concentrations (volume/weight) in the formal irrespective of the shape of the cavities and thermal conditions under which the experiment is performed. However, convective heat transfer performance is found to enhance with increasing concentration of nanoparticles in base fluids. Our strong opinion is that there is a critical concentration of nanoparticles to which the heat transfer—MCHs cooling via NFs—can be augmented. Therefore, such an optimum volume/weight concentration of NFs needs to be investigated in the future to ascertain the maximum thermal cooling that can be achieved using these transport media in MCHs.

Most published works used either empirical or experimentally obtained correlations to predict the thermophysical properties of NFs which cannot be used to effectively substitute the properties of the NFs being studied in terms of stability, size, and distribution, nanofluid synthesis, etc. It is most appropriate in experimental studies of this regard to experimentally obtain the critical thermophysical properties. The measurement of these properties and the subsequent result guarantee the accuracy and correctness of the study. The current trend of using green and hybrid NFs also calls for the experimental determination of the thermophysical properties of NFs.

There is, therefore, a need to examine the thermal performance of other types of NFs (new nanoparticles suspended in base fluids) in cavities and MCHs outside the ones currently published in the literature. The advent of hybrid and green NFs and the manipulative property of ferrofluid using magnetic field are future studies to be researched. Furthermore, inconsistency is observed in the reportage of uncertainty. Some publications report uncertainty in terms of bias, while some present based on precision, with very few stating both. We suggest that both (bias and precision) be presented for a better understanding of the random and system errors related to the experiment. In most of the literature studies particularly in convective heat transfer, the measurement uncertainty has not been carefully and completely analyzed. In order to fully analyze the measurement uncertainty, it is important that the accuracy of all the equipment used for the measurement of various parameters is clearly stated and accounted for.

Acknowledgments

This work was partially funded by Fundação para a Ciência e a Tecnologia (FCT), Portugal, through project PTDC/NAN-MAT/29989/2017.

References

Adio, S. A., M. Mehrabi, M. Sharifpur and J. P. Meyer. 2016a. Experimental investigation and model development for effective viscosity of MgO-ethylene glycol nanofluids by using dimensional analysis, FCM-ANFIS and GA-PNN techniques. *Int. Comm. Heat Mass Transf.* 72:71–83.

Adio, S. A., M. Sharifpur and J. P. Meyer. 2016b. Influence of ultrasonication energy on the dispersion consistency of Al_2O_3-glycerol nanofluid based on viscosity data, and model development for the required ultrasonication energy density. *J. Exp. Nanosci.* 11:630–649.

Ahmed, H. E., M. I. Ahmed, I. M. F. Seder and B. H. Salman. 2016. Experimental investigation for sequential triangular double-layered microchannel heat sink with nanofluids. *Int. Comm. Heat Mass Transf.* 77:104–115.

Ali, M., O. Zeitoun, S. Almotairi and H. Al-Ansary. 2013aa. The effect of alumina–water nanofluid on natural convection heat transfer inside vertical circular enclosures heated from above. *Heat Transf. Eng.* 34:1289–1299.

Ali, M., O. Zeitoun and S. Almotairi. 2013bb. Natural convection heat transfer inside vertical circular enclosure filled with water-based Al_2O_3 nanofluids. *Int. J. Therm. Sci.* 63:115–124.

Angayarkanni, S. A. and J. Philip. 2015. Review on thermal properties of nanofluids: Recent developments. *Adv. Colloid Interface Sci.* 225:146–176.

Assael, M. J, I. N. Metaxa, K. Kakosimos and D. Constantinou. 2006. Thermal conductivity of nanofluids-experimental and theoretical. *Int. J. Thermophys.* 27:999–1016.

Aybar, H. Ş., M. Sharifpur, M. R. Azizian, M. Mehrabi and J. P. Meyer. 2015. A review of thermal conductivity models for nanofluids. *Heat Transfer Eng.* 36:1085–1110.

Azizi, Z., A. Alamdari, M. R. Malayeri. 2016. Thermal performance and friction factor of a cylindrical microchannel heat sink cooled by Cu-water nanofluid. *Appl. Therm. Eng.* 99:970–978.

Babita, S. K. S. and S. M. Gupta. 2016. Preparation and evaluation of stable nanofluids for heat transfer application: A review. *Exp. Therm. Fluid Sci.* 79:202–212.

Chein, R. and G. Huang. 2005. Analysis of microchannel heat sink performance using nanofluids. *App. Therm. Eng.* 25:3104–3114.

Chen, H., Y. Ding, A. Lapkin and X. Fan. 2009. Rheological behaviour of ethylene glycol-titanate nanotube nanofluids. *J. Nanopart. Res.* 11:1513–1520.

Choi, S. U. S. 1995. Enhancing thermal conductivity of fluids with nanoparticles. *ASME FED* 231:99–105.

De la Peña, N. L. C., C. I. Rivera-Solorio, L. A. Payán-Rodríguez, A. J. García-Cuéllar and J. L. López-Salinas. 2017. Experimental analysis of natural convection in vertical annuli filled with AlN and TiO_2/mineral oil-based nanofluids. *Int. J. Therm. Sci.* 111:138–145.

Duangthongsuk, W. and S. Wongwises. 2017. An experimental investigation on the heat transfer and pressure drop characteristics of nanofluid flowing in microchannel heat sink with multiple zigzag flow channel structures. *Exp. Therm. Fluid Sci.* 87:30–39.

Eastman, J. A., S. U. S. Choi, S. Li and L. J. Thompson. 1997. Enhanced thermal conductivity through the development of nanofluids. *In Proceeding of Symposium Nanophase and Nanocomposite Materials II*, Boston, MA.

Eastman, J. A., S. U. S. Choi, S. Li, W. Yu and L. J. Thompson. 2001. Anomalously increased effective thermal conductivities of ethylene glycol-based nanofluids containing copper nanoparticles. *Appl. Phys. Lett.* 78:718–720.

Einstein, A. 1906. A new determination of molecular dimensions. *Annal. Physik* 4:37–62.

Gallego, M. J. P., L. Lugo, J. L. Legido and M. M. Pineiro. 2011. Enhancement of thermal conductivity and volumetric behavior of Fe_xO_y nanofluids. *J. Appl. Phys.* 110:014309.

Garbadeen, I. D., M. Sharifpur, J. M. Slabber and J. P. Meyer. 2017. Experimental study on natural convection of MWCNT-water nanofluids in a square enclosure. *Int. Comm. Heat Mass Transf.* 88:1–8.

Ghadimi, A., R. Saidur and H. S. C. Metselaar. 2011. A review of nanofluid stability properties and characterization in stationary conditions. *Int. J. Heat Mass Transf.* 54:4051–4068.

Ghazvini, M., M. A. Akhavan-Behabadi, E. Rasouli and M. Raisee. 2011. Heat transfer properties of nanodiamond-engine oil nanofluid in laminar flow. *Heat Transf. Eng.* 33:525–532.

Ghodsinezhad, H., M. Sharifpur and J. P. Meyer. 2016. Experimental investigation on cavity flow natural convection of Al_2O_3–water nanofluids. *Int. Comm. Heat Mass Transf.* 7:316–324.

Giwa, S. O., M. Shafirpur and J. P. Meyer. 2018. Heat transfer enhancement of dilute Al_2O_3-MWCNT water based hybrid nanofluids in a square cavity. *In Proceeding of 16th International Heat Transfer Conference*, August 10–15, 2018, Beijing, China.

Gui, N. G. J., C. Stanley, N. T. Nguyen and G. Rosengarten. 2018. Ferrofluids for heat transfer enhancement under an external magnetic field. *Int. J. Heat Mass Transf.* 123:110–121.

Ho, C. J., L. C. Wei and Z. W. Li. 2010. An experimental investigation of forced convective cooling performance of a microchannel heat sink with Al_2O_3/water nanofluid. *Appl. Therm. Eng.* 30:96–103.

Hong, K. S., T.-K. Hong and H.-S. Yang. 2006. Thermal conductivity of Fe nanofluids depending on the cluster size of nanoparticles. *Appl. Phys. Lett.* 88:031901.

Ilyas, S. U., R. Pendyala and M. Narahari. 2017. An experimental study on the natural convection heat transfer in rectangular enclosure using functionalized alumina-thermal oil-based nanofluids. *Appl. Therm. Eng.* 127:765–775.

Joubert, J. C., M. Sharifpur, A. B. Solomon and J. P. Meyer. 2017. Enhancement in heat transfer of a ferrofluid in a differentially heated square cavity through the use of permanent magnets. *J. Mag. Magnetic Mat.* 443:149–158.

Jung, J. Y., H. S. Oh and H. Y. Kwak. 2009. Forced convective heat transfer of nanofluids in microchannels. *Int. J. Heat Mass Transf.* 52:466–472.

Karami, E., M. Rahimi and N. Azimi. 2018. Convective heat transfer enhancement in a pitted microchannel by stimulation of magnetic nanoparticles. *Chem. Eng. Process- Process Intensif.* 126:156–167.

Kim, S. H., S. R. Choi and D. Kim. 2007. Thermal conductivity of metal-oxide nano-fluids: Particle size dependence and effect of laser irradiation. *J. Heat Transf.* 129:298–307.

Krieger, I. M. and T. Dougherty. 1959. A mechanism for non-Newtonian flow in suspensions of rigid spheres. *Trans. Soc. Rheol.* 3:137–152.

Kulkarni, D. P., P. K. Namburu and D. K. Das. 2007. Comparison of heat transfer rates of different nanofluids on the basis of the Mouromtseff number. *Electro. Cooling* 13(3):28–32.

Kulkarni, D. P., R. S. Vajjha, D. K. Das and D. Oliva. 2008. Application of aluminum oxide nanofluids in diesel electric generator as jacket water coolant. *Appl. Therm. Eng.* 28(14–15):1774–1781.

Kumaresan, V. and R. Velraj. 2012. Experimental investigation of the thermo-physical properties of water–ethylene glycol mixture based CNT nanofluids. *Thermo-chim Acta* 545(0):180–186.

Lee, S. and S. U. S. Choi. 1996. Application of metallic nanoparticle suspensions in advanced cooling systems. *In Proceedings of Recent Advances in Solids/Structures and Application of Metallic Materials*, ASME, PVP 342/MD-72, New York.

Lee, J. and I. Mudawar. 2007. Assessment of the effectiveness of nanofluids for single-phase and two-phase heat transfer in micro-channels. *Int. J. Heat Mass Transf.* 50:452–463.

Lee, S. W., S. D. Park, S. Kang, I. C. Bang and J. H. Kim. 2011. Investigation of viscosity and thermal conductivity of SiC nanofluids for heat transfer applications. *Int. J. Heat Mass Transf.* 54:433–438.

Li, H., Y. He, Y. Hu, B. Jiang and Y. Huang. 2015. Thermophysical and natural convection characteristics of ethylene glycol and water mixture based ZnO nanofluids. *Int. J. Heat Mass Transf.* 91:385–389.

Lomascolo, M., G. Colangelo, M. Milanese and A. de Risi. 2015. Review of heat transfer in nanofluids: Conductive, convective and radiative experimental results. *Renew. Sust. Energy Rev.* 43:1182–1198.

Mahian, O., A. Kianifar, S. Z. Heris and S. Wongwises. 2016. Natural convection of silica nanofluids in square and triangular enclosures: Theoretical and experimental study. *Int. J. Heat Mass Transf.* 99:792–804.

Mahrood, M. R. K., S. G. Etemad and R. Bagheri. 2011. Free convection heat transfer of non Newtonian nanofluids under constant heat flux condition. *Int. Comm. Heat Mass Transf.* 38:1449–1454.

Manay, E. and B. Sahin. 2016. The effect of microchannel height on performance of nanofluids. *Int. J. Heat Mass Transf.* 95:307–320.

Mary, E. E. J., K. S. Suganthi, S. Manikandan, N. Anusha and K. S. Rajan. 2015. Cerium oxide ethylene glycol nanofluids with improved transport properties: Preparation and elucidation of mechanism. *J. Taiwan Inst. Chem. Eng.* 49:183–191.

Meyer, J. P., S. A. Adio, M. Sharifpur and P. N. Nwosu. 2016. The viscosity of nanofluids: A review of the theoretical, empirical, and numerical models. *Heat Transf. Eng.* 37:387–421.

Moradi, H., B. Bazooyar, A. Moheb and S. G. Etemad. 2015. Optimization of natural convection heat transfer of Newtonian nanofluids in a cylindrical enclosure. *Chinese J. Chem. Eng.* 23:1266–1274.

Murshed, S. M. S. 2009. Correction and comment on "thermal conductance of nanofluids: Is the controversy over?" *J. Nanopart. Res.* 11:511–512.

Murshed, S. M. S. 2011. Determination of effective specific heat of nanofluids. *J. Exp. Nanosci.* 6:539–546.

Murshed, S. M. S. 2012. Simultaneous measurement of thermal conductivity, thermal diffusivity, and specific heat of nanofluids. *Heat Transf. Eng.* 33:722–731.

Murshed, S. M. S. and C. A. Nieto de Castro. 2014. *Nanofluids: Synthesis, Properties and Applications.* New York: Nova Science Publishers Inc.

Murshed, S. M. S. and P. Estellé. 2017. A state of the art review on viscosity of nanofluids. *Renew. Sust. Energy Rev.* 76:1134–1152.

Murshed, S. M. S., K. C. Leong and C. Yang. 2005. Enhanced thermal conductivity of TiO_2-water based nanofluids. *Int. J. Therm. Sci.* 44:367–373.

Murshed, S. M. S., K. C. Leong and C. Yang. 2006. Determination of the effective thermal diffusivity of nanofluids by the double hot-wire technique. *J. Phys. D-Appl. Phys.* 39:5316–5322.

Murshed, S. M. S., K. C. Leong and C. Yang. 2008a. Investigations of thermal conductivity and viscosity of nanofluids. *In. J. Therm. Sci.* 47:560–568.

Murshed, S. M. S., K. C. Leong and C. Yang. 2008b. Thermophysical and electrokinetic properties of nanofluids: A critical review. *Appl. Therm. Eng.* 28:2109–2125.

Murshed, S. M. S., K. C. Leong, C. Yang and N. T. Nguyen. 2008c. Convective heat transfer characteristics of aqueous TiO_2 nanofluids under laminar flow conditions. *Int. J. Nanosci.* 7:325–331.

Murshed, S. M. S., K. C. Leong and C. Yang. 2010. Thermophysical properties of nanofluids. In: *Handbook of Nanophysics: Nanoparticles and Quantum Dots*, ed. K. D. Sattler. Boca Raton, FL: Taylor & Francis, 32.-1–32-14.

Murshed, S. M. S., C. A. Nieto de Castro, M. J. V. Lourenço, M. Matos Lopes and F. J. V. Santos. 2011. A review of boiling and convective heat transfer with nanofluids. *Renew. Sust. Energy Rev.* 15:2342–2354.

Nielsen, L. E. 1970. Generalized equation for the elastic moduli of composite materials. *J. Appl. Phys.* 41:4626–4627.

Nelson, I. C., D. Banerjee and R. Ponnappan. 2009. Ponnappan Flow loop experiments using polyalphaolefin nanofluids. *J. Thermophys. Heat Transf.* 23(4):752–761.

Nieto de Castro, C. A., I. C. S. Vieira, M. J. V. Lourenço and S. M. S. Murshed. 2017. Understanding stability, measurements, and mechanisms of thermal conductivity of nanofluids. *J. Nanofluid.* 6:804–811.

Nnanna, A. G. 2007. Experimental model of temperature-driven nanofluid. *J. Heat Transfer.* 129:697–704.

Pak, B. C. and Y. I. Cho. 1998. Hydrodynamic and heat transfer study of dispersed fluids with submicron metallic oxide particles. *Exp. Heat Transfer* 11(2):151–170.

Pakdaman M. F., M. A. A. Behabadi and P. Razi. 2012. An experimental investigation on thermo-physical properties and overall performance of MWCNT/heat transfer oil nanofluid flow inside vertical helically coiled tubes. *Exp. Therm. Fluid. Sci.* 40(0):103–111.

Peyghambarzadeh, S. M., S. H. Hashemabadi, A. R. Chabi and M. Salimi. 2014. Performance of water based CuO and Al_2O_3 nanofluids in a Cu-Be alloy heat sink with rectangular microchannels. *Energy Convers. Manag.* 86:28–38.

Putra, N., W. Roetzel and S. K. Das. 2003. Natural convection of nano-fluids. *Heat Mass Transf.* 39:775–784.

Reddy, M. C. S. and V. V. Rao. 2013. Experimental studies on thermal conductivity of blends of ethylene glycol-water-based TiO_2 nanofluids. *Int. Comm. Heat Mass Transf.* 46:31–36.

Riazi, H., T. Murphy, G. B. Webber, R. Atkin, S. M. Tehrani and R. A. Taylor. 2016. Specific heat control of nanofluids: A critical review. *Int. J. Therm. Sci.* 107:25–38.

Rudyak, V. Y., S. V. Dimov, V. V. Kuznetsov and S. P. Bardakhanov. 2013. Measurement of the viscosity coefficient of an ethylene glycol-based nanofluid with silicon-dioxide particles. *Doklady Phys.* 58(5):173–176.

Saeedinia, M., M. A. A. Behabadi and M. Nasr. 2012. Experimental study on heat transfer and pressure drop of nanofluid flow in a horizontal coiled wire inserted tube under constant heat flux. *Exp. Therm. Fluid. Sci.* 36:158–168.

Selvakumar, P. and S. Suresh. 2012. Convective performance of CuO/water nanofluid in an electronic heat sink. *Exp. Therm. Fluid Sci.* 40:57–63.

Shahrul, I. M., I. M. Mahbubul, S. S. Khaleduzzaman, R. Saidur and M. F. M. Sabri. 2014. A comparative review on the specific heat of nanofluids for energy perspective. *Renew. Sust. Energy Rev.* 38:88–98.

Sharifpur, M., S. Yousefi and J. P. Meyer. 2016. A new model for density of nanofluids including nanolayer. *Int. Comm. Heat Mass Transf.* 78:168–174.

Sharifpur, M., N. Tshimanga, J. P. Meyer and O. Manca. 2017. Experimental investigation and model development for thermal conductivity of α-Al$_2$O$_3$-glycerol nanofluids. *Int. Comm. Heat Mass Transf.* 85:12–22.

Sharifpur, M., A. B. Solomon, T. L. Ottermann and J. P. Meyer. 2018. Optimum concentration of nanofluids for heat transfer enhancement under cavity flow natural convection with TiO$_2$– Water. *Int. Comm. Heat Mass Transf.* 98:297–303.

Shin, D. and D. Banerjee. 2014. Specific heat of nanofluids synthesized by dispersing alumina nanoparticles in alkali salt eutectic. *Int. J. Heat Mass Transf.* 74:210–214.

Sivakumar, A., N. Alagumurthi and T. Senthilvelan. 2017. Effect of serpentine grooves on heat transfer characteristics of microchannel heat sink with different nanofluids. *Heat Transf. Asian Res.* 46:201–217.

Solomon, A. B., J. van Rooyen, M. Rencken, M. Sharifpur and J. P. Meyer. 2017a. Experimental study on the influence of the aspect ratio of square cavity on natural convection heat transfer with Al$_2$O$_3$/water nanofluids. *Int. Comm. Heat Mass Transf.* 88:254–261.

Solomon, A. B., M. Sharifpur, J. P. Meyer, J. S. Ibrahim and B. Immanuel. 2017b. Convection heat transfer with water based mango bark nanofluids. *In 13th International Conference on Heat Transfer, Fluid Mechanics and Thermodynamics,* Portorož, Slovenia.

Solomon, A. B., M. Sharifpur, T. Ottermann, C. Grobler, M. Joubert and J. P. Meyer. 2017c. Natural convection enhancement in a porous cavity with Al$_2$O$_3$-ethylene glycol/water nanofluids. *Int. J. Heat Mass Transf.* 108:1324–1334.

Starace, A. K., J. C. Gomez, J. Wang, S. Pradhan, G. C. Glatzmaier and C. Greg. 2011. Nanofluid heat capacities. *J. Appl. Phys.* 110:1–5.

Sundar, L. S., K. V. Sharma, M. T. Naik and M. K. Singh. 2013. Empirical and theoretical correlations on viscosity of nanofluids: A review. *Ren. Sust. Ener. Rev.* 25:670–686.

Tshimanga, N., M. Sharifpur and J. P. Meyer. 2016. Experimental investigation and model development for thermal conductivity of glycerol–MgO nanofluids. *Heat Transf. Eng.* 37:1538–1553.

Tuckerman, D. B. and R. F. W. Pease. 1981. High-performance heat sinking for VLSI. *IEEE Elect. Dev. Lett.* 2:126–129.

Turgut, A., I. Tavman, M. Chirtoc, H. P. Schuchmann, C. Sauter and S. Tavman. 2009. Thermal conductivity and viscosity measurements of water-based TiO$_2$ nanofluids. *Int. J. Thermophys.* 30:1213–1226.

Vajjha, R. S. and D. K. Das. 2009a. Specific heat measurement of three nanofluids and development of new correlations. *J. Heat Transf.* 131(7):071601–071607.

Vajjha, R. S. and D. K. Das. 2009b. Experimental determination of thermal conductivity of three nanofluids and development of new correlations. *Int. J. Heat Mass Transf.* 52:4675–4682.

Wang, X., X. Xu and S. U. S. Choi. 1999. Thermal conductivity of nanoparticle–fluid mixture. *J. Thermophys. Heat Transf.* 13:474–480.

Wang, Y., T. S. Fisher, J. L. Davidson and L. Jiang. 2002. Thermal conductivity of nanoparticle suspensions. *In Proceedings of 8th AIAA/ASME Joint Thermophysics and Heat Transfer Conference*, St. Louis, MI.

Xie, H., J. Wang, T. Xi and Y. Liu. 2002. Thermal conductivity of suspensions containing nanosized SiC particles. *Int. J. Thermophys.* 23:571–580.

Xuan, Y., Q. Li and W. Hu. 2003. Aggregation structure and thermal conductivity of nanofluids. *AIChE J.* 49:1038–1043.

Yiamsawas, T., O. Mahian, A. S. Dalkilic, S. Kaewnai and S. Wongwises. 2013. Experimental studies on the viscosity of TiO_2 and Al_2O_3 nanoparticles suspended in a mixture of ethylene glycol and water for high temperature applications. *Appl. Energy* 111:40–45.

Yu, L. and D. Liu. 2013. Study of the thermal effectiveness of laminar forced convection of nanofluids for liquid cooling applications. *IEEE Trans. Compon. Packag. Manuf. Technol.* 3:1693–1704.

Zennifer, M. A., S. Manikandan, K. S. Sughanti, V. L. Vinodhan and K. S. Rajan. 2015. Development of CuO-ethylene glycol nanofluids for efficient energy management: Assessment of potential for energy recovery. *Ener. Conv. Manag.* 105:685–696.

Zhang, H., S. Shao, H. Xu and C. Tian. 2013. Heat transfer and flow features of Al_2O_3-water nanofluids flowing through a circular microchannel: Experimental results and correlations. *Appl. Therm. Eng.* 61:86–92.

Zhou, S.-Q. and R. Ni. 2008. Measurement of the specific heat capacity of water-based Al_2O_3 nanofluid. *Appl. Phys. Lett.* 92:1–3.

8

Determination of Energy Efficiency of Hot Water Boilers and Calculation of Measurement Uncertainties

Sotirios Karellas, Panagiotis Vourliotis, Platon Pallis, Ioannis-Alexandros Sofras, and Emmanuel Kakaras

National Technical University of Athens

CONTENTS

8.1 Introduction

The energy investigation of heat generation systems can be evaluated with two different approaches.

The first approach considers permanent conditions and uses the efficiency of the "boiler space heater."

Boiler space heater means a space heater that generates heat using the combustion of fossil fuels and/or biomass fuels.

"Space heater" means a device that:

a. provides heat to a water-based central heating system in order to reach and maintain at a desired level the indoor temperature of an enclosed space such as a building, a dwelling, or a room, and

b. Is equipped with a heat generator (burner unit), designed to transmit to water the heat released from burning.

For the purposes of this chapter, the following definitions are valid:

- *Boiler*: the combined boiler body–burner unit, designed to transmit to water the heat released from burning,
- *Appliance*:
  - the boiler body designed to have a burner fitted,
  - the burner designed to be fitted to a boiler body.

The efficiency (performance rate, η) is calculated:

a. By the direct method from the ratio of useful energy to consumed energy:
   η = useful energy/consumed energy, or alternatively

b. By the indirect method through heat losses of the system:
   η = (consumed energy - thermal losses)/consumed energy

The second approach uses energy balances for an extended period of time, usually one year, which takes into account losses during standby and operation times, which are often decisive for energy balances. Hence, the second approach uses the annual efficiency of the heating installation.

The relevant harmonized norms, the European Union Directives, and the regulations related to the test–certification of the boiler space heater are presented in References. Ensuring all the necessary conditions and parameters for the proper conduct of the test for the determination of the nominal useful heat output as well as the boiler efficiency is specified in the relevant European norms [1–20].

8.2 Procedure for Determining the Energy Efficiency of Hot Water Boiler

8.2.1 Determination of the Nominal Heat Output

As effective rated heat output of the boiler (expressed in kW) is defined, the useful heat output is determined during the boiler efficiency certification process which is being delivered during continuous operation with a specific fuel, with a useful efficiency as determined during the certification process.

The amount of useful heat output transmitted to the heat carrier (water) is measured. It can be determined in the boiler circuit or by means of a secondary heat exchanger.

The useful heat output transmitted to the water is determined by measuring either:

a. the mass flow of cold water entering the boiler circuit and the rise of temperature between the outlet water temperature and the inlet water temperature (bypass method), or

b. the mass flow of the water circulating in the boiler circuit and its temperature rise, or

c. the mass flow and the temperature rise over a secondary heat exchanger corrected by the heat loss of this secondary heat exchanger. The heat produced by the boiler is transferred to the cooling water by means of a secondary heat exchanger. The heat received by the latter is calculated from the mass flow and the temperature rise of the cooling water. The heat losses from the well-insulated connections between the boiler and the secondary heat exchanger and those of the secondary heat exchanger itself are determined either by preliminary tests or by calculation. The heat output of the boiler is the sum of the two amounts of heat.

The test for the determination of the nominal heat output shall be carried out at a firing rate such that the output is at least 100%, but does not exceed 105% of the nominal value, and the requirements concerning the nominal heat output shall be met. If the heat output exceeds 105%, a second test shall be carried out at a firing rate between 95% and 100% of the nominal heat output of the boiler. The actual value for the nominal heat output shall be determined by linear interpolation between the two test results. The nominal heat output shall be determined at a water rate that is adjusted to obtain a return water temperature of 60°C ± 1°C and a temperature difference between the flow and return water temperature of 20°C ± 2°C.

Note: The conditions for determination of the rated heat output in the former versions of EN 304 have been a mean flow temperature of between 80°C and 90°C, and the mean temperature difference between flow and return has been between 10 and 25 K. However, this is not in line with the Regulation EU 813/2013.

8.2.1.1 Description of the Hot Water Boiler Testing Plant (Bypass Method)

In Figure 8.1, a line diagram of the test rig of a typical hot water boiler testing plant (bypass method) is presented. From a large-capacity water tank (7) (with overflow control), the cold water supply (temperature T_E) is pumped to cool the boiler. The flow of cold water is regulated by a regulating valve (3). The flow water exits the boiler (temperature T_F), and a portion of water supply (equal to the cold makeup water) is led to a surge vessel (6) from which the overflow is led to the weighing container (10), after being cooled in the heat exchanger (13). The remaining water (from the exit boiler) returns to the boiler from the recirculation circuit (short circuit) by means of a circulator (2). There is a regulating valve (4) for regulating the supply of recirculation water

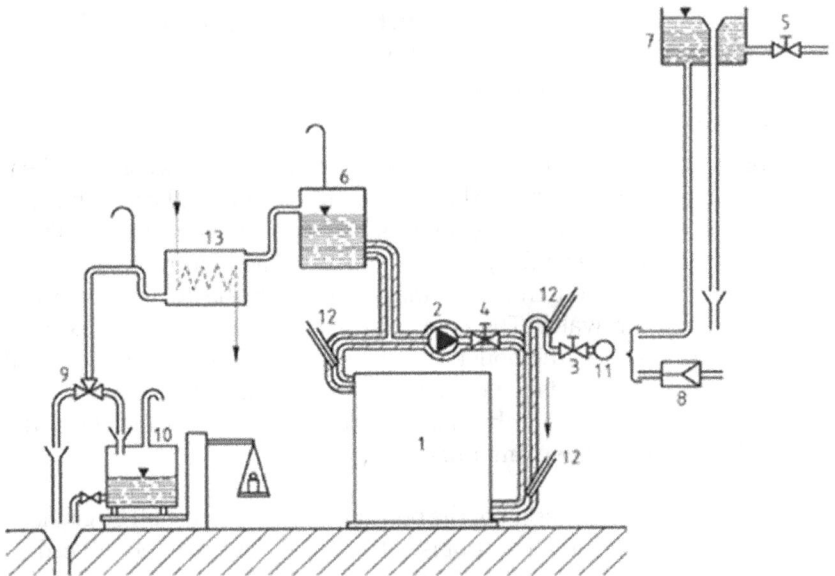

Key

1	boiler under test	8	constant pressure distribution pipe
2	circulating pump	9	three-way tap
3	control valve I	10	weighing vessel
4	control valve II	11	water meter
5	control valve III	12	temperature measurements
6	compensating tank	13	cooler
7	constant head tank		

FIGURE 8.1

Schematic diagram of the boiler testing plant (bypass method). Test rig with short circuit section and three possible arrangements for cold water supply.

in the pump depression. Thus, by means of the throttle valves (3) and (4), the supply of cold water entering the system (water for cooling the boiler) and the flow rate of water recirculation are regulated and finally the inlet and exit temperatures (T_R) and (T_F) of the water in the boiler, are regulated respectively. The cold water temperature (T_E) is mixed with exit water temperature (T_F) and gives the return water temperature in the boiler (T_R).

The useful heat output is determined by measuring the amount of the cold water flowing through the system and the corresponding temperature difference $(T_F - T_E)$. The water supply is measured either by weighing the passing quantity into the weighing vessel (10) after it passes through the balancing vessel (heat exchanger 13) (cooling the hot water exiting the boiler with cold water from the grid) so as to avoid mass losses of the water to be weighed due to evaporation or alternatively with modern flow mass meters, operating with the Coriolis principle. The test rig is complemented by the fuel supply circuit for liquid fuels, the fuel tank, and the corresponding supply and return pipes.

8.2.1.2 Calculation of the Nominal Heat Output $\dot{Q}_{out}$

The calculation of the nominal heat output $\dot{Q}_{out}$ is given by Eq. (8.1):

$$\dot{Q}_{out} = \frac{\dot{q}_{mw1}}{3600} \cdot \bar{c}_{pwFE} \cdot (T_F - T_E) + L, \text{ in kW} \qquad (8.1)$$

where:

$\dot{q}_{mw1}$: the measured mass flow of cold water entering the system, in kg/h; due to the continuity principle, it is identical to the hot water flow leaving the system.

T_E: the entering temperature of the supply cold water, in °C.

T_F: the exit temperature of the water (flow temperature), in °C.

T_R: the return temperature of the water entering to the boiler, in °C.

$\bar{c}_{pwFE}$: the specific heat capacity of water at temperature $T_{mwFE} = \dfrac{T_F + T_E}{2}$, in kJ/(kg·K).

$L = A_L + B_L (T_{mwL} - T_L)$: heat loss in kW from the hydraulic circuit of the test rig, taking into account losses from the hot walls of the pipes and the pump, and also the energy contribution of the circulator. The temperature T_{mwL} is:

$$T_{mwL} = \frac{T_F + T_R}{2} \text{ in } °C \qquad (8.2)$$

T_L: the ambient temperature, in °C.

The latent losses of the test rig circuit result from the circuit's loss correlation curve between the temperature of the circuit water and the temperature of ambient air. This correlation curve has been obtained after a series of tests, which are carried out according to EN 303.02 and EN 303.03.

Also, the calculation of the recirculation water supply $\dot{q}_{mw2}$ was as follows:

$$\dot{q}_{mw2} = \dot{q}_{mw1} \cdot \frac{(T_R - T_E)}{(T_F - T_E)}, \text{ in kg/h} \qquad (8.3)$$

Therefore, the calculation of the total flow of the water $\dot{q}_{mwtot}$ was as follows:

$$\dot{q}_{mwtot} = \dot{q}_{mw1} + \dot{q}_{mw2} = \frac{\dot{Q}_{out}}{\left(\bar{c}_{pwFR} \cdot (T_F - T_R)\right)} \cdot 3600, \text{ in kg/h} \qquad (8.4)$$

Measured values: $\dot{q}_w$, T_F, T_E, T_R, and T_L

The specific heat capacity of the water at the temperature T_{mwFE} is given by Eq. (8.5a):

$$\bar{c}_{pwFE} = A + B \cdot T_{mwFE} + C \cdot T_{mwFE}^2 + D \cdot T_{mwFE}^3 \text{ , in kJ/(kg·K)} \qquad (8.5a)$$

The specific heat capacity of the water at the temperature $T_{mwFR = TmwL}$ is given by Eq. (8.5b):

$$\bar{c}_{pwFR} = A + B \cdot T_{mwFR} + C \cdot T_{mwFR}^2 + D \cdot T_{mwFR}^3 \text{ , in kJ/(kg·K)} \qquad (8.5b)$$

where $A = 4.2121$; $B = -0.002421$; $C = 0.00005288 = 5.288E\text{-}05$;   and $D = -0.000000333 = -3.33E\text{-}07$.

8.2.2 Calculation of Heat Input $\dot{Q}_{in}$

8.2.2.1 Calculation of the Nominal Heat Input $\dot{Q}_{in}$, When Oil Fuel or Solid Fuel Is Used

The calculation of heat input $\dot{Q}_{in}$, when oil fuel or solid fuel is used, is given by Eq. (8.6):

$$\dot{Q}_{in} = \left(\frac{\dot{q}_{mf}}{3600}\right) \cdot H_{im} \text{ in kW,} \qquad (8.6)$$

where:

$\dot{q}_{mf}$: the mass flow rate of fuel in kg/h.

H_{im}: the imperior or lower calorific capacity or net calorific value (NCV) of the fuel in kJ/kg.

Measured values: $\dot{q}_{mf}$ and H_{im}

In general, the fuel heating calorific capacity expresses the energy released when a fuel is fully burnt under specific pressure and temperature conditions.

It is obvious that the combustion heat released varies with the pressure and temperature conditions at which the combustion takes place. Heat is measured when:

- A specific amount of fuel is burnt completely with dry air, the exhaust flue gas temperature drops to a reference temperature of 25°C, and the water vapor is condensed by yielding its heat. In this case, the total heat attributed is characterized as superior calorific H_s (superior or gross calorific capacity or value, GCV).

- However, if all of the fuel's chemical energy is released and the exhaust flue gas temperature drops to the reference temperature of 25°C, and assuming that the water vapor contained in the exhaust flue gases is not liquefied, then the total heat attributed is characterized as a lower calorific value H_{im} (imperior or net calorific capacity or value, NCV).

It is obvious that the two calorific capacities differ in the vapor latent heat (due to combustion of hydrogen and moisture contained in the fuel), that is, Eq. (8.7):

$$H_{im} = H_s - (8.936\ \gamma_h + \gamma_w)r \text{, in kJ/kg} \tag{8.7}$$

where
γ_h: the mass content of the fuel in hydrogen, in kg/kg.
γ_w: the mass content of the fuel in moisture, in kg/kg.
r: the heat of water vaporization at 25°C, in kJ/kg; at atmospheric pressure, it is equal to 2442 kJ/kg.

8.2.2.2 Calculation of Heat Input $\dot{Q}_{in,r}$, When Gas Fuel Is Used

The test method is performed according to the European standards, and the heat input is calculated from Eq. (8.8):

$$\dot{Q}_{in,r} = 0.278 \cdot \dot{V}_{Gas,r} \cdot H_{imv,r} \text{, in kW} \tag{8.8}$$

where:
r: the size status under standard reference conditions (gas volume at *15°C and 1013.25 mbar*).
0.278: unit conversion factor of the ratio = 1000/3600.
$\dot{Q}_{in,r}$: the heat input in kW, which is the product of the measured volume flow rate of the fuel gas over the net calorific value of the gas, under the same reference conditions.
$H_{imv,r}$: the net calorific value, expressed in MJ/m³ (gas volume at 15°C and 1013.25 mbar).
$\dot{V}_{Gas,r}$: fuel gas volume flow rate in m³/h reduced to standard conditions at 15°C and 1013.25 mbar, which is given by Eq. (8.9):

$$\dot{V}_{Gas,r} = \dot{V}_{Gas,m} \cdot \frac{(P_a + P_g - P_s)}{(P_n)} \cdot \frac{(T_n + T_r)}{(T_n + T_{gas})} \cdot \left(\frac{1}{K}\right) \text{, in m}^3/\text{h} \tag{8.9}$$

where:

$\dot{V}_{Gas,m}$: the measured fuel gas volume flow rate under test conditions, in m³/h.

P_a: the barometric atmospheric pressure during the test, in mbar.

P_g: the gas pressure in the meter, in mbar.

P_s: the partial vapor pressure = $\varphi \cdot p_k$, in mbar.

φ: the relative humidity (degree of water vapor saturation).

p_k: the saturation pressure of the water vapor at temperature T_{gas}, in mbar.

P_n: the normal pressure, which is equal to 1013.25 mbar).

T_n: the normal temperature, which is equal to 273.15 K.

T_r: the reference temperature, which is equal to 15 °C.

T_{gas}: the temperature of the fuel gas in the meter, in °C.

K: compressibility coefficient; $K = Z/Zn = 1.00-(P_a+P_g)/450$ for natural gas and $K = 1.00$ for atmospheric air.

Z: the deviation factor from the ideal state of the constituent gas equation.

Zn: the deviation factor for normal conditions.

Measured values: $\dot{V}_{Gas,m}$, $H_{imv,r}$, P_a, P_g, and T_{gas}

8.2.3 Efficiency of Boiler at Full Nominal Load

The efficiency of the boiler is determined according to the following alternative methods:

- *Direct method*: by determining the amount of heat absorbed by the heat carrier (water) during the test (nominal heat output) and the heat quantity supplied simultaneously with the fuel and air (heat input).
- *Indirect method*: by determining all thermal losses in the heat balance.

8.2.3.1 Efficiency Measurement with the Direct Method

The calculation of the hot water boiler efficiency based on the direct method is performed according to the procedures and requirements described in standards EN 304, EN 303.03 and EN 303.05 for liquid, gas, and solid fuel, respectively.

Direct boiler efficiency η_{Direct}: The term "useful efficiency" (expressed in %) describes the ratio of the heat output $P_{N = Qout}$ transmitted to the boiler water to the heat input $Q_{B = Qi}$ (which is the product of the net calorific value at constant fuel pressure and the consumption expressed as a quantity of fuel per time unit). Thus, in the general case, we define the efficiency of the boiler, with Eq. (8.9):

$$\eta_{Direct} = \eta_k = \dot{Q}_{out}/\dot{Q}_{in} \cdot 100, \text{ in\%} \qquad (8.10)$$

where:

η_k: the efficiency of the boiler, in %.

$\dot{Q}_{out}$: the heat output of the boiler transmitted to the boiler water, in kW.

$\dot{Q}_{in}$: the heat input of the boiler, in kW.

8.2.3.2 Efficiency Measurement with the Indirect Method of Thermal Losses

In a boiler, as in any thermal machine, the utilization of the thermal and chemical energy supplied is incomplete (during the transformation into useful thermal energy, a part is lost to the environment in the form of various thermal losses). Thus, the useful energy always falls short of what is attributed to those losses, so that the efficiency is always lower than the unit [21–23]. Hence, Eq. (8.9) is written as follows:

$$\eta_{\text{Indirect}} = \eta_k = \left(\dot{Q}_{\text{in}} - \sum Q_i\right)\Big/Q_{\text{in}} \qquad (8.11)$$

where:

$\sum Q_i$: the sum of all thermal losses of the boiler, which are discussed in detail below.

According to what has been mentioned above, Eq. (8.11) can be written as:

$$\eta_{\text{Indirect}} = \eta_k = \frac{\dot{q}_{\text{mf}} \cdot H_{\text{im}} - \dot{q}_{\text{mf}} \cdot \sum q_i}{\dot{q}_{\text{mf}} \cdot H_{\text{im}}} = \frac{H_{\text{im}} - \sum q_i}{H_{\text{im}}} = 1 - \frac{\sum q_i}{H_{\text{im}}} \qquad (8.12)$$

where:

$\sum q_i$: the sum of all heat losses that occur in the boiler during the combustion of the fuel mass unit, or in other words, the part of the chemical energy of the fuel mass unit which is not transferred to the fluid medium (water) but is lost either in the environment or in the ash of the fuel. These heat losses are allocated to the following categories:

q_d: heat losses due to the fuel that escapes from the gaps of the grid and falls into the ashtray, without burning.

q_t: heat losses from the fuel residues, located within the ash at the end of combustion.

q_f: heat losses due to soot and coke found on fly ash.

q_q: heat losses due to the temperature of solid or liquid combustion residues.

q_A: heat losses due to the (relatively) high temperature of the exhaust flue gas, with which it leaves the boiler and exits into the atmosphere.

q_u: heat losses due to the analyzed exhaust flue gas content of some combustible gases that have not burnt completely.

q_S: heat losses due to radiation and heat transfer from the walls of the external surfaces of the boiler to the environment.

If we characterize the ratio $q_i/H_{\text{im}} \cdot 100 = v_i$, then the efficiency can be written as follows:

$$\eta_{\text{Indirect}} = \eta_k = \left(100 - \sum v_i\right), \text{in } \% \qquad (8.13)$$

In this case, v_i will be the heat loss due to the i cause, reduced to H_{im} and expressed as %. In Figure 8.2, the flow of power and losses are illustrated graphically in a typical central heating system.

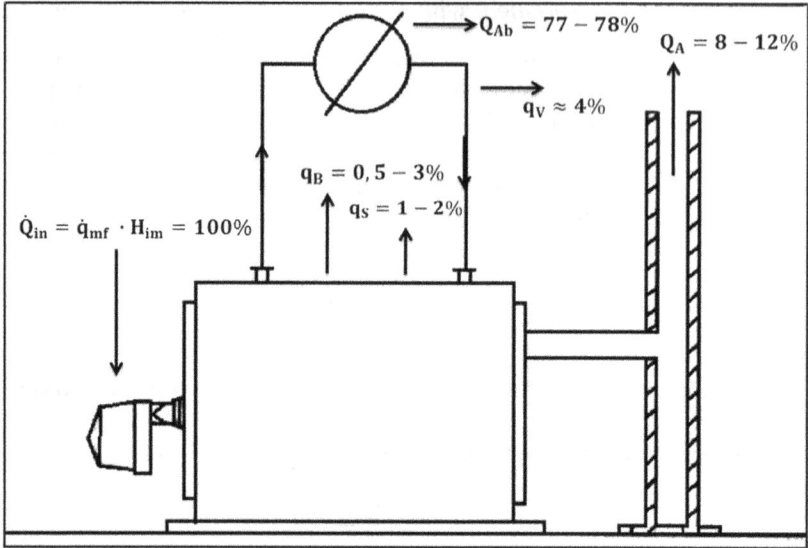

FIGURE 8.2
Flow of energy/losses in a typical central heating system.

where:
q_B: the standby losses of the boiler body.
q_V: the thermal losses of the distribution of the hydraulic circuit.
Q_{Ab}: the final percentage of useful thermal energy to meet the building's thermal requirements.

8.3 Measurement Accuracies and Uncertainties

All measurements shall be made with instruments calibrated in accordance with the manufacturer's instructions. External fixed or portable instruments shall be used unless it can be proven that the instruments' sensors installed on the boiler by its manufacturer have been located correctly and system's accuracy is certified.

The stated accuracy of the measurement devices for the following parameters shall not exceed the below respective limit:

  a. Atmospheric pressure = 50 Pa.
  b. Waterside pressure loss = 2% of the measured value.
  c. Water flow rate = 1% of the measured value.
  d. Air volume flow rate = 2% of the measured value.

e. Time
   1. up to 1 h: 0.2 s;
   2. beyond 1 h: 0.1% of the measured value
f. Auxiliary electrical energy = 2% of the measured value.
g. Temperatures:
   1. Ambient: 2 K;
   2. Water: 1 K;
   3. Combustion products: 2 K;
   4. Surface: 2 K;
h. CO, CO_2, O_2 NO_x, and C_xH_y:
   1. CO_2 content: 0.1% volume from the full scale
   2. O_2 content: 0.1% volume from the full scale
   3. CO content: 5 ml/m^3
   4. NO_x content: 5 ml/m^3
   5. C_xH_y content: 5 ml/m^3
i. Mass = 0.05% of the full scale.
j. Pressure flue gas:
   1. ≤ 60 Pa: 1 Pa;
   2. > 60 Pa: 2% of the measured value.

The full range of the measuring apparatus shall be chosen in such a way that it is suitable for the maximum anticipated value. The measurement accuracies indicated above concern individual measurements.

For measurements requiring a combination of individual measurements, the lower accuracies associated with individual measurements may be necessary to attain the total required uncertainty. The test rig shall be set up in such a way that the efficiency can be determined within an absolute expanded uncertainty of 2% points.

8.4 Procedures for Calculation of the Uncertainty Efficiency in a Hot Water Boiler

8.4.1 Evaluation Methods for Measuring Errors

Following a proposal by international working groups, in order to calculate the uncertainties of measurands, two different evaluation methods can be applied:

- Type A, and
- Type B.

The **Type A** evaluation can be applied when the uncertainty of a parameter is evaluated by a series of repeated observations under the same measurement conditions. This requires an appropriate statistical analysis of the method in order to allow observation of the measurement dispersion.

The **Type B** evaluation is not calculated from a statistical analysis of many observations, such as the Type A method, but its estimation is based on metrological justified criteria taking into account all existing information on value fluctuations. This information is basically taken from the measurement accuracy provided by the instruments, since it is not possible to take measurements with a standard deviation less than the internal measurement uncertainty of the instrument.

This description of procedures applies when the Type B procedure is used to assess the uncertainty of measuring the efficiency.

8.4.2 Definitions

The mean value of the measurement series is given by Eq. (8.14):

$$\bar{x} = \frac{\sum_{i=1}^{n} x_i}{n} \tag{8.14}$$

where:
$\bar{x}$: the arithmetic mean of n measurements.
x_i: the value of the ith measurement.
n: the number of measurements received.
The variability of a series of measurements is given by Eq. (8.15):

$$V(x) = \frac{\sum_{i=1}^{n} (x_i - \bar{x})^2}{n-1} \tag{8.15}$$

For Type B measurements, the variability of a series of measurements is equal to the instrument's measurement variability used for this measurement.

The standard deviation of a measurement is given by the variability (Eq. (8.15)) and is calculated from Eq. (8.16):

$$\sigma(x) = \sqrt{V(x)} \Rightarrow \sigma^2(x) = V(x) \tag{8.16}$$

where:
$\sigma(x)$: the standard deviation of the measured value.
The measurement uncertainty of a measurand is given within a confidence interval and depends on it. Usually, the uncertainty of a measurand is given in the form of a space in which the measured quantity is placed with a certain probability. The following values are defined as uncertainty:

1. Absolute uncertainty:

$$u = 2 \cdot \sigma = 2 \cdot \sqrt{V(x)} = \sqrt{\varepsilon_\mu^2 + S^2} \qquad (8.17)$$

2. Relative uncertainty:

$$u_r = 100 \cdot \frac{u}{\bar{x}} \qquad (8.18)$$

where:
  ε_μ: the maximum possible measurement error value.
  S: the systematic error of instrument measuring.
  The uncertainty u of a sample of a specific number of observations derived from a probabilistic set of normal distribution with mean value $\bar{x}$ and standard deviation σ is given:

- $u = \pm 1\sigma$ for a confidence interval of 68.3%
- $u = \pm 2\sigma$ for a confidence interval of 95%
- $u = \pm 3\sigma$ for a confidence interval of 99.8%

where:
  u: measurement error/uncertainty for a given confidence interval.
  σ: the standard deviation of the measurement sample.
  The efficiency of the boiler is given at 95% confidence interval. Also in the instruments used, the confidence interval is 95%, 2σ.

8.4.3 Uncertainty Calculation of Useful Heat Output

The calculation of the nominal heat output $\dot{Q}_{out}$ is given by Eq. (8.1). The specific heat capacity of the water $\bar{c}_{pwFE}$ at the temperature T_{mwFE} is given by Eq. (8.5a).
  The specific heat capacity of the water depends on its temperature. In order to measure the useful thermal power, the specific heat capacity is calculated at a temperature that is equal to the average inlet temperature of the supply water in the hydraulic circuit and the exit of the boiler water. The average water temperature in the circuit is given by the following equation:

$$T_{mwFE} = \frac{T_F + T_E}{2} \qquad (8.19)$$

In Figure 8.3, the calculation process of the useful heat output is illustrated.
  The variability of the useful heat output measurement is given by Eq. (8.20):

```
                    ┌─────────────────────────┐
                    │  Calculation of Nominal │
                    │  Heat Output Q̇_out in kW.│
                    └─────────────────────────┘
```

$$\boxed{\text{Calculation of Nominal Heat Output } \dot{Q}_{out} \text{ in kW.}}$$

$$\boxed{\text{Measurement of exit (flow) water temperature } T_F, \text{ in °C.}}$$

$$\boxed{\text{Measurement of the entering cold water temperature } T_E, \text{ in °C.}}$$

$$\boxed{T_{mwFE} = (T_F + T_E)/2}$$

$$\boxed{\text{Measurement of the entering cold water mass flow } \dot{q}_{mw1}, \text{ in kg/h.}}$$

$$\boxed{\text{Calculation of mean water temperature } T_{mwFE}, \text{ in °C.}}$$

$$\boxed{\text{Calculation of the water mean specific heat capacity } c_{pwFE} \text{ for the mean temperature } T_{mwFE}, \text{ in kJ/(kg·K)}}$$

$$\dot{Q}_{out} = \frac{\dot{q}_{mw1}}{3600} \cdot \overline{c}_{pwFE} \cdot (T_F - T_E) + L, \text{ in kW}$$

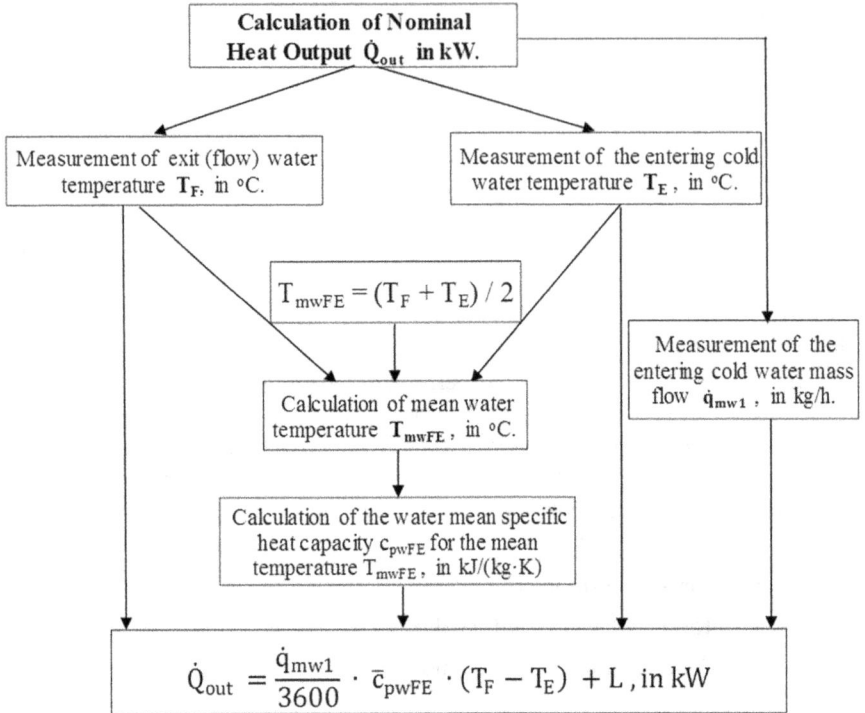

FIGURE 8.3
Calculation process of useful heat output.

$$V(\dot{Q}_{out}) =$$

$$\dot{Q}_{out}^2 \cdot \left[\left[\left(\frac{(\overline{c}_{pwFE} \cdot (T_F - T_E))}{(\dot{q}_{mw1} \cdot \overline{c}_{pwFE} \cdot (T_F - T_E) + L)} \right)^2 \right] \cdot V(\dot{q}_{mw1}) \right)$$

$$+ \left(\left(\frac{\overline{c}_{pwFE} \cdot q_w}{(\dot{q}_{mw1} \cdot \overline{c}_{pwFE} \cdot (T_F - T_E) + L)} \right)^2 \cdot (V(T_E) + V(T_F)) \right)$$

$$+ \left(\left(\frac{\dot{q}_{mw1} \cdot (T_F - T_E)}{(\dot{q}_{mw1} \cdot \overline{c}_{pwFE} \cdot (T_F - T_E) + L)} \right)^2 \cdot \left(V(\overline{c}_{pwFE}) \right) \right)$$

$$+ \left(\left(\frac{1}{(\dot{q}_{mw1} \cdot \overline{c}_{pwFE} \cdot (T_F - T_E) + L)} \right)^2 (V(L)) \right) \right] \qquad (8.20)$$

or by Eq. (8.21):

$$V(\dot{Q}_{out}) = \dot{Q}_{out}^2 \cdot \left[\left(A_{out} + B_{out} + C_{out} + D_{out} + E_{out} \right) \right] \tag{8.21}$$

where:

$$A_{out} = \left(\left(\frac{\overline{c}_{pwFE} \cdot (T_F - T_E)}{\left(\dot{q}_{mw1} \cdot \overline{c}_{pwFE} \cdot (T_F - T_E) + L \right)} \right)^2 \cdot \left(V(\dot{q}_{mw1}) \right) \right) \tag{8.22}$$

$$B_{out} = \left(\left(\frac{\overline{c}_{pwFE} \cdot \dot{q}_{mw1}}{\left(\dot{q}_{mw1} \cdot \overline{c}_{pwFE} \cdot (T_F - T_E) + L \right)} \right)^2 \cdot \left(V(T_E) \right) \right) \tag{8.23}$$

$$C_{out} = \left(\left(\frac{\overline{c}_{pwFE} \cdot \dot{q}_{mw1}}{\left(\dot{q}_{mw1} \cdot \overline{c}_{pwFE} \cdot (T_F - T_E) + L \right)} \right)^2 \cdot \left(V(T_F) \right) \right) \tag{8.24}$$

$$D_{out} = \left(\left(\frac{\dot{q}_{mw1} \cdot (T_F - T_E)}{\left(\dot{q}_{mw1} \cdot \overline{c}_{pwFE} \cdot (T_F - T_E) + L \right)} \right)^2 \cdot \left(V(\overline{c}_{pwFE}) \right) \right) \tag{8.25}$$

$$E_{out} = \left(\left(\frac{1}{\left(\dot{q}_{mw1} \cdot \overline{c}_{pwFE} \cdot (T_F - T_E) + L \right)} \right)^2 \cdot \left(V(L) \right) \right) \tag{8.26}$$

In Eq. (8.20), where there is a V symbol, such as $V(T_E)$ or $V(T_F)$, the measurement variability of the specified measurand is replaced. Where V is not present, the mean value obtained from the metering process is replaced and equals to the value recorded by the measuring instrument of the measurand.

The measurement variability of the specific heat capacity is calculated by knowing that the relative uncertainty of the value resulting from Eq. (8.5a) is 0.07% over a confidence interval of 95%. Therefore:

$$V\left(\overline{C}_{PmwFE} \right) = \left(\frac{0.07}{2 \cdot 100} \cdot \overline{C}_{PmwFE}^2 \right). \tag{8.27}$$

Thus, by replacing it in Eq. (8.20), the measurement variability of the useful heat output is calculated.

8.4.4 Uncertainty Calculation of Heat Input

The case of a hot water boiler fired with liquid fuel (heating oil) is considered. Hence, the calculation of the heat input $\dot{Q}_{in}$ is given by Eq. (8.28):

$$\dot{Q}_{in} = \dot{q}_{mf} \cdot H_{im} \tag{8.28}$$

where:

$\dot{Q}_{in}$: the heat input, in kW.
$\dot{q}_{mf}$: the mass flow of liquid fuel, in kg/s.
H_{im}: the net calorific value of the liquid fuel, in kJ/kg.
In Figure 8.4, the calculation process of the heat input is illustrated.
The variability of the heat input measurement is given by Eq. (8.29):

$$\frac{V(\dot{Q}_{in})}{\dot{Q}_{in}^2} = \frac{V(\dot{q}_{mf})}{\dot{q}_{mf}^2} + \frac{V(H_{im})}{H_{im}^2} \tag{8.29}$$

The variability of the calorific capacity of the fuel $V(H_{im})$ measurement is calculated based on the absolute uncertainty given by a certified laboratory conducting the fuel analysis.

The measurement of fuel mass flow can be performed in two possible ways:

1. *Direct*: measuring the flow rate using an appropriate flow meter.

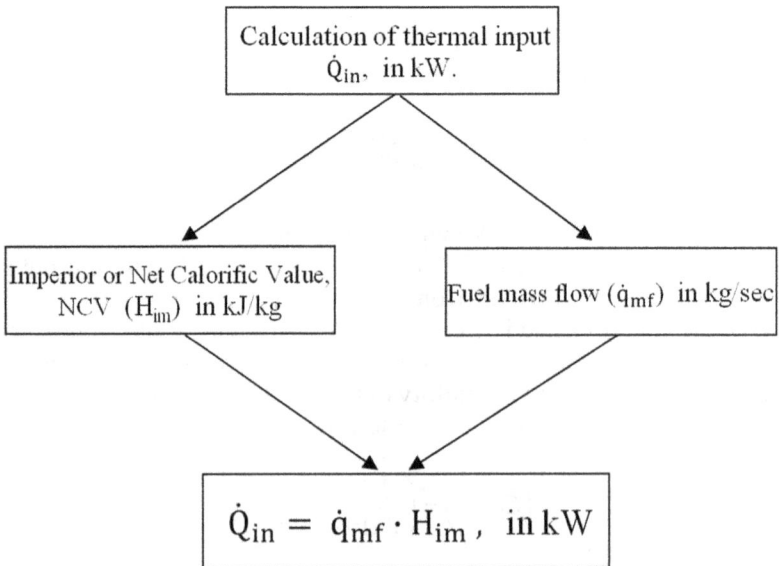

FIGURE 8.4
Measurement process of heat input.

2. *Indirect*: measuring the flow rate by weighing the fuel mass and measuring the time required to burn a certain amount of fuel (e.g., 200 g). For instance, the measurement begins when the scale shows that the weighing container contains, e.g., 13 kg of fuel and counting with the timer the amount of time required for the combustion of 200 g (the measurement sequence is therefore carried out at 12.8 kg, 12.6 kg, 12.4 kg, 12.2 kg and ends when the measuring balance contains 12 kg). This gives five time intervals during which the corresponding amount of 200 g of fuel is consumed, and therefore, five different fuel mass flow rates are derived and the fuel mass flow rate over the entire measurement interval is calculated as the average value of the calculated five mass flow rates.

In the case where the mass flow rate is calculated using a *flow meter*, the relative uncertainty of the measurement of the fuel mass flow rate is given by the manufacturer of the flow meter. If the fuel mass flow rate measurement error is given, the variability of this measurement is given as known using Eq. (8.16).

For the flow meter, applying Eqs. (8.16)–(8.18), the variability for measuring the mass flow rate is given by Eq. (8.30):

$$V(\dot{q}_{mf}) = \left(\frac{\alpha}{2 \cdot 100} \cdot \dot{q}_{mf} \right)^2 \tag{8.30}$$

where:

α: the accuracy percentage of measurement relative uncertainty of the mass flow meter, which is given by the instrument manufacturer for the given flow rate range.

In the case where the fuel mass flow is calculated by the indirect method (using a balance), the variability of this measurement is given by Eq. (8.31):

$$V(\dot{q}_{mf}) = \left[\frac{V(m_{f2})}{\Delta_{mf}^2} + \frac{V(m_{f1})}{\Delta_{mf}^2} + \frac{V(t)}{t^2} \right] \cdot \dot{q}_{mf}^2 \tag{8.31}$$

where:

$$V(m_{f1}) = \left(\frac{\alpha}{2 \cdot 100} \cdot \Delta_{mf} \right)^2 \tag{8.32}$$

$$V(m_{f2}) = \left(\frac{\alpha}{2 \cdot 100} \cdot \Delta_{mf} \right)^2 \tag{8.33}$$

α: the accuracy percentage of measurement relative uncertainty of the fuel mass flow balance, which for the given weighing range is given by the instrument manufacturer.

Δ_{mf}: the total amount of fuel consumed during the measurement, in kg.

m_{f1}, m_{f2}: the initial and final value of the weighing mass of the fuel in the measuring balance, respectively, in kg.

And:

$$V(t) = \left(\frac{\alpha}{2 \cdot 100} \cdot t \right)^2 \qquad (8.34)$$

where:

α: the accuracy percentage of measurement relative uncertainty of the timer, given by the instrument manufacturer.

t: the total time of the measurement, in s.

8.4.5 Variability of Efficiency Measurement

Boiler efficiency is calculated from Eq. (8.35) based on the values of the useful heat output and the heat input, calculated from Eqs. (8.20) and (8.29):

$$\eta = \frac{\dot{Q}_{out}}{\dot{Q}_{in}} \qquad (8.35)$$

The overall variability of efficiency "η" is given by Eq. (8.36):

$$\frac{V(\eta)}{\eta^2} = \frac{V(\dot{Q}_{in})}{\dot{Q}_{in}^2} + \frac{V(\dot{Q}_{out})}{\dot{Q}_{out}^2} \qquad (8.36)$$

where:

η: the efficiency, calculated from Eq. (8.35).

$V(\eta)$: the variability of efficiency measurement.

The efficiency standard deviation results from its variability (Eq. (8.37)):

$$\sigma(\eta) = \sqrt{V(\eta)} \qquad (8.37)$$

The efficiency is given with uncertainty over a 95% confidence interval. Therefore, for the range of efficiency within a 95% confidence interval, Eq. (8.38) applies:

$$\eta - 2\sigma \leq \eta \leq \eta + 2\sigma \qquad (8.38)$$

The calculation process for variability and uncertainty of boiler efficiency measurement is presented in Figure 8.5 as a flow diagram.

The relative % uncertainty of the measurement of the efficiency, for a 95% confidence interval, is given by Eq. (8.39):

$$u_r = 100 \cdot \frac{2 \cdot \sigma}{\eta} \qquad (8.39)$$

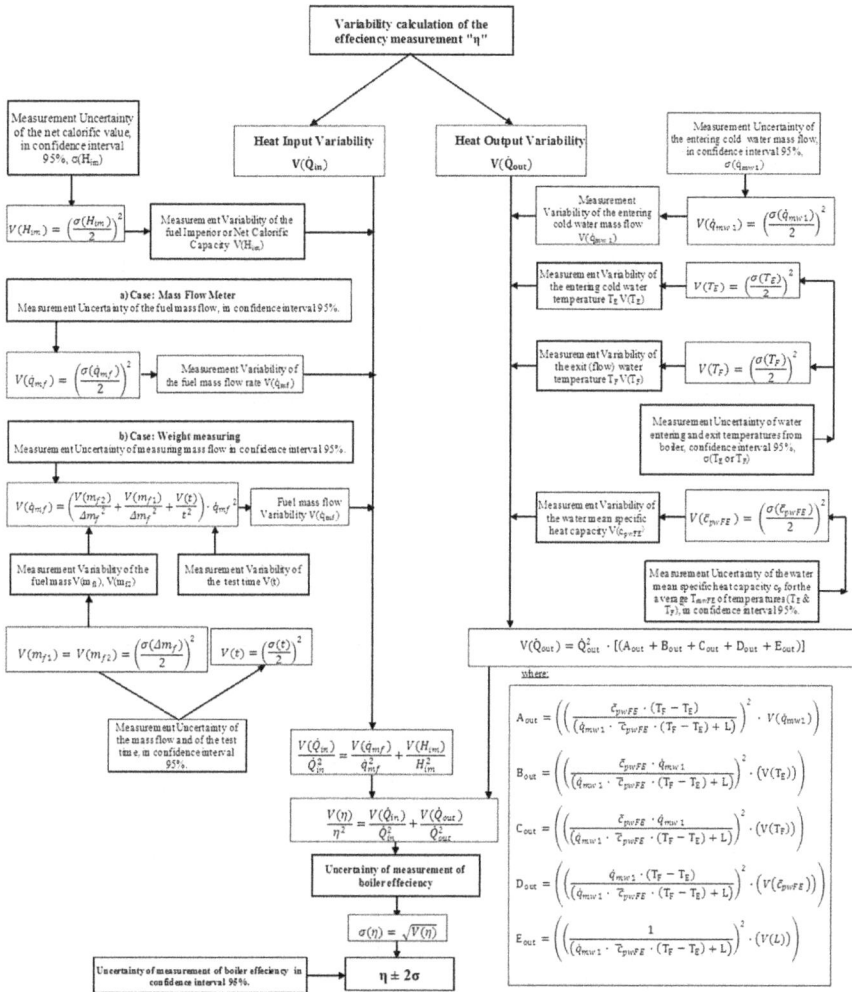

Variability calculation of the effeciency measurement "η"

FIGURE 8.5
Flow diagram of the calculation process for variability and uncertainty of boiler efficiency measurement, at a 95% confidence interval and 2σ.

8.5 Parametric Analysis and Graphical Depiction of the Effect of the Measured Values of the Q_{in}, Q_{out}, and "η" Calculations

Application Test Case in a Fuel Liquid Boiler with Useful Nominal Heat Output 40 kW

The measurands are those that affect the determination of the measurement uncertainty of the following calculated values:

1. The heat input, $\dot{Q}_{in}$.
2. The heat output, $\dot{Q}_{out}$.
3. The efficiency of hot water boiler, "η".

An example of the uncertainty calculation of measuring the efficiency of a hot water boiler fired with liquid fuel (heating oil) is then considered. Table 8.1 gives the measured and calculated values of the measurands required in order to calculate the efficiency and the related uncertainty of its measurement. In this context, a parametric analysis for the effect of the measurands on the measurement uncertainty is presented in Figures 8.6–8.14 A&B, and the quantified effect of measurands on the $\dot{Q}_{in}$ and $\dot{Q}_{out}$ calculations is given in Figure 8.14a and b, respectively.

TABLE 8.1

The Measured and Calculated Values of the Measurands That Calculate the Efficiency and the Uncertainty of Its Measurement

Measurand Description	Symbol	Unit	Value
Ambient conditions			
Ambient temperature	T_L	°C	23.1
Barometric atmospheric pressure	P_a	mbar	997
Liquid fuel (light oil)			
Difference of liquid fuel mass	Δmf	kg	1
Initial fuel mass	m_{f1}	kg	14
Final fuel mass	m_{f2}	kg	13
Time of testing	t	s	1010.54
Fuel mass flow	$\dot{q}_{mf}$	kg/h	3.562
Supply water			
Entering cold water temperature	T_E	°C	23.4
Return water temperature	T_R	°C	60.3
Exit (flow) water temperature	T_F	°C	80.4
Entering cold water mass flow	$\dot{q}_{mw1}$	kg/h	601.518
Heat output			
Heat losses from the test rig	L	kW	0.3942
Useful heat output	$\dot{Q}_{out}$	kW	40.227
Heat input			
Gross calorific capacity	H_s	MJ/kg	45.74
Lower calorific capacity or net calorific value	H_{im}	MJ/kg	43.05
Heat input	$\dot{Q}_{in}$	kW	42.596
Efficiency	η	%	94.44
Absolute expanded uncertainty (2σ) of the efficiency $[U(n)]$, due to the expressed value of the efficiency in %, in confidence interval 95% and $k = 2$		%	0.6

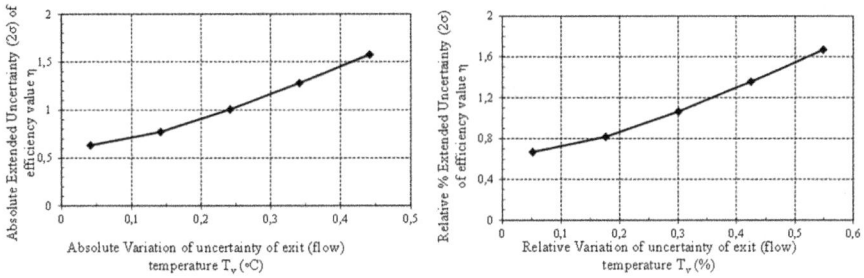

FIGURE 8.6
Absolute and relative effect of exit (flow) water temperature $(T_{v=TF})$ uncertainty on efficiency "η".

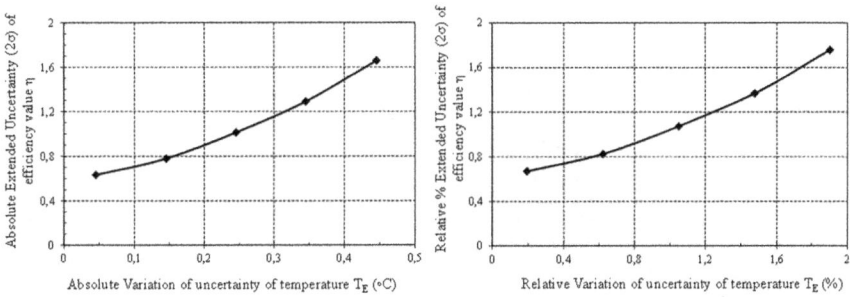

FIGURE 8.7
Absolute and relative effect of water supply temperature (T_E) uncertainty on efficiency "η".

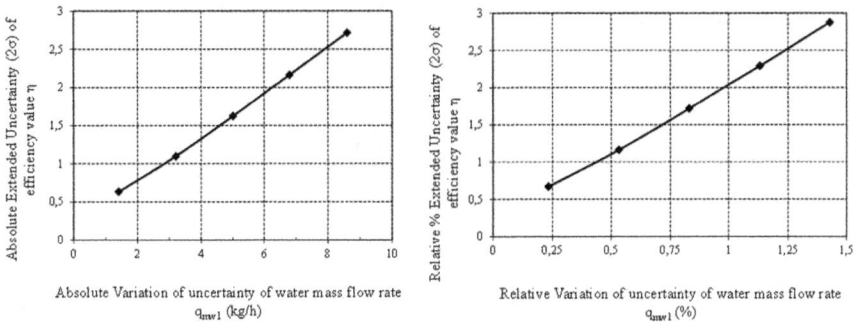

FIGURE 8.8
Absolute and relative effect of water mass flow rate $(\dot{q}_{mw1})$ uncertainty on efficiency "η".

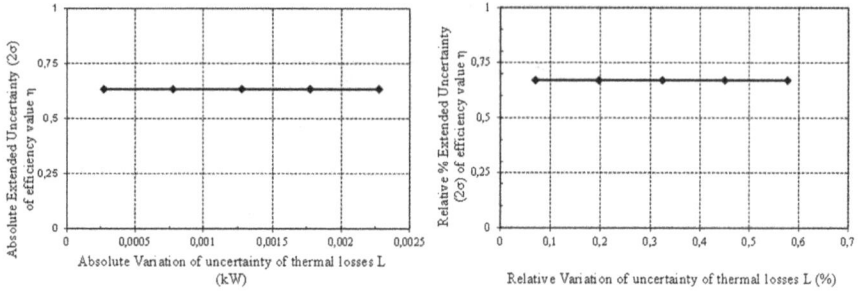

FIGURE 8.9
Absolute and relative effect of thermal losses L of test rig circuit on efficiency "η".

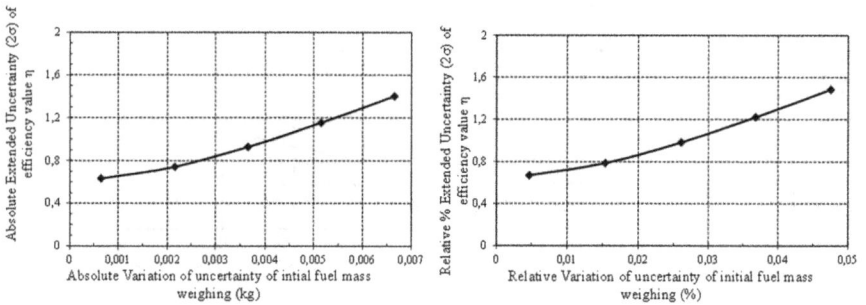

FIGURE 8.10
Absolute and relative effect of initial fuel mass weighing ($mf1$) uncertainty on efficiency "η".

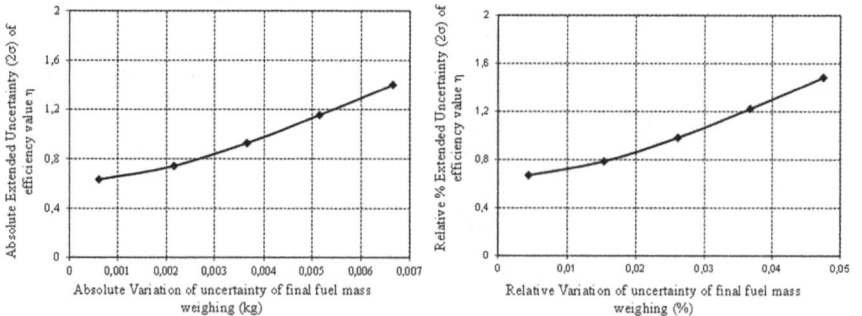

FIGURE 8.11
Absolute and relative effect of final fuel mass weighing ($mf2$) uncertainty on efficiency "η".

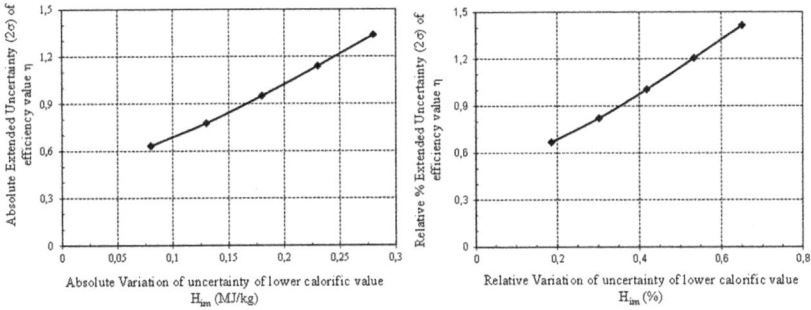

FIGURE 8.12
Absolute and relative effect of lower calorific value (*Him*) uncertainty on efficiency "η".

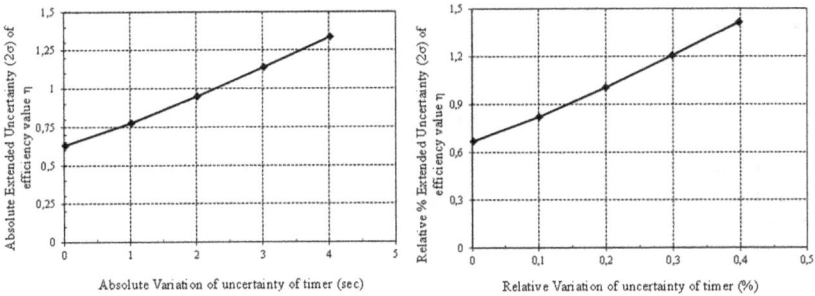

FIGURE 8.13
Absolute and relative effect of cycle test time uncertainty on efficiency "η".

Effect of the measured values on Q_{in} calculation

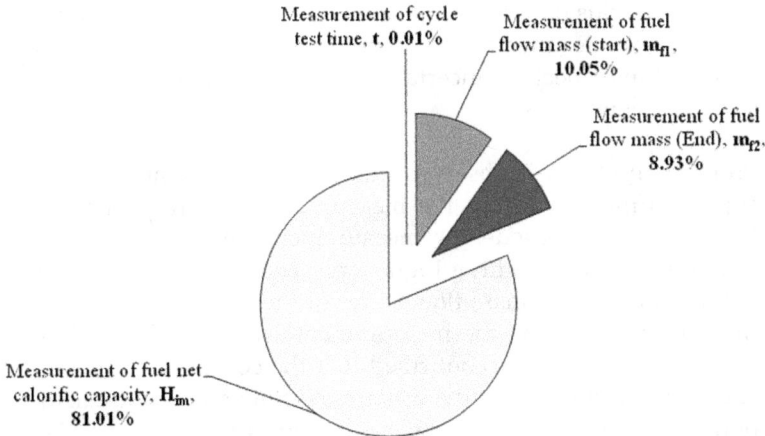

FIGURE 8.14A
Quantified effect of measurands on the $\dot{Q}_{in}$ calculations and hence the measurement uncertainty of efficiency "η".

Effect of the measured values on Q_out calculation

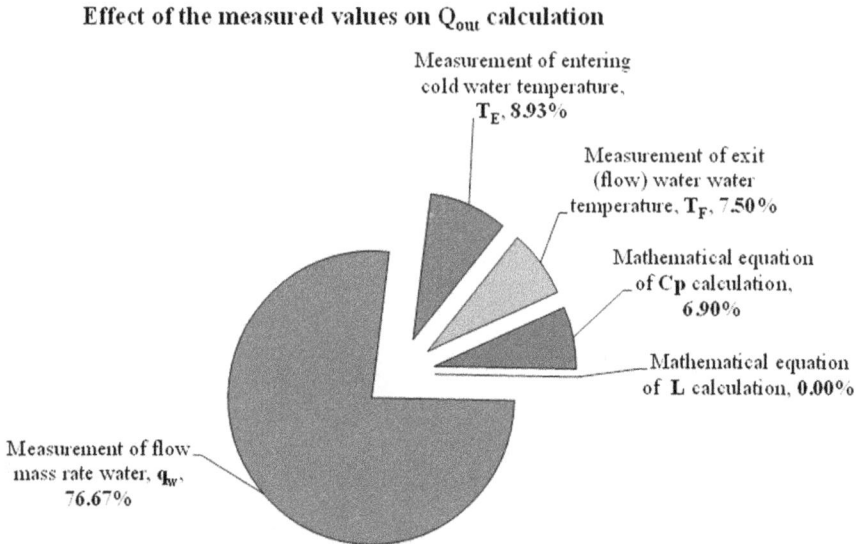

FIGURE 8.14B
Quantified effect of measurands on the $\dot{Q}_{out}$ calculations and hence the measurement uncertainty of efficiency "η".

8.6 General Conclusions from Parametric Analysis of the Absolute and Relative % Uncertainty of the Boiler Efficiency

The general conclusions resulting from the parametric analysis for the relative and absolute uncertainty of the hot water boiler value efficiency, in relation to the respective uncertainty of each measuring instrument, are summarized below:

- From Figure 8.8, it is observed that as the uncertainty of the water supply is increased—i.e., the measurement accuracy of the water flow rate is decreased—the measurement uncertainty of the efficiency is increased with a high slope. That is, it is concluded that the accuracy of the mass flow meter of the entering cold water has the greatest effect on the measurement accuracy of the efficiency. From Figure 8.14b, it is confirmed that the contribution of the measurement uncertainty of the water mass flow rate in the calculation of the useful heat output is the most important.

- From Figures 8.6 and 8.7, it is observed that a very slight change in the uncertainty of measuring the temperature of the exit or entering cold water results in a marked increase in the uncertainty of measurement efficiency. It can be concluded that the accuracy of the water temperature sensors has a significant effect on the uncertainty of calculating the boiler efficiency.

- From Figure 8.12, it is observed that the variation in the uncertainty of the lower calorific capacity must be considerably large to cause a comparable variation in the efficiency uncertainty in relation to the changes caused by other measurands such as water mass flow meter, fuel mass flow by measuring weight, timer, and thermoresistances. Nevertheless, the influence of the measuring accuracy of the fuel lower calorific capacity on the uncertainty of measuring efficiency is also important. From Figure 8.14a, it is confirmed that the contribution of the fuel lower calorific capacity uncertainty in calculating the uncertainty of the heat input is the most important.

- The effect of the accuracy of calculating the water mean specific heat capacity and thermal loss L (Figure 8.9) on the uncertainty of calculating the boiler efficiency is constant, since their uncertainty originates not from the measurement procedure but from a mathematical relationship.

References

1. Commission Regulation (EU) No 813/2013 of 2 August 2013 implementing Directive 2009/125/EC of the European Parliament and of the Council with regard to ecodesign requirements for space heaters and combined heaters.
2. Directive 92/42/EEC (21/05/1992): Performance requirements for new hot water boilers fired with liquid or gaseous fuels.
3. Directive 2009/125/EC of the European Parliament and of the Council of 21 October 2009 establishing a framework for the setting of ecodesign requirements for energy related products (recast).
4. Directive 2010/30/EU of the European Parliament and of the Council of 19 May 2010 on the indication by labeling and standard product information of the consumption of energy and other resources by energy-related products (recast).
5. Commission Regulation No 811/2013 of 18 February 2013 supplementing Directive 2010/30/EU of the European Parliament and of the Council with regard to energy labeling of space heaters, combined heaters, space heater assemblies, temperature regulators and solar apparatus, as well as the Combined Heater, Thermostat, and Solar Unit assemblies.
6. Directive 2002/91/EC of the European Parliament and of the Council of 16th December 2002 on "Energy efficiency in buildings". EEEK L.1, 4.1.2003, (2003).
7. Directive 2010/31/EC of the European Parliament and of the Council of 19 May 2010 on the energy performance of buildings (recast). EEEK L.153, 18.6.2010, (2010).

8. EN 304:2017: "Heating boilers - Test regulations for heating boilers with boilers atomizing oil."

9. EN 303.01:2017: "Heating boilers - Part 1: Heating boilers with forced draft burners - Terminology, general requirements, testing and marking".

10. EN 303.02:2017: "Heating boilers - Part 2: Heating boilers with forced draft burners - Special requirements for boilers with boilers atomizing oil".

11. EN 303.03:1998 (together with its attachments): "Heating boilers - Part 3: Gas central heating boilers - A complex comprising a boiler and a burner with forced air supply".

12. EN 303-4: 1999: "Heating boilers - Part 4: Heating boilers with forced draft burners - Special requirements for forced draft burners with a payload of up to 70 kW and a maximum operating pressure of 3 bar - Terminology, special requirements, testing and marking".

13. EN 303.05: "Heating boilers - Part 5: Heating boilers for solid fuels, manually or automatically powered, with a maximum nominal thermal output of up to 300 kW - Terminology, requirements, testing and marking".

14. EN 303.06:2000: "Heating boilers - Part 6: Heating boilers with forced draft burners - Particular requirements for dual-function domestic hot water hot water boilers with mechanical fuel oil burners, nominal thermal input up to 70 kW".

15. EN 303-7:2011: "Heating boilers - Part 7: Central gas boilers with forced draft burners with nominal thermal output up to 1000 kW".

16. EN 15502-2-1 + A1:2017: "Gas-fired central heating boilers - Part 2-1: Specific standard for Type C devices and Type B2, B3 and B5 nominal thermal inputs up to 1000 kW".

17. EN 15502-1 + A1:13-08-2015: "Central heating gas boilers - Part 1: General requirements and testing".

18. EN 15502-2-2:2014: "Gas-fired central heating boilers - Part 2-2: Specific standard for B1 type devices".

19. EN 656:1999 (together with its attachments): "Central heating gas boilers - Type B boilers with a nominal thermal input from 70 kW to 300 kW".

20. EN 267 + A1: 12-03-2012: "Monoblock oil spraying burners".

21. Papageorgiou, N. 1991. *Steam Generators I, General Principles*. Athens: Simeon Publishing.

22. Kakaras, E. and Karellas, S. 2015. *Decentralized Thermal Systems*. Athens: Tsiotras Publishing.

23. Gumz, W. 2013. *Handbuch der Brennstoff und Feuerungstechnik*. Berlin: Verlag Springer.

9

Psychrometric Performance Testing for HVAC&R Components and Equipment

Orkan Kurtulus

Purdue University

Christian K. Bach, Romit Maulik, and Omer San

Oklahoma State University

Davide Ziviani

Purdue University

Craig R. Bradshaw

Oklahoma State University

Eckhard A. Groll

Purdue University

CONTENTS

9.1 Introduction

The *HVAC&R* industry and researchers are conducting psychrometric performance tests to determine component or system performance. Performance
tests involve basic measurements such as temperature, humidity, pressure,
flow rate, and electrical power. Tests are usually conducted in an environment
where air properties are precisely controlled. Psychrometric performance
testing is generally conducted using either psychrometric chambers or psychrometric conditioning setups. In both cases, temperature and humidity
are precisely controlled by dedicated sensors. Researchers are using fixed
temperature operation to execute testing plans for performance testing and
model validation at component or system level. Alternatively, the dynamic
behavior of the equipment can be evaluated, e.g., using hardware-in-the-loop
approach such as load-based testing.

9.2 Psychrometric Conditioning

Psychrometric rooms can provide various temperature and humidity
conditions to simulate different climate zones and indoor environments to
study the performance and behavior of the tested component or equipment.
Testing *HVAC&R* components such as heat exchanger coils, air handling
units, outdoor units, and heat pump water heaters requires temperature and
humidity to be stable and within tight limits.

The psychrometric conditioning unit (PCU) of the psychrometric room
provides a counterload to cancel the heating and cooling effect generated by
the tested equipment, the air leakage, and other loads. In addition, a humidity control unit can be utilized to manage latent loads.

A PCU generally includes circulation fan(s), electric heating element(s),
cooling coil(s), refrigeration rack(s) or chilled water source, humidification

system, and (optionally) a dedicated dehumidification system. Figure 9.1 shows a simplified representation for a PCU with cooling coil, fan, reheat coil, and humidifier. PCUs are instrumented to control temperature and humidity accurately. Relative humidity and temperature probes are the main sensor types utilized for controls instrumentation. Operating tolerance of the room dry-bulb and wet-bulb temperature is generally specified by the testing standard with a typical value of 0.5 K (1 F). Stricter tolerances may apply for the average value of these properties during the test (e.g., test condition tolerance). The operator should refer to currently applicable values to the applicable testing standards, e.g., for the USA often (ASHRAE 37(RA-2019), 2009).

Psychrometric chart: Operating psychrometric rooms requires expertise in psychrometry and psychrometric chart due to their complex processes. The psychrometric chart allows to draw and illustrate psychrometric processes. The state of the moist air can be calculated based on two properties at a specified pressure using thermodynamic relations.

Presentation of moist air states in psychrometric charts allows for a quicker understanding of the processes. These charts include various property lines, including dry-bulb temperature (T_{db}) (°C), wet-bulb temperature (T_{wb}) (°C), dew-point temperature (T_{dp}) (°C), relative humidity (φ) (°C), enthalpy (h) (kJ/kg), specific volume (v) (m3/kg), and humidity ratio (w) ($kg_{water}/kg_{dry\ air}$). An advantage of the psychrometric chart is that only two independent intensive properties need to be known to find all other properties in the chart. A simple representation of the psychrometric chart is shown in Figure 9.2.

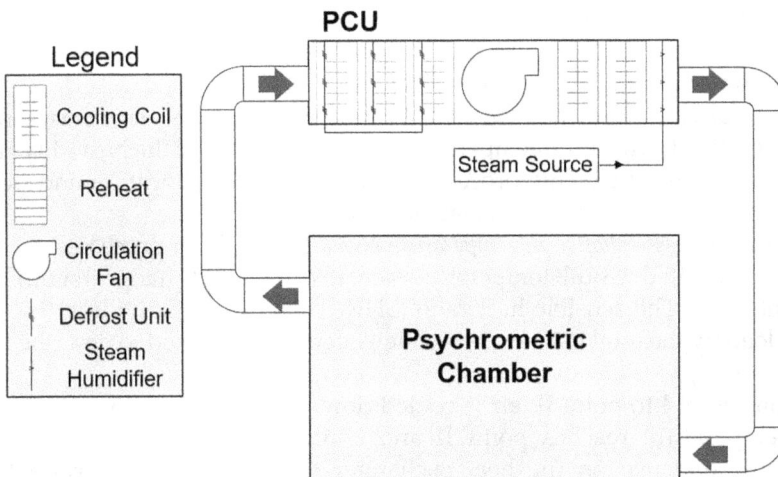

FIGURE 9.1
Schematic representation of a psychrometric chamber air distribution system.

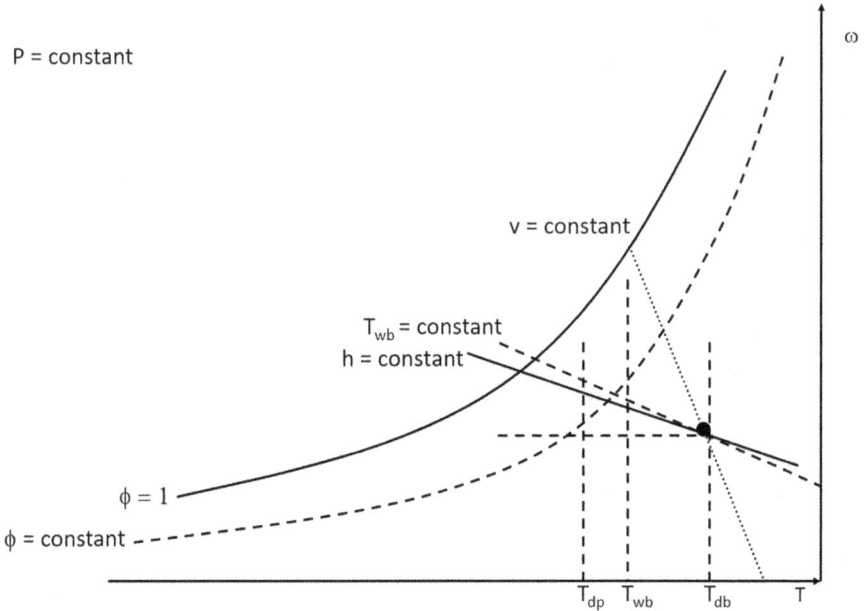

FIGURE 9.2
Representation of the psychrometric chart.

Sensible process—change in dry-bulb temperature: By moving horizontally across the chart, only the dry-bulb temperature changes. This represents heating (moving to the right) or cooling (moving to the left). Since no humidity is added or removed, it is called a (pure) sensible process of cooling or heating, caused by a sensible load. Figure 9.3 represents (pure) sensible load between points I and II and III and IV.

Latent process—change in humidity: By moving the state point between IV and I vertically, dry-bulb temperature stays constant, but the humidity ratio changes (Figure 9.3). This process demonstrates humidification and dehumidification caused by a (pure) latent load.

Mixed process—change in temperature and humidity: Mix processes involve both change in dry-bulb temperature (sensible load) and change in humidity (latent load). The sensible heat ratio (SHR), calculated by dividing the sensible load by the total load (e.g., sensible+latent load), is used to describe load behavior.

From point I to point II, air is cooled down using the cooling coil. When air temperature reaches point II and continues to drop, moisture will start to condense on the heat exchanger fin surface once surface temperature is below the dew point of the entering air. A cooling coil can be used as the sole method to remove humidity in PCUs without a dedicated

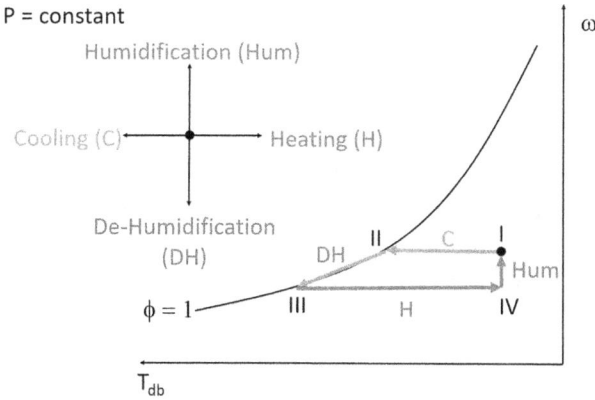

FIGURE 9.3
Representation of heating cooling humidification and dehumidification processes.

dehumidification unit. In this case, run-time will be limited by frost formation on the fins for low-ambient-temperature tests. Frost formation will occur quickly in high-humidity conditions and it may initially increase the heat transfer. However, airflow rate and heat transfer start to reduce, which impacts the heat exchanger's performance and the achievable conditions in the room. Continuous operation can be achieved by using two fully redundant coils: allowing one to be in operation while the other one is on standby or in defrost mode.

Adding a dedicated adsorption- or absorption-type dehumidifier may improve the humidity control and reduce defrost cycles of the coils. These systems are advantageous to keep the humidity within the control tolerances, and may not need as many – if any – defrost cycles.

The most common way to add sensible heating is using an electrical heater. Other alternatives are steam heater, gas furnace, and a heat pump. However, solid-state relay (SCR)-controlled electric heaters provide the most precise temperature control. Refer to the section on safety mechanisms (Section 7.2) for recommendations for electric heater operation.

Steam humidification is the most common process used in PCUs. It utilizes a steam source, steam flow control valve, and injection distributor. A jacketed injection distributor can be used for extending the operation range below the freezing temperature. The temperature of atmospheric steam is 100°C, causing a minimal change in the air temperature on completion of the mixing process. Adding high-temperature steam increases the dry air temperature, leading to cross-dependencies in temperature and humidity controls of the PCU. Alternatively, air–water spray injection nozzles can simultaneously lead to a reduction in air temperature. However, spray-nozzle humidification does not allow for operation below the freezing point of water.

9.3 Airflow Distribution

Airflow distribution within psychrometric facilities has a substantial effect on measurement results. Velocity, temperature, and humidity maldistribution can cause discrepancies in performance of split system indoor units and packaged heat exchanger coils. Additionally, recirculation between air outlet and inlet of the tested unit(s) can result in non-uniformity of the inlet conditions. Some testing facilities try to mitigate the issue by rerouting air from the flow measurement apparatus directly to the facilities' PCU. However, recirculation may be typical in field applications for outdoor units. This section outlines different aspects of airflow distribution.

9.3.1 Common Practices and Limitations

Depending on the facility and space requirements, the PCU can be installed inside or outside of the chambers. One common practice for air distribution is to draw air from a location at the top of the psychrometric room, condition it in the PCU, and then return it to the room at the floor level. One common concept is using a raised floor to handle air distribution throughout the floor level.

If overall floor space is limited, it is preferred to install the PCU inside the chamber. The advantage of this configuration is a reduction in heat transfer to the ambient. However, the facility is losing useful floor area in the chamber.

A psychrometric chamber may utilize air ducts where air pulled closer from the ceiling and returned to the floor level. Air is then distributed uniformly across the floor as shown in Figure 9.1.

One of the air delivery options is using ducts as distributing the air under the floor (**case 1**). Conditioned air enters the room from the main duct system; then, it branches out to small diffusers to deliver airflow evenly. Alternatively, a raised perforated flooring can be used, where the space below the raised floor (**case 2**) functions as a large plenum. Both air delivery systems need additional commissioning work to achieve uniform air velocities across the floor. Finding air velocity maldistribution across the floor is common upon initial commissioning of rooms, and it needs to be addressed to obtain repeatable testing results.

Depending on the requirements, different floor systems can be designed. If a heavy equipment needs to be tested, raised floors may not be an option. Large equipment psychrometric rooms often employ strong concrete or steel floors. Alternatively, a top/bottom design with the indoor room located below the outdoor room can be utilized. The top/bottom layout is particularly advantageous for testing rooftop AC units.

The main objective of reducing air velocity non-uniformity is to achieve a more uniform testing environment that provides:

- Precise and stable control for inlet air temperature and humidity for the tested equipment,
- Reduced temperature control-related issues due to differences between tested equipment instrumentation and reference temperature sensor of the controlled chamber, and
- Repeatable test conditions.

9.3.2 Experimental Evaluation of Airflow

Airflow across the floor can be measured with air capture hoods, typically sold for air balancing of diffusers in office environments. A capture hood measures the volumetric airflow of a rectangular or round area. It is recommended to divide the floor to generate a grid that allows for fast post-processing and visualization of the uniformity of the airflow across the floor. Once the airflow distribution is mapped, it is easier to identify areas with high or low flow and to take corrective action.

In Figure 9.4, airflow measurement of the previously introduced cases 1 and 2 is shown. Airflow is not evenly distributed across the floor in both cases. The next two paragraphs discuss the cases and suggest appropriate corrective actions.

Case 1 has a high flow rate in the southeast corner because of unbalanced branches as shown in Figure 9.4. In case 1, the far corner of the (southeast) diffuser receives more flow rate than the diffusers closer to the supply header. Due to the high airflow velocity in the southeast corner, turbulence

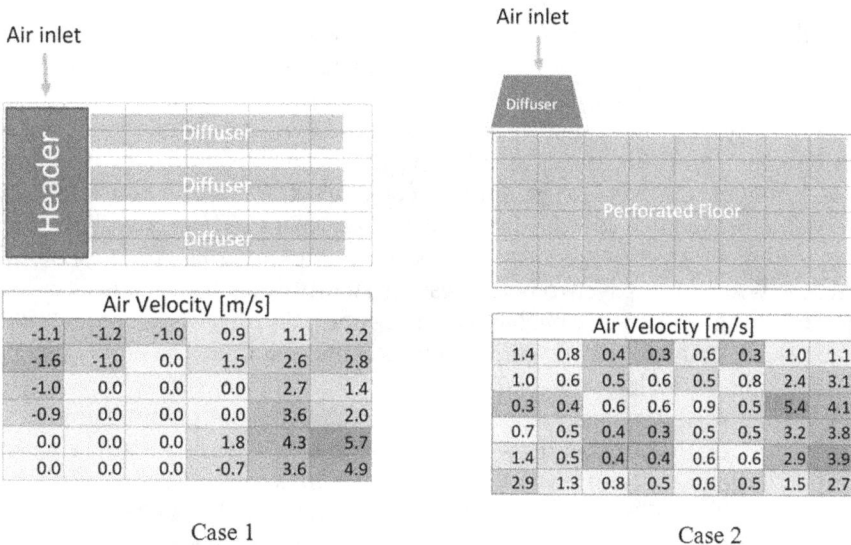

Case 1

Air Velocity [m/s]					
-1.1	-1.2	-1.0	0.9	1.1	2.2
-1.6	-1.0	0.0	1.5	2.6	2.8
-1.0	0.0	0.0	0.0	2.7	1.4
-0.9	0.0	0.0	0.0	3.6	2.0
0.0	0.0	0.0	1.8	4.3	5.7
0.0	0.0	0.0	-0.7	3.6	4.9

Case 2

Air Velocity [m/s]							
1.4	0.8	0.4	0.3	0.6	0.3	1.0	1.1
1.0	0.6	0.5	0.6	0.5	0.8	2.4	3.1
0.3	0.4	0.6	0.6	0.9	0.5	5.4	4.1
0.7	0.5	0.4	0.3	0.5	0.5	3.2	3.8
1.4	0.5	0.4	0.4	0.6	0.6	2.9	3.9
2.9	1.3	0.8	0.5	0.6	0.5	1.5	2.7

FIGURE 9.4
Cases of airflow measurement results.

is introduced to room that causes negative airflow velocity at the northwest corner. It is clear that additional changes are required. For instance, adjustable dampers can be installed on each diffuser to distribute airflow evenly in case 1. Another alternative can be installing fixed orifices. Numerical investigation may help reduce the required time to determine the orifice sizes or damper opening.

In *case 2*, airflow in this example is always in positive direction but not evenly distributed as shown in Figure 9.4. This case uses a floor plenum for air supply. One option to improve airflow velocity uniformity can be improved by adding series of air vanes to direct airflow or installing perforated plates of various opening areas.

Both cases can be addressed iteratively. Although uniform airflow distribution can be achieved, it can be quite costly. CFD analysis may support a reduction of the required number of adjustments.

9.3.3 Application of CFD and Effects onto Equipment Tests

Experimental evaluation of airflow is generally time-consuming and can only be done after installation is complete. Air recirculation for outdoor units and flow uniformity requirements for heat exchanger coil testing are examples of where airflow plays a critical role for the experiment. For design and layout purposes, computational fluid dynamics (CFD) appears to offer a fast and easy way to predict airflow. However, a CFD model is only a limited and incomplete representation of the modeled physical environment. Several areas, as outlined below, need to be taken into consideration.

Modeling simplifications: In general, a geometric model used in CFD simulations has much less detail than the actual setup. This is done to decrease the required meshing time and to allow a larger minimum mesh element size. Similarly, environmental conditions are simplified to allow them to be determined and input with a reasonable amount of effort. Instead of complex flow boundary conditions that may have flow profiles and non-uniform temperature, simplified constant temperature, constant flow, constant pressure, and wall assumptions are made. These simplifications can lead to a loss in accuracy and, in extreme cases, miss key physics. If bulk airflow within the psychrometric testing facility rooms—as caused by the conditioning system—is neglected, then simulation results for, e.g., tested equipment inlet ductwork can be much different than the flow velocity profiles measured later on.

Meshing simplifications: In order to limit required computation time, the goal is generally to maximize mesh element size to reduce mesh element count. If done without care, this can result in large errors, convergence issues, and loss of mass conservation. To resolve flow behavior near walls, mesh element size should be fine enough to allow $y+$ values near 1 and much lower than 5. Additionally, the growth of mesh elements should be limited to no more than 20%–30%, e.g., growth factor of 1.2–1.3. Furthermore, it is

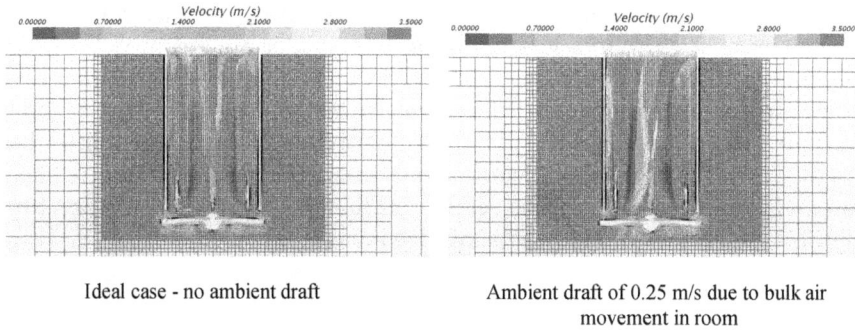

Ideal case - no ambient draft   Ambient draft of 0.25 m/s due to bulk air
                                       movement in room

FIGURE 9.5
CFD simulation predicts substantial sensitivity of flow field nonuniformity in an inlet duct section used for unitary equipment testing.

generally recommended to refine the mesh in areas that are expected to have large velocity gradients.

Turbulence model: The k ω-SST model that is well suited for vorticity-dominated flows (Menter, 1994), such as caused by typical redirection of air-flows within psychrometric coil testing facilities. It uses a blending function to combine wall treatment and bulk airflow in an appropriate fashion.

Mesh/grid independence: Any CFD result should always be checked for solution consistency using at least three different mesh sizes to check for issues caused by insufficient mesh detail.

CFD application example—inlet duct for unitary equipment testing: Figure 9.5 shows example simulations for a piece of air inlet ductwork used for unitary equipment testing. Air enters the duct by passing an air sampling tree and an inlet damper. The exit of the duct will be connected to an air handling unit. The specific objective here was to investigate whether bulk airflow in the surrounding domain has an effect on the air velocity distribution at the outlet of the duct. The results clearly indicate that this is the case, suggesting that bulk airflow in the surroundings of the actual experimental setups should be limited.

9.4 Psychrometric Measurements

One of the unique features of psychrometric testing is having moist air in the process. Typically, AC and refrigeration equipment lead to a dehu-midification (e.g., latent load) process for coolers and evaporators, or to a humidification process for systems that employ water evaporation for cool-ing (e.g., swamp/adiabatic coolers, evaporative pre-cooling of air before con-densers, cooling towers). This removal or addition of moisture changes the air properties and needs to be considered for accurate capacity calculations of the test articles. While it is tempting to consider air incompressible for psychrometric testing, it is generally required to compensate for differences in absolute pressure between different measurement locations. This includes accurate determination of the pressures not only at the equipment inlet, but also at the flow measurement equipment's inlet and the psychrometric measurement device location. This section outlines best practices for psychrometric measurements that help minimize measurement errors.

9.4.1 Air Sampling

Air sampling devices are generally used when an average airflow humidity measurement is required. These devices generally employ a sampling device with a number of holes divided into branches of smaller diameter and a trunk to collect sampled air. A fan is then used to pull air through the sampling holes, past a humidity measurement device, and then exhaust it either into a psychrometric room or into the duct downstream of the measurement position. Some guidelines, e.g., US 10 CFR, Part 430, Subpart B, Appendix M, Section 2.14.1, exist for the design of air samplers (US Department of Energy, 2017).

Air sampling causes pressure drop, which leads to measurement error for the measured wet-bulb temperature in the sampling device. Wet-bulb humidity sensors (psychrometers) are most sensitive to this effect. Figure 9.6

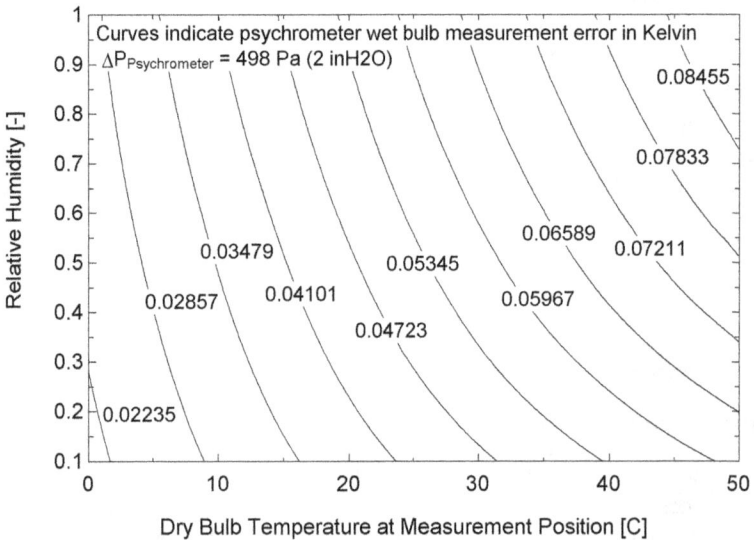

FIGURE 9.6
Wet-bulb temperature measurement error for psychrometer at fixed sampling device pressure drop.

shows that an error of up to 0.1 K occurs if the pressure is not measured directly at the psychrometer with a sampling apparatus that has approximately 500 Pa pressure drop. This deviation is of the same order of magnitude as required sensor accuracy in testing standards. Static pressure for high-accuracy psychrometric measurements should therefore be measured at the humidity measurement device, and not at the sampling tree.

In some cases, air sampling and measurement are spatially separated; this includes remotely located dew-point meters and psychrometers connected to the sampled location via an air duct. In the case of psychrometers, the heat gains or losses lead to a measurement error that

- Decreases with increasing airflow rate and air duct insulation, and
- Increases with increasing duct length and temperature difference between duct and surroundings.

For remotely mounted dew-point meters, heat gains to the connection tube do not affect the measurement. However, heat losses from the generally non-insulated sampling tube can cause even more severe measurement problems due to the risk of condensation within the sampling tube. In addition, hygroscopic sampling tube materials need to be avoided.

9.4.2 Temperature Sensors

In practice, thermocouples and more accurate reference temperature sensors are utilized for temperature measurements. The following sections will describe both temperature measurement devices and give practical tips on how to use and install them.

9.4.2.1 Thermocouples (TCs)

TCs consist of two dissimilar metals joined together to form a junction. The junction generates a millivolt signal corresponding to the temperature. The millivolt signal can be interpret as temperature using a data acquisition system (DAQ). DAQ systems often have integrated cold junctions, which directly determine the cold junction temperature value. Advantages of using TCs are as follows:

- Low cost,
- Simple to use,
- Widely available, and
- Capable of accurately reading temperature.

The T-type TC is the most commonly used TC type for temperature sensing in HVAC&R applications. One of the dissimilar metals is copper (Cu), and the other one is constantan (Cu-Ni). Typical T-type TCs' measuring range is 250°C–350°C with 0.5°C accuracy.

Although TCs are widely available and easy to work with, the following considerations need to be taken into account before they are implemented to a measurement system.

Wiring: It is crucial to connect TC wires directly to the DAQ using thermocouple extension-grade wire. For extended connections greater than 3 m or extensions around high electrical noise levels, shielded cable is recommended to reduce electromagnetic interference.

Creating Junctions: In most cases, users are buying their TC wires in bulk. If pre-manufactured TC temperature probes are not used, the user needs to create his/her junction using thermocouple-grade wire. The following options are the most practical ones to make thermocouple junctions:

- *Electronic s type rosin core solder*: The user creates junction using low-temperature electronics type rosin core solder requiring some heat-up and cooldown time during the process.

- *TC welder*: This device is specifically designed for creating junctions for TCs and directly welds the two wires of the thermocouple without the need for additional materials. TC welds are fast due to the small thermal mass of the weld point and the good conductivity of the wires that quickly cool down the weld.

Regardless of the option, TCs should be calibrated before they are installed on the test stand for accurate and repeatable measurements. In situ calibration, e.g., calibrating the entire measurement chain including TC extension wire and DAQ, further increases the achievable accuracy.

9.4.2.2 Resistance Temperature Devices

Resistance temperature devices (RTDs) contain a resistor with specified nominal resistance at 0°C. Typical nominal resistances are 100, 200, 500, and 1,000 Ω. The resistance value changes based on the temperature. RTDs are offered for wide temperature measurement ranges starting from − 200°C to +over +750 C°C.

RTDs are enclosed as a probe in a sheath, which is usually manufactured from 316 stainless steel to protect the fragile sensor. RTD accuracy, especially for three- and four-wire versions, is better than TCs' accuracy and listed between 0.005°C and 0.3°C at 0°C. Accuracy often is specified as a function of measured temperature for temperatures other than 0°C (Dally et al., 1984).

In contrast to TCs, RTD sensors require excitation power to measure temperature. This excitation voltage or current leads to a self-heating of the sensing element. Therefore, excitation power needs to be carefully selected to limit measurement error.

Wiring: RTDs do not require a specific wire type. However, wire length may lead to unacceptable measurement error if using two-wire configurations without in situ calibration.

Application: RTDs are well suited for controlling PCUs. They are also available as averaging temperature sensor versions for spatially averaged measurements. Figure 9.8 shows an example installation.

9.4.2.3 Temperature Grids and Air Mixers

Temperature grids are an important tool for ensuring accurate psychrometric measurements since they allow to detect non-uniform air temperatures. Air mixers reduce the temperature non-uniformities at the expense of pressure drop and additional duct length before the measurement location. Details for the installation of air mixers and temperature grids are provided in the standard(s) applicable for the target sales region. Europe follows ISO, DIN, and EU standards; the USA follows ASHRAE and ANSI standards; Japan follows JRA; and Russia follows GOST-R standards. Each measurement standard provides detailed information about the methodology for each measurement type. As an example, ASHRAE 41.1 (2013) provides detailed information on how to create temperature grids, details of air mixer design, and other methodology-related information. Note that ASHRAE standard 41.1 (2013) covers only dry-bulb temperature measurement. Wet-bulb temperature measurements are covered under ASHRAE standard 41.6 (2014).

The temperature grid should be located where the airflow is uniform for accurate measurements. If air temperature distribution is not uniform across the measurement plane, an air mixer should be installed to reduce temperature non-uniformity. A rule of thumb for the location of the measurement plane should be five hydraulic diameters (D_h) away from the mixer.

9.4.2.4 Recommendation for Selecting Temperature Sensors

Let's consider two different systems. System 1 is a testing equipment instrumented for standard performance testing. System 2 is a psychrometric chamber's PCU where the air is conditioned based on the testing requirement.

System 1: Standard performance tests require rapid instrumentation since facility time in the psychrometric rooms is expensive. It is preferred to use instrumentation over and over again. TCs can be the best option for this type of installation since they can be quickly installed, are robust, and easily repaired in-house with a thermocouple welder. Figure 9.7 shows a typical temperature grid installed on the outdoor unit.

FIGURE 9.7
Thermocouple grid installed on outdoor unit.

RTDs are not preferred in this application for various reasons, including the following:

- RTDs require three- or four-wire connections for accurate measurements leading to cables with much larger diameters than thermocouple wire.
- RTDs are manufactured with specific probe length, length of the thermally sensitive portion (e.g. RTD element) and diameter.
- TC wires can be reused for another test, or a new TC junction can be made in-house easily and quickly whereas RTD sensors are specific for the application.
- RTD probes will make temperature grid installation bulky and heavy.

System 2: Consider a PCU where heating coil face temperature is needed. Instead of creating a temperature grid, it would be a great option to install an averaging temperature sensor. Figure 9.8 shows representation of the averaging temperature sensor in front of electrical heater.

Benefits of RTDs for this application include the following:

- They have better accuracy and larger temperature measurement range than thermocouples. Life span of the RTDs is longer than that of TCs since the sensing element is well encapsulated.

FIGURE 9.8
RTD averaging sensor in front of electrical heater.

- RTDs' sheath will protect the sensor from the environment, with waterproof cable connections available to increase robustness against condensation.

9.4.3 Psychrometers

Psychrometers employ a comparison between a dry surface thermometer's temperature (e.g., "dry bulb") and a thermometer fitted with a wet sock (e.g., "wet bulb"). An early mention of psychrometers is Ernst Ferdinand August, who is credited with patenting the term psychrometer in 1818 (Teague and Gallicchio, 2017). August's early version did not include a requirement of forced convection; modern psychrometers as used for equipment and component testing are required to maintain a certain airflow speed for increased accuracy.

The airspeed requirement for psychrometers reduces measurement error caused by radiative heat exchange with the environment by the sock while also not leading to frictional heating or blowout of the humidity from portions of the sock. ASHRAE 41.6 (ASHRAE 41.6, 2014), for example, requires an airflow speed range for the wet-bulb sensor of 700–2,000 fpm (3.5–10 m/s). The US federal register unified test method for air conditioners and heat pumps (US Department of Energy, 2017) suggests a narrower air velocity range of 800–1,200 ft/min (4–6 m/s). Note that the required air velocity depends on a wide range of parameters.

ASHRAE RP-1460 provided a detailed update that resulted in an updated ASHRAE Standard 41.6 (ASHRAE 41.6, 2014). The results lead to design recommendations that extend far beyond velocity ranges and most significantly include the use of radiation shields but are limited to temperature sensors of 6.35 mm (0.25 in) diameter.

9.4.4 Relative Humidity Sensors

There are a variety of instruments using different sensing methods for relative humidity (RH). One of the common types are polymer film electronic hygrometers. These sensors are manufactured using a polymer material that is sensitive to the water content of the air. The polymer, if used as a capacitor's insulation, leads to a moisture-dependent capacitance, which is converted by an electronic circuit into a relative humidity output signal.

These sensors are of relatively low cost and have a fast response to humidity changes. However, they become less accurate for high humidity levels and cannot be used for measuring more than 95% of relative humidity. Additionally, exposure to high humidity and condensing conditions may result in loss of calibration. Independently of these more severe issues, recalibration is recommended every six months by using a more accurate reference sensor (e.g., dew-point meter) since the polymer film circuit output can drift over time.

Sensor accuracy and response time depend on the technology used. RH can be measured within the 2%–3% range, and response time can be in the range of 1–120 s for 63% changes in RH (ASHRAE standard 41.6, 2014).

9.4.5 Dew-Point Meters

Dew-point monitors, also called chilled mirror sensors, determine the dew point of air by detecting visible condensation of moisture on a temperature-controlled cooled surface and are considered a precision measurement device. An illustration of a chilled mirror measurement is shown in Figure 9.9.

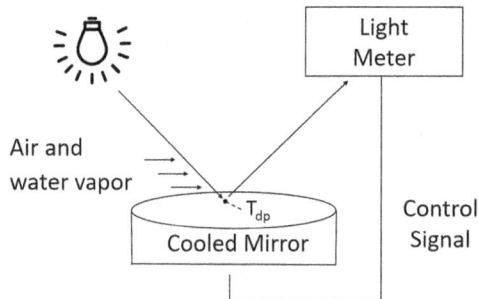

FIGURE 9.9
Illustration of chilled mirror measurement.

The chilled mirror surface is artificially cooled to a temperature point where visible condensation occurs. A light sensor determines the condensation forming and controls the surface temperature. Surface temperature is controlled to create a stable form of water droplet. In other words, water droplet will not change in size by time. The required mirror temperature to maintain droplet size is the dew point of the air passing through the instrument (ASHRAE standard 41.6, 2014).

Dew-point meters are considered a precise and repeatable measurement device. Their temperature accuracy is below 0.2°C. They are, however, sensitive to dust, and it is important to keep the mirror surface clean. Figure 9.10 shows the chilled mirror in a clean and a dirty condition.

If the mirror is flooded, as can happen during fast changes in air moisture, the dew-point meter switches to (or is manually switched, based on configuration) into a self-cleaning cycle. Once in the cleaning cycle, the controller rapidly increases the surface temperature attempting to evaporate all moisture from the surface. This rapid evaporation may also remove some solid impurities from the surface.

The following points are essential for every user when working with dew-point meters:

- The Chilled mirror needs to be cleaned regularly to remove any dirt from the surface as per the manufacturer's cleaning instructions.
- Air sampling is done by creating a measurement ring around the duct where it is sampled with at least four sample points. The connection is similar to the piezometer ring connection shown in ASHRAE standard 41.2 (2018).

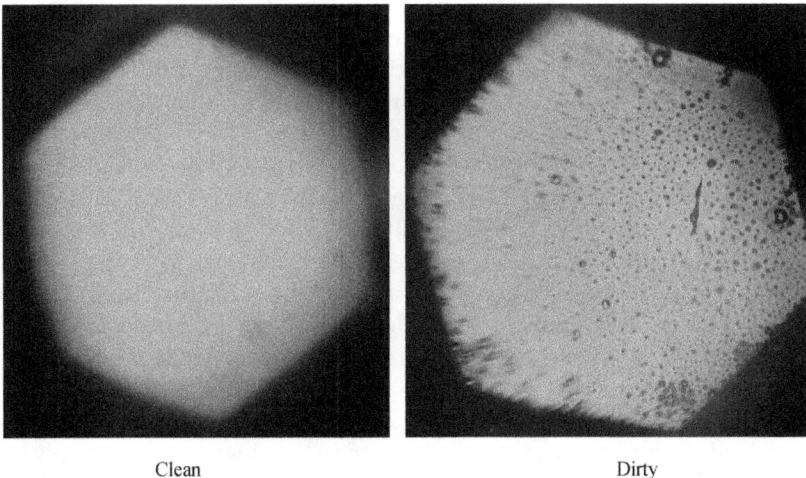

Clean   Dirty

FIGURE 9.10
Cooled mirror condition of dew-point monitor.

Usually, sampling tubes are made of plastic. The users need to be careful with the following:

- All sampling tube lengths need to be identical.
- During the tests, PVC tube may absorb water vapor, biasing dew-point measurements. The users needs to select tube materials that are compatible to work in the psychrometric chamber temperature range and do not absorb water. For permanent installations it may be preferable to change tube and fitting material to non-corroding metals such as stainless steel.
- If the test condition requires high-humidity conditions, the users may need to thermally isolate sampling tubes and/or install a parallel rope heater to avoid condensation inside the tubes.

9.4.6 Other Recommendations

Given the significant cost of psychrometric facilities, it is recommended to use redundant psychrometric measurements. Two dry-bulb temperature sensors combined with a wet-bulb and a relative humidity sensor allow detecting instrumentation issues by comparing dry-bulb temperature sensors as well as humidity measurements.

9.5 Airflow Measurements

Airflow measurements are generally required to test the airside capacity of components during psychrometric performance testing. While critical flow nozzles are used as a reference measurement (e.g., Collins et al., 2016), they generally do not find application in practical applications due to excessive cost for the required re-compression of the airflow. The de-facto standard is the use of standardized airflow measurement nozzles with well-defined geometries. Anemometers and Pitot tubes are used in some applications but have lower accuracy than nozzle-based measurements for typical HVAC applications. Their usage is generally limited to field applications and confirmation of flow uniformity at the inlet to test articles. This section introduces the most common measurement devices and gives recommendations for their use.

Note: Typical measurements include actual airflow rate (e.g., CFM) as well as standard airflow rate (e.g., SCFM). Standard airflow rate is an equivalent airflow rate for a predefined standard density. It is obtained by first calculating the mass flow using density and measured airflow and then dividing by the standard density.

9.5.1 Code Tester/Nozzle Box

Airflow rate measurements can be obtained using one or multiple flow nozzles. If multiple nozzles are used, then the resulting measurement device is typically called a code tester or a nozzle box. Accurate results can be obtained with low-turbulence inlet, and vortices at the nozzle outlet need to be avoided. Additionally, influence from adjacent nozzles and code tester walls needs to be limited. As a result, design standards such as ASHRAE standard 41.2, 2018 provide detailed guidelines for minimum distances between the inlet to the code tester, settling means, nozzle plane, outlet settling means, and distance to exit duct. Minimum distances are also specified between nozzles as well as between nozzles and code tester walls.

Code testers do not directly measure the flow rate. Instead, flow rate is calculated with a set of equations using pressure, temperature, humidity, and differential pressure in combination with the nominal nozzle dimensions as an input. This set of equations was found to be accurate if the code tester is designed according to the standard, capable of achieving a accuracy of 0.4% of reading of the measured flow rate (Collins et al., 2016).

Code testers typically include a booster fan. This allows compensating for the pressure drop across settling means, nozzles, and connection ductwork and enables control of the flow rate and/or static pressure at the exit of the tested equipment or component. While the measurement range of the code tester can be adjusted over a wide range by covering nozzles, the fan may set limits to the actual usable range. Multiple parallel variable-speed fans can allow overcoming this issue at the expense of greater complexity of the airflow booster system.

For unitary equipment testing, booster fan speed needs to be carefully controlled to maintain the correct external static pressure on the equipment. Excessive positive or negative static pressure will lead to air leakage, in turn causing large differences between airside capacity and refrigerant-side capacity.

One pitfall of (remotely) installed code testers is that pressure, temperature, and flow rate need to be calculated at the correct position. The pressure and temperature at the piece of equipment will be different at the code tester than at the test object. Therefore, appropriate corrections for the density need to be applied.

9.5.2 Anemometers

For airflow distribution and uniformity measurements, the most commonly used anemometers are hot-wire (or heated thermistor-based) anemometers and vane anemometers.

Hot-wire and heated thermistor anemometers are single-point measurement devices. Therefore, they need to be installed to create a grid to capture flow characteristics in the installed plane.

Vane anemometers usually come as handheld devices. They can be used as a secondary measurement device to confirm the airflow. Similar to hot-wire anemometers, airflow needs to be measured in a grid to determine airspeed.

Regardless of which anemometer is used, it is crucial to measure the cross-sectional area as a grid to obtain accurate average airflow values.

9.5.3 Pitot Tubes

Pitot tubes are the most common method for measuring the airflow. Pitot tubes are designed to measure total pressure (P_T) as well as static pressure (P_S) with the difference between the two being dynamic pressure (P_D). Equations (9.1) and (9.2) show how airflow is calculated using Pitot tubes:

$$P_D = P_T - P_S. \tag{9.1}$$

Air velocity at the measurement plane can be obtained from the following:

$$V = \frac{2 \times P_D}{\rho}, \tag{9.2}$$

where:
- P_D = dynamic pressure (Pa),
- P_T = total pressure (Pa),
- P_S = static pressure (Pa),
- ρ = airflow density (kg/m$_3$), and
- V = air velocity (m/s),

At the bottom of the Pitot tube, there are two connection ports. These ports can be attached to the measurement device HI (P_T) and LOW (P_S) port using PVC tubes.

When the Pitot tube is used, measured air velocity range and accuracy are dependent on the attached differential pressure transducer. In general, it is difficult to measure low velocities with Pitot tubes due to the low value of the dynamic pressure P_D.

The following bullet points need to be considered when Pitot tubes are used in the field:

- Pitot tube measurement nose needs to be perpendicular to the measured plane.
- While measuring the airflow, additional measurements such as air temperature and atmospheric pressure need to be collected to

calculate the density of the air. Changing $10°$ angle during the measurement will cause non-repeatable measurement and decrease the accuracy.

Note: Pitot tubes are generally used as a secondary measurement method, for the determination of flow distribution, or as a means of calibration.

9.5.4 Leakage Testing

Psychrometric measurements of equipment and component capacities generally require flow rate measurements as well as conditioning measurements. For accurate measurements, leakage of both air and heat needs to be sufficiently small relative to the airflow rate or condition difference between the inlet and outlet of the component.

Airflow leakage should typically be below 0.25% of the measured value for laboratory measurements and 1% of the measured value for field measurements (ASHRAE standard 41.2, 2018). When testing airflow leakage, the setup is blocked off on one side with the other side being equipped with a small fan to pull air out or push air into the duct setup. It is essential to minimize the modifications to reduce the risk of accidentally introducing new leakage paths when reconnecting the setup to the code tester or airflow measurement device.

The best practice is to equip the code tester with means to use it directly for leakage measurements. This may include a small nozzle that covers approximately 1% of the target flow rate. That approach also allows to include leakage within the code tester itself. Alternatively, the entire code tester may be included in the leakage measurement. Air leakage may be affected by static pressure, e.g., leakage pathways increasing in size due to (elastic) deformation of ductwork. Therefore, static pressure during leakage testing should be comparable to the static pressure during actual performance testing or include both positive and negative static pressure to specifically determine leakage into and out of the piece of equipment.

9.6 Controls of Psychrometric Testing Facilities

Psychrometric testing facilities generally include a large number of control loops, some of which are at the system level (e.g., chiller capacity, steam generator output, booster fan frequency) and some are at a supervisory level (e.g., air temperature, air humidity, static pressure). The interdependence of variables, e.g., changes in temperature, also affects the humidity, generally requiring automatic controls.

Typically, bundled temperature and RH sensor or psychrometer is used for controlling the temperature and humidity level of the room. Since the sensing instrument is single-point measurement sensor, the location of the sensor has a substantial impact on the room temperature and RH control.

The conditioned air is generally cooled and dehumidified below the target set point. Electric reheat and steam re-humidification is then used to allow stable operation and tight control. Most importantly, that approach allows relatively quick changes between different operating conditions to increase the throughput of expensive psychrometric facilities. With respect to cost, application of redundant safety mechanisms on electric heaters is a requirement for psychrometric testing facility safety. This section gives an introduction to the controls of psychrometric facilities and additionally gives recommendations for safety features to prevent the need of costly repairs.

9.6.1 Steady-State Operation

HVAC&R testing requires the user to provide certain stable conditions to test equipment. When stable condition is provided to a testing equipment, it will reach steady-state operation. Test conditions can be provided with support equipment such as chillers with thermal buffer tanks, direct expansion systems (DX), electrical heater, constant temperature water source, geothermal wells, steam, and burner. The mentioned support equipment allows to control humidity and temperature.

9.6.2 Safety Mechanisms

One of the critical problems commonly observed in initial installations of psychrometric test facilities is engineering mistakes on safety loops. These issues primarily relate to issues with electric heaters and typically result in major cost for repairs on newly completed facilities. Electric heaters are capable of achieving temperatures far above the melting points of typical insulation materials used for insulation, wiring, and lighting fixtures. Failure to carefully design fully redundant safety mechanisms will, sooner or later, result in merciless accidents during times where the laboratory is unoccupied.

The following steps are recommended to prevent damage to the facilities:

1. Software temperature limits "hard-coded" into the control software that limits temperature close to the heaters as well as in the facility itself.
2. Software logic and redundant hardwired airflow switches to prevent heater operation without airflow.
3. Software limitations according to the manufacturer limitations on heater output for low airflow rates to prevent heater burnout. These limits should be set based on a mapping of measured minimum

local air velocity at the heater and not on assumed uniform airflow distribution and nominal fan performance.

4. Hardwired temperature switches to shut off heaters in case airflow switch and other circuits fail. This switch should be a manual reset switch and located in a fashion that allows it to sense hot air raising up from the heating elements. The manual reset is required to prevent failure of the contactors due to repeated switching—as can occur if all other above safety mechanisms fail. Automatic reset switches at a lower temperature can be used in addition to limit the number of times an electrician needs to be called to reset the manual reset switch. These automatic reset switches should only be used to deactivate SCRs but not to switch contactors that supply mains power.

5. Melting fuses or manual reset bimetallic switches that should be used in series with the heating elements as a last defense. These devices should be selected at a temperature higher than the above safety mechanisms to limit their activation to when it is actually needed.

While typically less severe, failure of humidification controls can also lead to problems. It is recommended to use a limiting humidistat to shut off humidification above the room's safe maximum humidity limits. This limiting humidistat should be located at a location where it can sense the humidity addition without fans in operation. As a passive and secondary safety mechanism, ductwork should be sloped with a drain at the lowest point to prevent filling ductwork with condensate until mechanical failure.

While not necessarily related to control-based safety issues, drains used for coils, steam injection condensate, and other purposes should have a minimum internal diameter of 20 mm and be equipped with an inlet screen to limit the risk of blockage. The screen should be checked for blockage by construction debris after system modifications, as well as on regular intervals.

References

ASHRAE 37 (RA-2019) (2009). *ANSI/ASHRAE Standard 37-2009(ra-2019) - Methods of Testing for Rating Electrically Driven Unitary Air-Conditioning and Heat Pump Equipment*. ASHRAE, Atlanta.

ASHRAE 41.1 (2013). *ASHRAE 41.1-2013 - Standard Methods for Temperature Measurements*. ASHRAE, Atlanta.

ASHRAE 41.2 (2018). *ASHRAE 41.2-2018 - Standard Methods for Air Velocity and Airflow Measurements*. ASHRAE, Atlanta.

ASHRAE 41.6 (2014). *ASHRAE 41.6-2014 - Standard Methods for Humidity Measurements.* ASHRAE, Atlanta.

Collins, P. E., Beck, T., Schaefer, J. (2016). Verification of the Accuracy of Air Flow Measurement Using the Multi- Nozzle Chamber Method. 2016 *ASHRAE Annual Conference—Papers.*

Dally, J. W., Riley, W. F., McConnell, K. G. (1984). Instrumentation for Engineering Measurements. John Wiley & Sons.

Menter, F. R. (1994). Two-equation eddy-viscosity turbulence models for engineering applications. *AIAA Journal*, 32: 1598–1605.

Teague, K. A., Gallicchio, N. (2017). The Evolution of Meteorology: A Look into the Past, Present, and Future of Weather Forecasting. John Wiley & Sons, Oxford.

US Department of Energy (2017). 10 CFR Ch. II (1-1-2017 edition). Subchapter D-Energy Conservation - Appendix M to Subpart B of Part 430 Uniform Test Method for Measuring the Energy Consumption of Central Air Conditioners and Heat Pumps. Subsection 2.14.2. U.S. Government Publishing Office.

Part C

Heat Flux Measurements, Optical Techniques, and Infrared Thermography

10

Surface Temperature Measurement on Complex Topology by Infrared Thermography

Maximilian Elfner and Hans-Jörg Bauer

Institute of Thermal Turbomachinery (ITS), Karlsruhe Institute of Technology

Achmed Schulz

formerly ITS

CONTENTS

10.1 Introduction

The determination of an object's temperature, especially its surface temperature, is of great importance for many applications. Examples can be found in civil engineering, mechanical engineering, electrical engineering, and even simple household application. The utilization of radiation emitted by the object is one if not the oldest method to determine its temperature. In times where a quantitative temperature measurement was not possible (way before a quantitative measure of temperature even was conceived), visible radiation in the form of object color was used to determine the correct temperature for, e.g., blacksmithing.

Even thousands of years later, optical methods are advantageous if other, now available methods for measuring the temperature are not feasible. This might occur when the object of interest is too remote, too fragile, or simply too hot. Even if a conventional method is feasible, optical methods often prove to be better suited. They are fast, since no physical object (the probe itself) has to be heated. They do not influence the thermal balance of the object, since only negligible amounts of heat are transferred to the probe. They do not disturb the surrounding of the object itself, which is of great importance when (forced) convection has to be considered. And finally, optical methods are robust against interferences from magnetic or electrical fields.

However, the measured quantity, the object's surface radiation, is only an indicator for its temperature. To derive a physical temperature, complex calibration procedures are required. If not calibrated properly, the measurement uncertainty can easily be worse than the judgment of an experienced blacksmith.

The following chapter will present procedures, which, when applied correctly, allow a quantitative determination of an object's surface temperature with very low measurement uncertainty. To gain a better understanding of the underlying theoretical considerations, Section 10.2 deals with the basics of radiation theory, why infrared thermography is used in many applications, and the differences to visible light. Section 10.3 will cover basic detector and camera types, their applications, and also implications during the use under different conditions. Section 10.4 will then cover the camera calibration. This includes everything that is needed to produce a clean, low-noise infrared recording even under adverse conditions. Section 10.5 describes the process of calibrating such an infrared recording to a physical temperature, again considering different situations and calibration techniques. Finally, in Section 10.6, the residual measurement error for different calibration methods will be considered.

Not every step will be applicable to every type of radiation detector or every type of process. You have to choose the steps that are applicable to your specific system. In general, the process is designed for modern FPA-style infrared cameras. When using, e.g., a point pyrometer, the camera calibration will most likely be of lesser importance. However, temperature calibration has to be performed with equal care. On the other hand, there might be situations when a qualitative radiation measurement is sufficient, and adding a temperature calibration then only adds uncertainty.

10.2 Basic Theoretical Considerations

Every object with a temperature above 0 K emits radiation. For the sake of clarity, we will assume that our objects have negligible translucency in the interesting range of wavelengths. This holds true for most technical applications; however, there are some exceptions. Window materials are translucent in specific wavelength bands only. They also emit small amounts of radiation. This will be covered later in Sections 10.3 and 10.5. Other materials may behave against intuition: One example is certain polymers (e.g., PTFE, PEI), which are opaque in the optical wavelength band but quite translucent for longer wavelengths.

The assumption of vanishing translucency yields two major simplifications:

1. The radiation energy exchange is constrained to an infinitesimally small surface layer. Energy transport in the object itself is only occurring by means of conduction.

2. In a stationary state, Kirchhoff's laws are applicable. They then state twofold:

   a. Good emitters are good absorbers, or more exactly, a surface's emissivity ε equals its absorption α:

$$\varepsilon = \alpha. \tag{10.1}$$

   b. Incident radiation can either be absorbed or be reflected. The amount of reflected radiation is defined by a surface's reflectivity ρ (with a value between 0 and 1). Thus:

$$\alpha + \rho = \varepsilon + \rho = 1. \tag{10.2}$$

While many different approaches existed to describe the amount of emitted radiation before, it was Max Planck who finally derived a formulation that was applicable to all wavelengths (or frequencies) of radiation, paving the road for modern quantum physics along the way. He found that the radiation emitted by a surface, the specific (per area and steradian) spectral (per wavelength) radiance of an ideal body $W_{bb,P}$ can be described by:

$$W_{bb,P}\left(\lambda,T_S\right)=\frac{2hc^2}{\lambda^5\left(e^{hc/\lambda kT_S}-1\right)} \tag{10.3}$$

where c describes the speed of light, h is the Planck constant, λ is the wavelength, k is the Boltzmann constant, and T_S is the surface temperature. This law is shown for a practical range of values in Figure 10.1. Each curve at a constant surface temperature T_S shows a clear maximum of emitted radiation, while a decrease in both directions away from this maximum is visible (yielding a bounded amount of emitted energy when integrating over all wavelengths). This maximum shifts from UV (short wavelength compared to optical) at very high temperatures above 5,000 K (the sun has an effective surface temperature of 5,778 K) over the optical range at ~3,000 K (a typical value for a light bulb) to the infrared range at temperatures around 2,000 K and below. While the emitted radiation is always a mix of different wavelengths, the response and precision of a detector are best when working close to the maximum of radiance.

This is the main reason why infrared thermography at wavelengths from ≈1 to ≈15 μm is the most common way of optical temperature measurement in technical applications. The temperature at the radiation maximum for those wavelengths ranges from ≈300 to ≈2,000 K, which covers the most common applications in science and engineering. However, there is no optimal detector covering the full range of the infrared spectrum. Different detector types

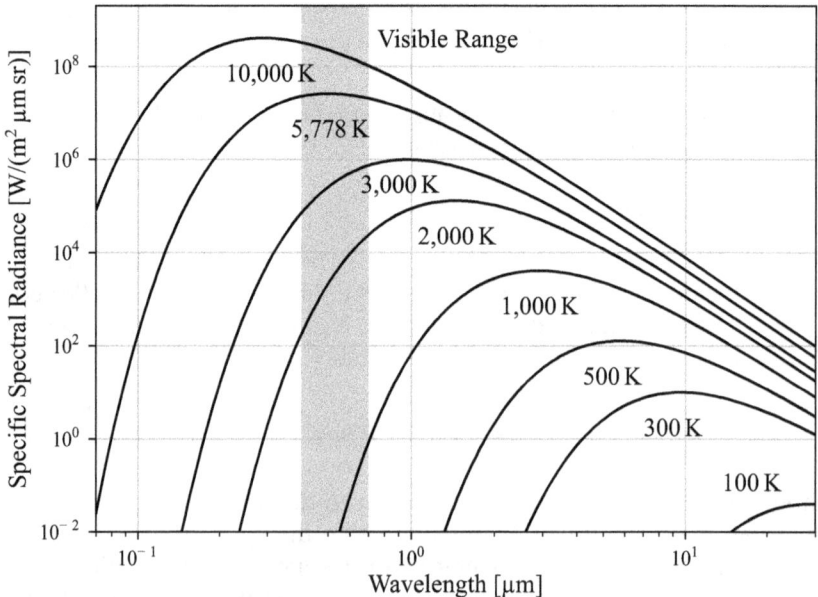

FIGURE 10.1
Radiance as computed by Planck's law.

are used for different wavelength bands. Those will be briefly described in Section 10.3.

Above, we introduced the emissivity without further clarification. We will now consider the effect of the emissivity. Planck's law only describes ideal surfaces (in general bodies, in our simplified case surfaces) with an emissivity of unity. Those surfaces absorb all incoming radiation and reflect none:

$$\alpha = \varepsilon = 1 \quad \rho = 0 \tag{10.4}$$

Such a surface with a constant emissivity of one is termed the surface of a black body (considering temperatures below the threshold of visible radiation, it would appear as perfect black and only emit thermal radiation). This black body surface emits the maximum radiation possible and thus follows Planck's law (Eq. (10.3)). For real bodies, a changing emissivity has to be considered. In general, the emissivity ε (and thus also the reflectivity) is a function of the surface temperature T_S, the wavelength λ, and the polar angle of view θ:

$$\varepsilon = f(T, \lambda, \theta) \tag{10.5}$$

Particularly for metal surfaces, the surface condition also plays a major role. While the emissivity of, e.g., polished steel ranges between ≈ 0.05 and ≈ 0.2, oxidized steel can achieve emissivities as high as ≈ 0.8 (Gubareff et al. 1960). This spatial (and often also temporal) variation is almost impossible to consider during calibration. Thus, for a low measurement uncertainty, **we recommend to coat the surface with special infrared coatings**.

For most non-conductive materials and coatings, simplifications can be made. Using narrow-band detectors, the wavelength influence is negligible. Also, for temperature ranges resolvable with a single detector system, the temperature dependence usually is small. Lastly, with angles of view below $50°$ from normal incidence, the remaining dependency vanishes and the emissivity can be considered constant in many technical applications. A surface with such a constant emissivity (sometimes only the vanishing wavelength dependence) is termed a gray body surface. Exceptions may occur with curved surfaces of complex geometries. They will be discussed in Section 10.5.3.1. Planck's law can then be rewritten including the emissivity (the most general case without the above simplifications shown):

$$W_{\mathrm{real},P}\left(\lambda, T_S, \varepsilon\right) = \varepsilon\left(T_S, \lambda, \theta\right) W_{bb,P}\left(\lambda, T_S\right) \tag{10.6}$$

The last important law when dealing with radiation is the Stefan–Boltzmann law. It states that the total specific radiance from a surface can simply be computed as follows:

$$W_{\mathrm{real},SB}\left(T_S, \varepsilon\right) = \varepsilon(T, \lambda, \theta) W_{bb,SB}\left(T_S\right) = \varepsilon(T, \lambda, \theta)\sigma T_S^4 \tag{10.7}$$

where σ is the Stefan–Boltzmann constant. The Stefan–Boltzmann law can also be deduced by integrating Planck's law over all wavelengths and a semi-infinite sphere space. This results in the exact definition of $\sigma = 2\pi^5 k^4/15h^3c^2$. The major implication from the Stefan–Boltzmann law is the T^4 proportionality for radiation power. This will be of major concern for the calibration procedures presented in Sections 10.4 and 10.5.

10.3 Camera Types and Practical Application

After the introduction of the basic theoretical foundations, it is clear that infrared thermography is well suited for technical temperature measurements. The next question to consider is how the radiation can be detected and digitized in different experimental setups. We will focus on modern systems that allow a quantification of the radiation signal.

10.3.1 Camera Types

In general, there are three major parts to a radiation detection system: The physical radiation detector, the detector geometry, and finally, the optical system. Those parts can be combined (almost independently) to build different systems.

The radiation detector describes the smallest physical unit detecting the incoming radiation. Two different types are usually found in infrared cameras: Shorter wavelength radiation up to $\approx$5 μm is commonly detected using CCD (charge-coupled device, increasing charge with incident photons) or CMOS (complementary metal–oxide–semiconductor, decreasing charge with incident photons) photon detectors made from indium–antimony (InSb) or similar materials. Those detectors are cooled to $\approx$20 K to reduce noise. The sensitivity to shorter wavelengths leads to a better signal-to-noise ratio at elevated surface temperatures of $\geq$420 K. Another major advantage when using semiconductor detectors is a well-controlled state of charge. This allows the user to control the integration time t_I. With this additional setting, the infrared system's response and dynamic range can be calibrated and customized to the application (and also digitally increased; see Section 10.4.3).

For lower object temperatures, longer wavelengths produce better signal-to-noise ratios due to the shift in the radiance peak. Those longer wavelengths up to 15 μm are commonly detected using bolometers. They are not true photon detectors based on semiconducting material. In fact, each detector consists of an absorber, a thermal capacity, and a defined thermal resistance to a heat sink. The incident radiation is absorbed and heats up the detector element. Finally, the temperature of the detector is determined and digitized.

Those small-scale detector elements are combined to form larger arrays. The current state of technology are FPA detector types with sensors made from many single pixel (px) elements. They range from 0.3 to 2 Mpx and are similar to those of standard optical cameras.

Another commonly used detector element is the pyrometer, especially for industrial application in demanding surroundings. They do not include a FPA with discrete detector elements. Instead, a small (typically a few millimeters) sensing element records the irradiance and outputs a single mean value over its sensing surface.

Finally, the optical system allows the customization of the detector system and completes the camera system. For FPA-style detectors, the optical system is a common lens built from multiple elements with high transmissivity in the infrared range. Several different focal lengths are available. For pyrometric devices, the optical system usually is a simple focus lens. Special application devices exist, e.g., waveguides for boroscopic access.

10.3.2 Practical Application

Two main different setups have to be considered in practical application: open measurement situations on the one hand, and closed measurement situations with boundaries of higher temperature than the test surface on the other hand. Closed measurement situations with low temperature boundaries are similar to open setups, and only the window effects have to be considered. Open setups typically arise when determining the temperature of hot parts, e.g., power electronic cooling and failure analysis or heater air flow analysis. Closed setups are commonly found in cooling research, e.g., convective cooling of gas turbine parts, where the object of interest is actively cooled and thus of lower temperature. Those two scenarios are shown in Figure 10.2. They have a direct influence on the choice of camera system and calibration procedure (the latter will be discussed in Section 10.5.1). In general, an object's surface emits the specific radiance W_S. This radiation then passes several obstructions with different transmissivities. In simple open setups, those obstructions are the atmosphere and the lens system. This situation is shown in Figure 10.2a.

In closed setups, additional effects have to be considered, e.g., different media or windows. A complex closed setup is shown in Figure 10.2b. Those additional effects can be summarized as the product of all transmissivities $\tau_{tot} = \prod_i \tau_i(\lambda, T_i)$, and thus, the radiation from the surface is (with the simplifications made in Section 10.2) as follows:

$$E_S(T_S) \propto \tau_{tot} \cdot W_S(T_S) \qquad (10.8)$$

Any object present in the optical path may reflect or emit radiation. This radiance also reaches the detector and generates a signal, which is not connected to the actual test specimen's surface temperature. Those effects include the

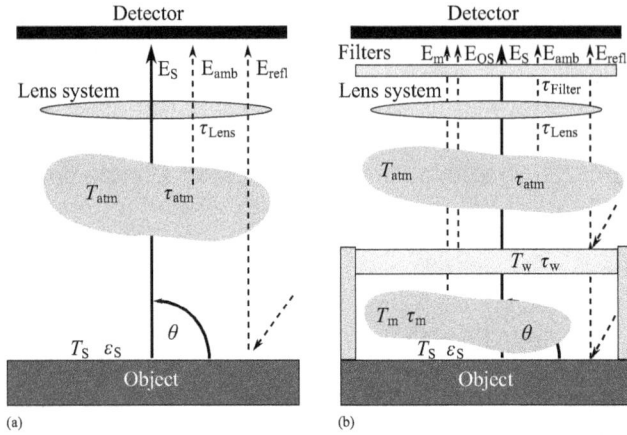

FIGURE 10.2
Different technical measurement setups. (a) A simple, open measurement setup and (b) a complex, open measurement setup.

ambient radiance E_{amb}, media emitted radiance E_m, radiance from the optical system and windows E_{OS}, and reflected radiation from any surfaces E_{refl}. We will sum up all those effects and denote this radiance as offset radiance E_{off}. Thus, the incident radiance at the detector E_D is composed of those two parts:

$$E_D = E_S(T_S, \varepsilon_S) + E_{off}(T_i, \tau_i, \varepsilon_i, \rho_i, ...) \tag{10.9}$$

This relation is different for every optical setup and measurement setup. Great care has to be taken when calibrating the optical system and, finally, when deducting a temperature from a radiation signal. Those two main calibration steps and different procedures are covered in Sections 10.4 and 10.5.

10.3.2.1 Window Materials

When using closed setups, windows are needed to allow optical access to the test specimen's surface. The window material plays a major role, since its transmissivity has to be as high as possible in the camera's spectral range.

For short-to-medium wavelength, there are a multitude of available materials. Standard fused silica (*quartz*) is transmissive for infrared up to ≈3 μm; however, its transmissivity is fairly low and temperature-dependent (Wood 1960). Different fluoride compounds (*magnesium, calcium, barium*) are comparably cheap options; however, their material strength is low, especially at elevated temperatures. For complex applications, *sapphire* is often used. It shows high transmissivity in the infrared and optical spectrum, has high mechanical strength, and is especially resilient against scratches (Thomas et al. 1988b). However, it is expensive and machining is challenging. Beyond ≈5 μm, the transmissivity reduces drastically. The exact wavelength of this reduction is temperature-dependent (Thomas et al. 1998a). To overcome this constraint, Ochs et al. (2009) propose the utilization of a 4.1 μm cutoff filter for window temperatures up to ≈600 K. This cutoff wavelength needs to be adapted to lower wavelengths if the window temperature further increases. For high-quality optical elements, such as lenses and filters, *germanium* can be used. However, its processing and machining are highly specialized.

Finally, if temperatures are comparably low, and low price combined with good machinability is required, *polyetherimide* may be used up to medium wavelengths (Philipp et al. 1989).

For medium-to-long wavelengths (usually detected with bolometers), the transmissivity above ≈8 μm has to be considered. Few materials can be used. *Calcium fluoride* has been discussed above, and *zinc selenide* is machinable and comparably cheap. Its mechanical strength is sufficient for most applications. However, zinc selenide reacts with water and is poisonous; thus, handling of the windows is challenging.

An extensive overview of different materials and their application range, especially considering their mechanical strength and high temperature resistance, can be found in Harris (1998).

10.4 Camera Calibration

Assume there is a certain irradiance E_D reaching the detector of a camera. The detector will then produce an output signal U that is proportional to its exposure time and irradiance weighted by the detector's spectral responsivity $r(\lambda)$:

$$U = g\left\{t_I \cdot \int_0^\infty r(\lambda) \cdot E_D(\lambda)d\lambda\right\} = g\left\{t_I \cdot I^*\right\}. \tag{10.10}$$

Considering a general detector, this response g is not linear and, in case of FPA detectors, not equal for each pixel. This leads to the conclusion that any calibration performed is also dependent on this response function and the integration time, thus a very specific set of parameters. To use a generalized calibration for each pixel and a multitude of measurement situations, several corrections have to be performed.

To perform those corrections, a calibration body is needed. This body (or a simple surface) needs to produce an even radiation signal, covering the field of view of the camera system. It is recommended to use black bodies, reducing the risk of inhomogeneous, reflected radiation. The calibration temperature needs to be adjustable in a range comparable to the expected measurement temperatures. Typical calibration bodies are black body cavities or solid heated copper blocks with coated, high-emissive surfaces.

10.4.1 Linearization

There are two major effects of detector nonlinearity: a nonlinearity with detector loading and a nonlinearity with changing integration time t_I even at optimal detector loading. Since they are both strongly connected to the integration time, those nonlinearities are specifically important for semiconductor detectors.

The first effect is shown in Figure 10.3a. The homogeneous calibration surface is recorded at a constant temperature with a wide range of integration times, and the normalized detector signal $\tilde{U}$ is evaluated at the detector center. A fairly linear range of response is clearly visible for a wide range of normalized integration times $\tilde{t}_1$; however, at very small and very large detector values, the response changes. This nonlinearity is hard to correct and, as a result, it is recommended to use integration times that produce detector responses from $\tilde{U} \approx 0.20$ to $\tilde{U} \approx 0.85$ only. These values differ for each camera system and may be taken as an orientation.

The second effect of nonlinearity is shown in Figure 10.3b. A random, single pixel is chosen, and the ratio of integration times $R_{i,j} = t_{I,i}/t_{I,j}$ and the ratio of response values $\tilde{U}_{R,i,j} = \tilde{U}_i/\tilde{U}_j$ are computed and plotted for each combination of recordings from the previous series, considering images in the linear range only (the only difference between two successive recordings being the integration time t_I). The line of identity is also shown. The response for each ratio is changing.

However, as discussed earlier, a linear and consistent response is crucial for calibration. Ochs et al. (2010) developed a linearization technique applicable to infrared detectors. They show that a polynomial correction of second degree can be applied to each pixel, linearizing its response. The linearization is temperature-independent and inherent to the detector, thus independent of the optical setup. If camera drift is ignored (which is a valid assumption for modern cameras), a single detector linearization can be applied to each measurement.

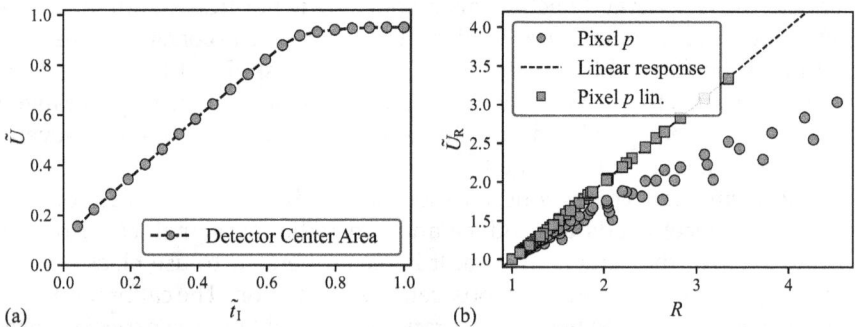

(a)   (b)

FIGURE 10.3
Detector nonlinearity. (a) Nonlinearity over detector loading and (b) nonlinearity rations.

To derive the correction constants, Ochs et al. (2010) again used the series of Q recorded images of the homogeneous, stationary calibration surface at different integration times. They then defined the target function for each pixel p which has to be minimized:

$$\epsilon_p = \sum_{q=1}^{Q-1}\sum_{k=1}^{Q-q}\left[\sum_{n=0}^{N}c_{p,n}\cdot\tilde{U}_{p,q}^n - R_{p,q+k}\cdot\sum_{n=0}^{N}c_{p,n}\cdot\tilde{U}_{p,q+k}^n\right]^2 \qquad (10.11)$$

where q and k are certain images and $c_{p,n}$ are the pixel-wise polynomial coefficients. Simply speaking, for each possible combination of integration times $R_{q,q+k}$, the ratio of corresponding, corrected detector responses $\left[c_{p,N}\cdot\tilde{U}_p^n\right]_{q,q+k}$ needs to be equal to $R_{q,q+k}$. Since this correction can only be performed to scale, $c_{p,N}=1-\sum_{n=0}^{N-1}c_{p,n}$ has to be fulfilled. Two degrees of freedom remain per pixel. The use of more images and thus more ratios is recommended. Usually, approximately ten images covering a wide range of linear response yield good results. Finally, the scaled irradiance for each pixel I_p can be computed:

$$I_p = \frac{1}{t_I}\sum_{n=0}^{N}c_{p,n}\tilde{U}_p^n. \qquad (10.12)$$

The ratio of this scaled and linearized irradiance is shown in Figure 10.3b. The scaled irradiance then is independent of the integration time and directly proportional to a physical radiance. The integration time can be used to adapt the camera system to the scene. However, each pixel has its own, individual scale a_p:

$$I_P = a_P \cdot I_P^*. \qquad (10.13)$$

Thus, when using this linearization procedure, a non-uniformity correction has to be performed to equalize the scale for all pixels.

10.4.2 NUC: Non-Uniformity Correction

The system's non-uniformity arises from different sources: As every optical lens system does, infrared lenses suffer from radially changing optical characteristics and vignetting. The latter is worse compared to visible wavelengths since the lens is usually colder than the object of interest and thus absorbs parts of its emitted radiation. Finally, with the linearization approach presented before, each pixel has an independent scale and thus needs to be rescaled.

A typical shape of non-uniformity is shown in Figure 10.4a: The linearized and scaled and normalized detector value is shown in arbitrary units [A.U.] (which is proportional to the irradiance but of no physical magnitude).

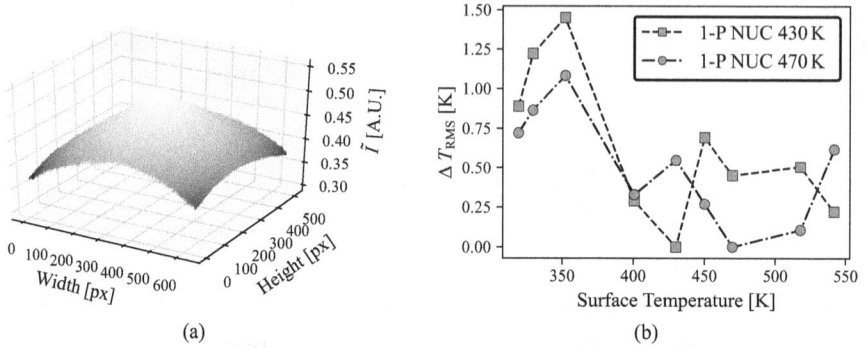

FIGURE 10.4

Detector non-uniformity. (a) Typical non-uniformity and (b) single-point NUC RMS error at different test surface temperatures.

A high value of irradiance at the center combined with an almost radially symmetric drop of to the image edges. The non-uniformity depends on the optical system, and its correction has to be performed anew for each optical configuration, independent of the process used. In the following, the most common methods are presented: a simple single-point correction, a two-point correction, and finally, the linearized two-point correction, which has proven beneficial in many applications. The uncertainty of those methods is computed as the RMS value of uniformity in temperature (different calibration techniques will be presented in Section 10.5). As we will see later, this measure of uncertainty is zero at the irradiation level of correction and increases for deviating irradiation levels.

10.4.2.1 Single-Point NUC

The most simple approach is based on a single recording of the calibration surface at constant conditions. A scalar correction $\Xi_{\mathrm{off},p}$ is determined for each pixel, defined by the pixel's deviation from the mean value $\bar{I}$:

$$\Xi_{\mathrm{off},p} = \bar{I} - I_p \tag{10.14}$$

This correction is easy to implement, uses small amounts of memory, and is fast. It can be applied to the linearized signal I or an arbitrary response U. However, large uncertainties arise when the irradiance level changes from the calibration point, which obviously happens during use when non-uniform, real surfaces are analyzed. Thus, such a simple correction can only be feasible when the target temperature can be estimated and varies only by small amounts. The residual error when the surface temperature is changing from the calibration level is shown in Figure 10.4b.

10.4.2.2 Two-Point NUC

The two-point NUC is the most common type of non-uniformity correction. The results are good, and the effort is comparably low. Instead of only one radiation level, two levels are used, and a linear correction function is derived for each pixel. Several approaches exist: Basic linear and quadratic corrections are described by M. Schulz and Caldwell (1995). Those procedures are further refined by Shi et al. (2005), including a model for the detector nonlinearity in the NUC.

In the following, only the newest method by Ochs et al. (2010) will be discussed in detail. This method, in combination with a *linearized* detector response I, performs superior for a multitude of complex situations. The same linear non-uniformity correction may be used for a non-linearized detector signal U. However, the integration times for the two correction levels have to be equal and the final NUC is only valid for this single integration time. The process is then comparable to the one of M. Schulz and Caldwell (1995).

If all pixels $p \in P$ have equal scale $a_p = \text{const.} \ \forall \ P$, the scaled irradiance I_p is uniform in case of uniform illumination. This is, however, not the case for the image I. The lens optic's mapping τ_p changes the object's surface radiance I^*_{obj} inhomogeneously, and a spatially varying offset $I^*_{\text{NUC, offset},p}$[1] needs to be considered. Thus, the pixel-wise irradiance can be expressed by:

$$I^*_p = \tau_p \cdot I^*_{\text{obj}} + I^*_{\text{NUC,Offset},p} \qquad (10.15)$$

With two images from the calibration surface at different, homogeneous radiation levels (e.g., different temperatures), a linear correction can be performed. The integration time can be chosen to match the radiation level of the surface; however, the generated detector signal must be covered by the range of the linearization. The two recordings then can be described by:

$$I_{p,1} = \frac{1}{t_{I,1}} \sum_{n=0}^{N} c_{p,n} \tilde{U}^n_{p,1} = a_p \cdot \left(\tau_p \cdot I^*_{\text{obj},1} + I^*_{\text{NUC,offset},p} \right) \qquad (10.16a)$$

$$I_{p,2} = \frac{1}{t_{I,2}} \sum_{n=0}^{N} c_{p,n} \tilde{U}^n_{p,2} = a_p \cdot \left(\tau_p \cdot I^*_{\text{obj},2} + I^*_{\text{NUC,offset},p} \right) \qquad (10.16b)$$

with their arithmetic mean values I_1 and I_2 over all pixels for a single readout. A linear NUC from those two recordings must satisfy:

$$I^{\text{NUC}}_p = \text{NUC}_{a,p} \cdot I_p + \text{NUC}_{b,p}. \qquad (10.17)$$

---

[1] This NUC offset $I^*_{\text{NUC,offset}}$ is superimposed to any object or scene radiation signal by the optical system of the camera. It must not be confused with the scene's offset radiation E_{off}.

Solving for the two NUC correction tables yields:

$$\text{NUC}_{a,p} = \frac{\bar{I}_2 - \bar{I}_1}{I_{p,2} - I_{p,1}}$$ (10.18a)

$$\text{NUC}_{b,p} = \bar{I}_1 - \frac{\bar{I}_2 - \bar{I}_1}{I_{p,2} - I_{p,1}} I_{p,1}$$ (10.18b)

Those two tables are each the size of the FPA's amount of pixels. They can be used for any integration time for a single optical setup. When changing the optical setup, e.g., changing the lens, filters, or extension tubes, a new NUC has to be performed. This can be confirmed by incorporating Eqs. (10.13) and (10.15) into Eq. (10.17):

$$I_p^{\text{NUC}} = \overline{a \cdot \tau} \cdot I_{\text{obj}}^* + \overline{a \cdot I_{\text{NUC,offset},p}^*}.$$ (10.19)

The non-uniformity-corrected irradiation thus contains the lens effects τ and is linear in object radiance and superimposed by a constant irradiance. This superimposed irradiance depends on the temperature of the camera system. When running long tests with high ambient temperatures and thus changing camera temperatures, the NUC quality can be further improved by recording an ISO surface at the new camera temperature level and updating the constant part of the NUC table $\text{NUC}_{b,p}$.

The residual error using this procedure is shown in Figure 10.5. The calibration surface is recorded at several different temperature levels. The images are linearized. Several non-uniformity corrections are computed using different pairs of images. Finally, the NUCs are applied to the images themselves and the residual error is determined for every image and for every NUC. As we have seen with the single-point NUC, the error is zero at the chosen levels of correction and increases moving to different radiation levels. The error is approximately one order of magnitude lower than the error in case of the single-point NUC. Interestingly, as it can be observed in Figure 10.5, the error, in some cases, can be reduced by extrapolation instead of using a wider temperature span for the NUC calculation and a following interpolation. The amount of feasible extrapolation is defined by the gradient of the error curves, which will depend on the specific optical setup. We advise to perform multiple recordings and build a comparable data set to decide on the NUC values. However, the error is bounded when interpolating, which is not the case for an extrapolation. Thus, extrapolation should only be considered when the range of temperatures is known in advance.

For the following sections, we will drop the superscript $I := I^{\text{NUC}}$ and assume that the data used for the following calibration steps is linearized and non-uniformity-corrected.

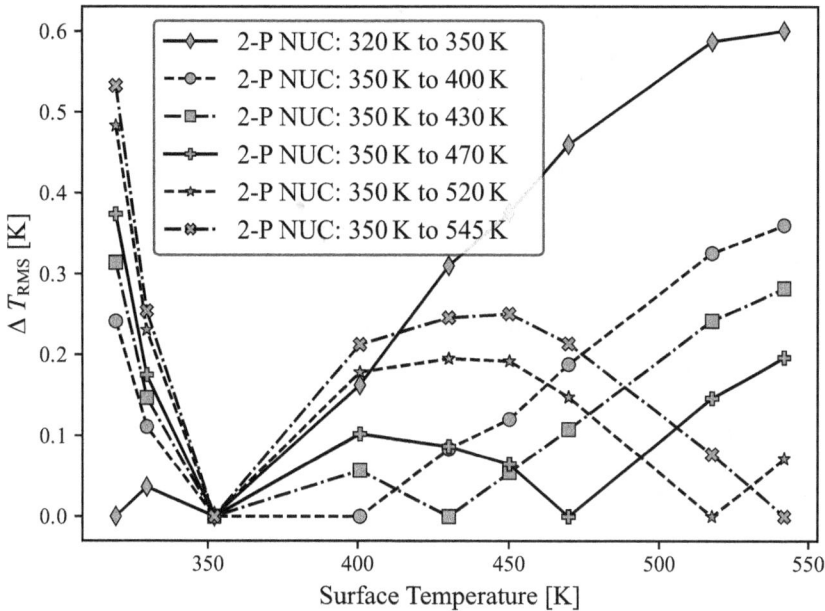

FIGURE 10.5
Two-point NUC RMS error at different test surface temperatures.

10.4.3 High Dynamic Range (HDR)

In Section 10.2 (Eq. (10.7)), we showed that when integrating over a larger range of wavelengths, the emitted radiance from a surface W_{real} is proportional to the fourth power of surface temperature, $\propto T_S^4$. If infrared thermography is applied to resolve surface temperatures of objects with larger spread of those temperatures, the radiation signal's dynamic range surpasses the capabilities of modern detectors. Those situations may occur with cooled components in hot environments, e.g., component cooling in mechanical engineering, or vice versa when examining hot components in cold environments, e.g., IC cooling in electrical engineering.

If a camera system with a CCD detector and thus a well-defined integration time t_i is available, HDR image generation can solve that issue. If and only if the linearization and non-uniformity correction as proposed in Section 10.4.1 have been performed correctly, HDR image generation is trivial.

Several images of the same scene are recorded in quick succession with different integration times and thus different exposures. The image with the lowest integration time needs to resolve the highest irradiance, and the image with the highest integration time needs to resolve the lowest irradiance. Several images in between can be recorded. A good starting point for most applications is four images. For each image with linearized and uniform radiation data I_i, a mask M_i is computed pointing to valid pixels:

$$M_i = I_i > I_{\min} \wedge I_i < I_{\max} \tag{10.20}$$

The final image can then be assembled, using data from a valid image at each pixel. If multiple images are valid at a single position, a single image can be chosen arbitrarily. If no image is valid, the integration time has to be adjusted or more images have to be recorded. With a well-calibrated camera, no seams will be visible.

The result of a HDR recording is shown in Figure 10.6: A light bulb was recorded using four images and the HDR technique. The full light bulb can be resolved, even though the temperature ranges from almost ambient temperature to ≈3,500 K for the glowing filament.

10.4.4 Position Estimation

For some of the advanced calibration procedures, the exact knowledge of the camera position relative to the object of interest is crucial. Those advanced methods will be presented in Section 10.5.3. In general, with more complex surface topologies and more complex experimental setups, simple 1D calibration approaches fail to produce exact results.

The topic of camera position estimation is too wide to be covered in this context; later on, we will assume we have determined the *optical center* $\vec{C}$ of the camera system correctly. Many different approaches for its determination with one, two, or even more cameras are discussed in detail by Hartley and Zisserman (2003). However, in most of the practical cases, only one infrared camera system is available per field of view. Those *single camera position estimation* problems are especially challenging to solve. An algorithm suited for infrared systems was developed by Elfner et al. (2018) and Elfner (2019).

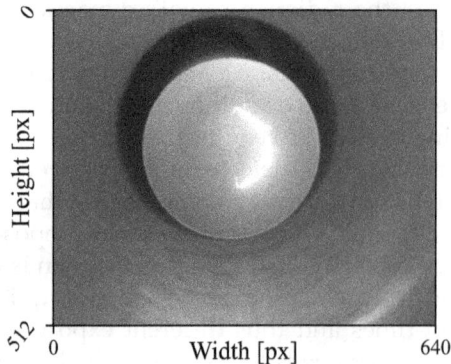

FIGURE 10.6
HDR infrared recording of a light bulb.

10.4.5 Focus Stacking

Infrared thermography can be used to resolve even the smallest detail. With a macrolens or extension tubes, high reproduction scales β can be achieved. However, with increasing reproduction scale, the depth of field decreases. Thus, focus stacking techniques are needed and can be applied to infrared recordings. The process will be roughly described in the following, and for further details, important references are given.

- A series of images of the scene are taken. The plane of focus is moved by moving either the camera or the lens focus throw.
- The images are aligned to each other. Since thermography images are often subject to patterns, global registration measures are superior to keypoint methods (e.g., Lowe 2004). An algorithm that was revealed to be especially advantageous to infrared images is the *ECC* algorithm (Evangelidis and Psarakis 2008).
- Sharp regions in every image are identified by image filters, e.g., Sobel filters (Sobel 2014).
- The final image is composed by using the sharp regions from every sub-image.

A process especially designed for infrared data has been developed by Elfner (2019). Figure 10.7 shows the result if the technique is applied correctly.

On the left (Figure 10.7a), a recording of a PCB is shown which was taken with a high reproduction ratio. Large out-of-focus areas are visible. On the right (Figure 10.7b), the same scene is shown with focus stacking from eight images, reducing out-of-focus areas and allowing the evaluation of data using the full field of view.

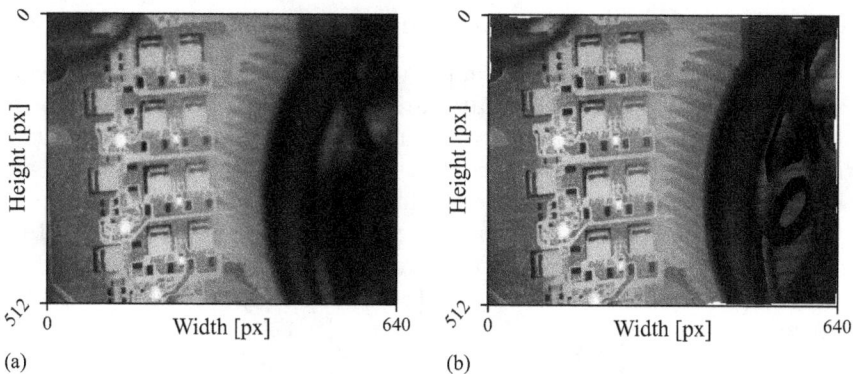

FIGURE 10.7
Comparison of basic and focus-stacked recording. (a) Single recording and (b) focus stacking of eight images.

10.5 Temperature Calibration

The camera calibration is only one part of the full calibration procedure. When performed correctly, the scaled irradiance signal I is linearly proportional to the detector irradiance E_D. As shown in Section 10.3.2, this irradiance needs to be calibrated to a certain temperature, since an analytical computation is, in general, not possible. The calibration procedures can be applied to FPA cameras as well as pyrometric devices.

10.5.1 The Semiempirical Calibration Function

A well-suited method for the calibration was proposed by Martiny et al. (1996). He defined a semiempirical relation between the irradiance and the surface temperature by inverting Eq. (10.3):

$$T_S = \frac{b}{\ln\left(\dfrac{r}{I_D} + f\right)} \tag{10.21}$$

This equation has three free parameters, r, b, and f, and thus needs to be calibrated. The most common approach is an in situ calibration technique. Several discrete thermal sensors (e.g., NTCs, thermocouples) are embedded in the test surface. After calibrating the recorded images following the procedures in Section 10.4, each thermal sensor location provides a pair of values $\{I_{D,i}, T_i\}$ that can be used for a LSQ fit method. At least three probes need to be implemented in every field of view to allow correct determination of the free parameters r, b, and f. The validity of this calibration technique was shown by A. Schulz (2000).

10.5.2 The Improved Calibration Technique

Ochs et al. (2009) showed that this approach can yield high uncertainties when large ranges of surface temperatures are present. They show that choosing different subsets of calibration pairs substantially changes the calibration, possibly introducing large uncertainties. Following their linearization and NUC approach, they propose a new calibration method.

First, they split the detector irradiance I_D according to Eq. (10.9) and the assumption of a linear system. Afterward, they build the calibration function depending on the surface's irradiance I_S, thus explicitly subtracting the offset radiance:

$$T_S = \frac{b}{\ln\left(\dfrac{r}{I_D - I_{off}} + f\right)} \tag{10.22}$$

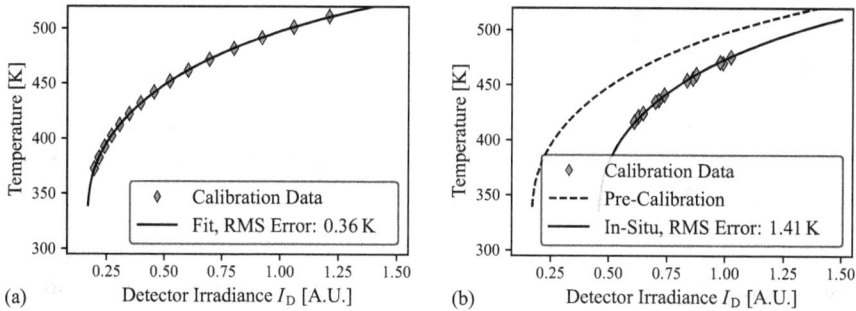

FIGURE 10.8
Different temperature calibrations. (a) Pre-calibrations according to Eq. (10.22) and (b) in situ offset calibrations at elevated temperatures.

introducing a fourth calibration parameter I_{off}. However, for many applications, certain calibration parameters can be determined in advance. This is of major importance: Using a calibration specimen with controllable temperature, a wide range of calibration pairs can be used, reducing the need for extrapolation. For open measurement applications as shown in Figure 10.2a, usually all four parameters can be predetermined. When windows are present (closed setup with cold boundaries), those windows have to be implemented during the pre-calibration. For closed channels, sketched in Figure 10.2b, different scenarios were described by Ochs et al. (2009). If the optical situation is imitated during calibration (the same lens system, the same windows), and the transmissivity of those obstructions is temperature-independent (e.g., for sapphire, this is achieved with the previously described 4.1 µm cutoff filter up to ≈650 K), the three parameters r, b, and f can be determined in advance. Only one parameter, the offset irradiance I_{off}, remains to be determined using an in situ calibration. A single temperature probe is sufficient, and multiple probes allow LSQ techniques. If the pre-calibration is performed correctly, only interpolation and no extrapolation has to be performed, greatly reducing the measurement uncertainty. An exemplary pre-calibration (Figure 10.8a) and its in situ offset calibration (Figure 10.8b) are shown in Figure 10.8. The residual RMS uncertainty of 1.41 K at a mean temperature of 445 K results in a final uncertainty of only ≈0.3%.

10.5.3 Adapting the Calibration to Local Effects

The parameters in the calibration function (Eq. (10.22)) are determined as scalar values. This is a feasible approach for many situations. However, with more complex surface topologies and test configurations, this approach may fail. In general, the pre-calibration parameters r, b, and f can still be assumed constant if the constraints described before are met. However, the radiation emitted from the surface (or better, the surface's emissivity ε_s) and the determined offset radiance have to be reconsidered. Two examples for those quantities follow.

10.5.3.1 Surface Curvature

Using special surface coatings and small band detectors, the surface emissivity is technically independent of the surface temperature and the wavelength. However, the angle of view θ still has to be considered. As a rule of thumb, angles of view above $\theta \geq 50°$ need more advanced correction procedures. Those angles may occur with curved surfaces of modern 3D geometries or by tilting the camera to the object of interest to allow different fields of view.

This change in emissivity with angle of view is different for each surface material and has to be determined in advance. For high-emissivity coatings, this change may be modeled with basic laws of electrodynamics; however, it is also strongly dependent on the surface's condition, e.g., paint thickness or paint roughness. For high accuracy at high angles of view, we recommend to determine a correction function for the specific surface or surface coating applied in the same batch with the same process as the actual test specimen. Based on Wen and Mudawar (2006), Elfner et al. (2017) proposed a function to model the calibration data for a normalized emissivity $\tilde{\varepsilon} = \dfrac{\varepsilon}{\varepsilon_\perp}$ (with $\varepsilon_\perp$ being the emissivity at $\theta = 0°$) using three constants C_0, C_1, and C_2:

$$\tilde{\varepsilon}(\theta) = 1 - \left(C_0 \frac{\pi}{\cos(C_1\theta)} + C_2 \right). \tag{10.23}$$

A simple way for determination of this function is moving the camera relative to the calibration specimen, as shown in Figure 10.9a. The acquired data and fits based on Eq. (10.23) are shown in Figure 10.9b for two object

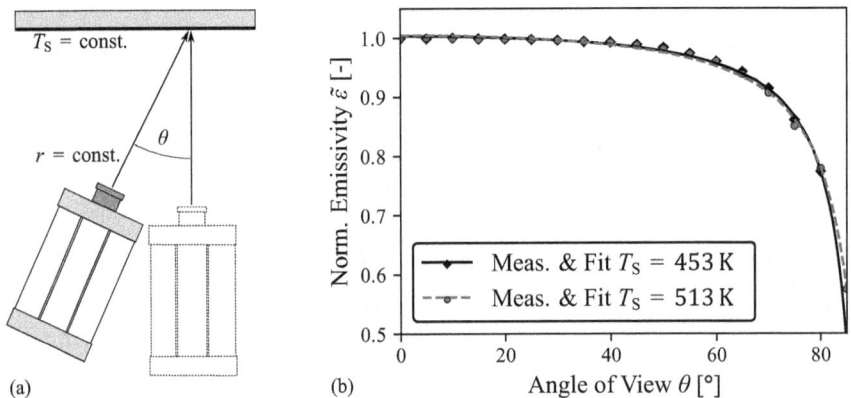

(a)   (b)

FIGURE 10.9
Determination of emissivity as a function of angle of view. (a) Test setup and (b) measurement and model fit for *Nextel Velvet Coating* at different temperatures.

temperatures. As stated before, this calibration is technically independent of the objects temperature[2] and thus has to be performed only once.

The change in emissivity then has to be included in the calibration procedure. Its effect is twofold: The emitted radiation from the test surface reduces due to the reduction in emissivity. The reflected radiation (the main contributor to I_{off}), however, increases due to $\varepsilon + \rho = 1$. Those effects may be included by using the inverse relations:

$$I_{D,corr} = I_D \cdot \tilde{\varepsilon}(\theta)^{-1} \text{ and } I_{off,corr} = I_{off}\tilde{\varepsilon}(\theta) \tag{10.24}$$

To avoid performing multiple in situ calibrations, the offset radiance I_{off} is determined as a virtual value at an angle of view of $\theta = 0°$ using the calibration pairs $\{I_{off, corr,i}, T_i\}$. The data calibration can then be performed similar to Eq. (10.22) by:

$$T_S(\theta) = \frac{b}{\ln\left(\dfrac{r}{I_{D,corr}(\theta) - I_{off}\tilde{\varepsilon}^{-1}(\theta)} + f\right)} \tag{10.25}$$

The spatially resolved nature of this procedure is given by $\theta = f(\vec{X})$, where θ can be computed with knowledge of the camera position $\vec{C}$ and the local surface normals of the test surface.

10.5.3.2 Non-Constant Offset Radiation

Infrared surface coatings are characterized by a technically perfect diffuse reflectivity (Dury et al. 2007). Thus, any hot structure reflected from the object of interest's surface, when using those coating, will be low-pass-filtered. This justifies the assumption of a scalar, spatially non-varying offset radiance I_{off} in most technical applications.

However, with complex test rigs, especially in closed setups, the need for a spatial correction may arise when external structures imprint on the test surface. A feasible procedure is the computation of a ray tracing for the test setup, solving for the irradiance at the test surface E_{inc}. This irradiance can then be normalized yielding a correction factor $\mathbb{C}$:

$$\mathbb{C} = \frac{E_{inc}(X)}{E_{inc,max}} \tag{10.26}$$

---

[2] As described below, the reflectivity changes with changing emissivity. Thus, this independence is only true when the emitted object radiation is much larger than the reflected radiation (usually the surroundings at ambient temperature). Therefore, we advise to perform this calibration at elevated object temperatures to mitigate effects from ambient radiation.

with reduced ray-tracing model error. This correction factor can be applied to the calibration function by:

$$T_S(\vec{X}) = \frac{b}{\ln\left(\dfrac{r}{I_D(\vec{X}) - \mathbb{C}(\vec{X})I_{\text{off}}} + f\right)}. \tag{10.27}$$

This ray-tracing approach, the correction procedure, and the resulting increase in measurement accuracy are described in more detail in Elfner et al. (2019).

10.6 Measurement Uncertainty

With all the presented corrections applied, a high-resolution, high-accuracy temperature map of an object's surface can be computed. In this final section, we will look at the residual error incorporated in the final data. Two main scenarios are analyzed: an open measurement and a closed measurement, considering small ($\theta = 20°$) and large ($\theta = 70°$) angles of view for the closed setup. We assume the calibration procedure with a linearization and a non-uniformity correction is performed. Using HDR has no influence on the measurement uncertainty. The temperature calibration is performed using the procedures presented in Sections 10.5.2 and 10.5.3. For all calculations, we assume a residual non-uniformity of $\Delta T_{\text{NUC}} = 0.3$ K and an uncertainty in temperature measurement for the in situ temperature probes of $\Delta T_{Pr} = 0.75$ K. The parameters used in the calibration functions were derived for a short-wavelength infrared system; thus, a temperature range of 340–600 K is chosen. This covers many technical applications.

The case with open measurement setup is analyzed first. The pre-calibration performed does not have to be scaled in situ and thus can be used as final calibration. The final error in measured surface temperature ΔT_S is computed as follows:

$$\Delta T_{S,\text{open}} = \frac{\partial T_s}{\partial I_D} \cdot \Delta I_D + \Delta T_{\text{NUC}} + \Delta T_{Pr} \tag{10.28}$$

With the uncertainty in detected irradiance of $\Delta I_D = 1\%$. Two different ranges of calibrations were computed and are displayed in Figure 10.10a: one for a low range of surface temperatures and one for a larger range and starting from a higher temperature.

The measurement uncertainty is, in general, low, usually below 1%. Two main features have to be mentioned: A calibration performed for a certain

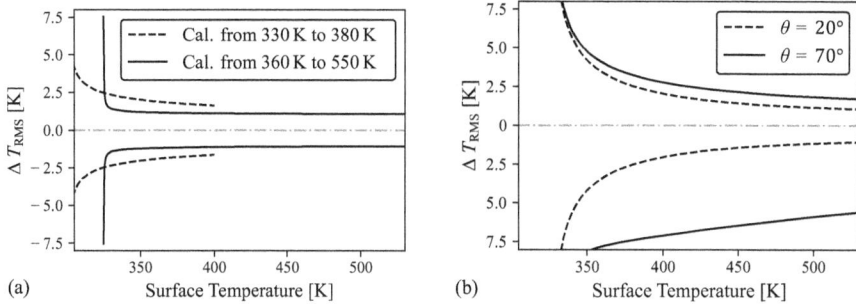

FIGURE 10.10
Different temperature calibration errors.

range of temperatures will perform best in this range. And, even more important, extrapolating to lower temperatures leads to large measurement uncertainty due to the high gradient of the calibration function in this range. Thus, it is recommended to build a calibration function covering the full range of expected temperatures and avoid extrapolating, even if the mean error may be marginally higher.

For the closed application (and hot boundaries) with angle of view dependence, the error can be computed in a comparable manner. In addition to the simple case, the uncertainty in offset radiation ΔI_{off} and the uncertainty in emissivity correction $\Delta \tilde{\varepsilon}$ have to be considered:

$$\Delta T_{S,\text{closed}} = \Delta T_{S,\text{open}} + \frac{\partial T_s}{\partial I_{\text{off}}} \cdot \Delta I_{\text{off}} + \Delta I_{\text{off}} + \frac{\partial T_s}{\partial \tilde{\varepsilon}} \Delta \tilde{\varepsilon}. \tag{10.29}$$

The uncertainty in offset radiation ΔI_{off} was determined from a multitude of calibrations from different experimental setups to a maximum of $\Delta I_{\text{off}} = 6\%$ (this value will change, most likely decrease, for a single specific setup), and the uncertainty in the emissivity correction $\Delta \tilde{\varepsilon}$ can be computed by using either the differential of Eq. (10.23) with an angular uncertainty of 2° when applying the correction or the absolute deviation $1 - \tilde{\varepsilon}$ when not applying the proposed correction.

Figure 10.10b shows the computed uncertainty for two angles of view (low: dashed line; and high: solid line), with and without correction (upper and lower plot area) for a calibration performed from 340 to 550 K.

For low angles of view (dashed lines), the error is higher compared to the open setup, especially at lower surface temperatures. At those temperatures, the ratio of offset radiation to object radiation strongly increases due to the T^4 relation, thus amplifying the effect of uncertainty in offset radiation. At higher surface temperatures, the uncertainties are comparable.

The effect of angle of view is clearly visible in the uncertainty plots for a high angle of view (solid lines). Without correction (lower area of Figure 10.10b), the uncertainty is higher (and, which is not shown here, spatially varying). With the angular correction applied (upper area of Figure 10.10 b), the uncertainty is only small amounts larger than for small angles of view. This reduces the global error, removes the spatial dependency from the measurement uncertainty, and thus allows the determination of surface temperatures on arbitrary surfaces. Due to the high nonlinearity in the correction model (Eq. (10.23)), this effect will strongly increase with angle of view: A curve for $\theta = 75°$ (without angular emissivity correction) already cannot be displayed with the axis ranges of Figure 10.10b.

10.7 Conclusion

Even under difficult conditions, infrared thermography can be used to determine high-resolution, high-accuracy surface temperature maps with errors below 2% for large ranges of surface temperatures. Using advanced techniques, even complex geometries can be analyzed. The calibration, however, has to be performed with great care. It is recommended to build or buy a calibration surface. Only then, the calibration may be performed for a specific surface condition, yielding the best measurement accuracy.

Many of the calibration steps can be conducted in advance and the calibration data stored in camera or for post-processing. With small drift in modern camera sensors, the time-consuming calibration procedures have to be repeated only in large time intervals, making the technique applicable and feasible in many situations.

References

Dury, M. R., Theocharous, T., Harrison, N., et al. 2007. Common black coatings: Reflectance and ageing characteristics in the 0.32–14.3μm wavelength range. *Optics Communications* Vol. 270, pp. 262–272. doi: 10.1016/j.optcom.2006.08.038.

Elfner, M. 2019. Assessment of new technologies for gas turbine rotor blade cooling. *Dissertation. KIT* 140 p. doi: 10.30819/4838.

Elfner, M., Glasenapp, T., Schulz, A., et al. 2019. A spatially resolved in-situ calibration applied to infrared thermography. *Measurement Science and Technology* Vol. 30, 9 p. doi: 10.1088/1361-6501/ab1db5.

Elfner, M., Schulz, A., Bauer, H.-J., et al. 2017. A novel test rig for assessing advanced rotor blade cooling concepts, measurement technique and first results. *In Proceedings of ASME Turbo Expo*, GT2017-64539. doi: 10.1115/GT2017-64539 Charlotte, NC, USA.

Elfner, M., Schulz, A., Bauer, H.-J., et al. 2018. Comparative experimental investigation of leading edge cooling concepts of turbine rotor blades. *In Proceedings of ASME Turbo Expo*, GT2018-75360 Oslo, Norway.

Evangelidis, G. D. and Psarakis, E. Z. 2008. Parametric image alignment using enhanced correlation coefficient maximization. *IEEE Transactions on Pattern Analysis and Machine Intelligence* Vol. 30(10), pp. 1858–1865. doi: 10.1109/TPAMI.2008.113.

Gubareff, G. G., Janssen, J. E., and Torborg, R. H. 1960. *Thermal Radiation Properties Survey: A Review of the Literature.* Minneapolis, MN: Honeywell Research Center, 293 p.

Harris, D. C. 1998. Durable 3–5 μm transmitting infrared window materials. *Infrared Physics and Technology* Vol. 39, pp. 185–201. doi: 10.1016/S1350-4495(98)00006-1.

Hartley, R. and Zisserman, A. 2003. *Multiple View Geometry in Computer Vision,* 2nd ed. New York: Cambridge University Press. doi: 10.1016/S0143-8166(01)00145-2.

Lowe, D. G. 2004. Distinctive image features from scale invariant keypoints. *International Journal of Computer Vision* Vol. 60, pp. 91–110. doi: 10.1023/B:VISI.0000029664.99615.94.

Martiny, M., Schiele, R., Gritsch, M., et al. 1996. *In Situ Calibration for Quantitative Infrared Thermography.* QIRT Eurotherm Series 50, Vol. 96.

Ochs, M., Horbach, T., Schulz, A., et al. 2009. A novel calibration method for an infrared thermography system applied to heat transfer experiments. *Measurement Science and Technology* Vol. 20(7), 9 p. doi: 10.1088/0957-0233/20/7/075103.

Ochs, M., Schulz, A., and Bauer, H.-J. 2010. High dynamic range infrared thermography by pixelwise radiometric self calibration. *Infrared Physics and Technology* Vol. 53, pp. 112–119. doi: 10.1016/j.infrared.2009.10.002.

Philipp, H. R., Le Grand, D. G., Cole, H. S., et al. 1989. The optical properties of a polyetherimide. *Polymer Engineering and Science* Vol. 29(22). doi: 10.1002/pen.760292205.

Schulz, A. 2000. Infrared thermography as applied to film cooling of gas turbine components. *Measurement Science and Technology* Vol. 11(7), 9 p.

Schulz, M. and Caldwell, L. 1995. Nonuniformity correction and correctability of infrared focal plane arrays. *Infrared Physics and Technology* Vol. 36, pp. 763–777. doi: 10.1016/1350-4495(94)00002-3.

Shi, Y., Zhang, T., Cao, Z., et al. 2005. A feasible approach for non-uniformity correction in IRFPA with nonlinear response. *Infrared Physics and Technology* Vol. 46, pp. 329–337. doi: 10.1016/j.infrared.2004.05.003.

Sobel, I. 2014. History and Definition of the So-Called Sobel Operator, More Appropriately Named the Sobel-Feldman Operator (Edition June 2015). In: Presentation at Stanford A.I. Project.

Thomas, M., Andersson, S., Sova, R., et al. 1998a. Frequency and temperature dependence of the refractive index of sapphire. *Infrared Physics and Technology* Vol. 39, pp. 235–249. doi: 10.1016/S1350-4495(98)00010-3.

Thomas, M., Joseph, R., and Tropf, W. 1988b. Infrared transmission properties of Sapphire, Spinel, Yttria, and ALON as a function of temperature and frequency. *Applied Optics* Vol. 27, pp. 239–245. doi: 10.1364/AO.27.000239.

Wen, C.-D. and Mudawar, I. 2006. Modeling the effects of surface roughness on the emissivity of aluminum alloys. *International Journal of Heat and Mass Transfer* Vol. 49, pp. 4279–4289. doi: 10.1016/j.ijheatmasstransfer.2006.04.037.

Wood, D. L. 1960. Infrared absorption of defects in quartz. *Journal of Physics and Chemistry of Solids* Vol. 13, pp. 326–336. doi: 10.1016/0022-3697(60)90017-2.

11

Optical Measurements for Phase Change Heat Transfer

Jungho Kim
University of Maryland

Iztok Golobič
University of Ljubljana

Janez Štrancar
Jožef Stefan Institute

CONTENTS

11.1 Introduction

Definitive understanding of phase change heat transfer mechanisms requires reliable local information for model testing and evaluation. Low-cost techniques to measure local heat transfer coefficient distributions are thus needed. Optical methods of temperature measurement have significant advantages over thermocouples, resistance thermometers, micro-heater arrays, and other point measurement techniques when optical access to the surface is available at the wavelengths of interest. Advantages include the ability to quantitatively measure the temperature distribution over large areas noninvasively at high speeds over a large temperature range without exposing the measuring instrument to harsh environments if measurements can be made through a window. Thermochromic liquid crystals have been used as temperature indicators and can accurately indicate whether or not a temperature threshold has been reached, but are not considered here due to their limited thermal response, limited temperature range, and degradation when exposed to UV light. In the sections below, a brief introduction to a technique commonly used in phase change heat transfer, infrared (IR) thermometry, is provided along with some references for readers who may be interested in this technique. A discussion of visible-light-based technologies and their advantages is then provided.

11.1.1 Infrared Thermography

A commonly used thermography technique in phase change heat transfer is based on IR in which the electromagnetic energy emitted by a substrate is used to measure temperature. Wavelength bands that exploit long-wave IR (LWIR, 7–14 μm), mid-wave IR (MWIR, 3–5 μm), and short-wave IR (SWIR, 0.4–1.7 μm) are available. LWIR cameras are generally uncooled and use bolometer arrays for imaging, but are noisier and less sensitive than cooled cameras and operate at low frame rates (10^1 Hz). MWIR cameras use semiconductor-based detectors that incorporate a cryocooler to minimize thermal noise. Spatial resolution, sensitivity (~20 mK), and frame rates (10^3 Hz) are much higher than for uncooled cameras, but their cost is much higher. SWIR cameras generally have uncooled but temperature-stabilized detectors and are useful for imaging at higher temperatures (>200°C). The majority of work in phase change heat transfer has used LWIR or MWIR cameras. Theofanous et al. (2002a,b) used a high-speed IR camera to measure the temperature distribution of a heated metallic film deposited on a sapphire surface during nucleate boiling and were perhaps the first to directly image wall heat transfer distributions during heater burnout.

IR temperature measurements in cases where the heated surface can be accessed directly without any intervening medium are fairly straightforward once the camera response has been calibrated. Typically, the in situ, pixel-by-pixel response for the camera is measured over the temperature range of interest. Golobic et al. (2009) and Stephan and co-workers (e.g., Schweizer and Stephan, 2009) measured the heat transfer distribution under single nucleating bubbles as they grew on thin metal foils. Stephan has also used a thicker CaF substrate in place of the thin film in order to increase the heat capacity of the substrate, so it is more representative of real surfaces (e.g., Fischer et al., 2012). Gerardi et al. (2010) used a high-speed IR camera in conjunction with a video camera to measure the bubble behavior on an ITO heated sapphire substrate—the IR camera measured the temperature distribution at the ITO surface, while the video camera was used to visualize the fluid behavior. Krebs et al. (2010), Shen et al. (2010), and Mani et al. (2011) used IR thermography to study flow boiling in micro-channels, droplet evaporation, and jet impingement, respectively. In these studies, an IR camera was used to view through a silicon substrate to visualize the temperature distribution at the silicon–water interface. Sefiane et al. (2008) visualized the spontaneously occurring hydrothermal waves within evaporating methanol, ethanol, and FC-72 droplets. Solotych et al. (2014, 2016) measured heat transfer distributions within plate heat exchanger geometries produced by machining chevron patterns into IR-transparent CaF_2 plates. Jung and Kim (2014) combined IR thermography, total internal reflection, and interferometry to study how a surface dries out as the heat flux is increased during pool boiling. These studies all required the use of IR-transparent substrates (e.g., CaF_2 and sapphire) in the wavelength used when direct optical access to the surface was not available.

If quantitative measurements are to be made in cases where one or more media are interposed between the heated surface and the IR camera, emission, reflection, transmission, and absorption of the media must be accounted for. Kim et al. (2012) described a coupled transient conduction/radiation technique whereby transient temperature and heat transfer distributions at the wall could be determined given the optical properties of the media. The technique was demonstrated for drop evaporation, pool boiling, and flow boiling, and a technique to measure the liquid film thickness in a tube was described. Bucci et al. (2016) extended this technique to account for the spectral dependence of the media and validated it using transient pool boiling experiments.

11.1.2 Optical Techniques (Visible Light)

Optical techniques based on visible light (0.4–0.7 μm), such as thermoreflectance and luminescence, enable the use of substrates such as glass and plastics that may block IR but are transparent to visible light and allow the use of widely available cameras and microscopes. The spatial resolution is much higher than with IR due to the use of visible light. Thermoreflectance has been used to measure the temperature and thermal properties of thin films. Since the setup required can be quite complicated and expensive and is generally not applicable to the complex geometries that can arise in boiling heat transfer research, this technique will not be further considered in this paper.

Luminescence-based temperature measurements exploit the temperature-sensitive emission of light when a material is excited by the absorption of photons. Luminescence relies on using a pulsed or continuous excitation source to stimulate emission of light from molecular probes deposited within or onto the surface of a substrate. Excitation can occur through the absorption of a single high-energy photon that results in transition of an electron in its ground state to an excited state, followed by emission of a photon at a lower energy (longer wavelength) as the electron decays back to the ground state. Excitation can also occur through the absorption of two lower-energy photons followed by emission at a higher energy (shorter wavelength). The latter can be used to prevent damage to the luminescent material or substrate by lowering the energy of the absorbed photons, or to confine the emission region to smaller spatial scales by sharper focusing of the excitation beam due to nonlinear excitation. The luminescence lifetime is the average time a molecule remains in an excited state before emission and can vary from nanoseconds to hours. Allison and Gillies (1997) refer to fluorescence as a type of luminescence whose duration is typically ~10^{-9} to 10^{-3}s, while the duration of phosphorescence is longer, ~10^{-3} to 10^{3}s. Since thermal energy can also cause the excited electrons to decay back to the ground state without emission of a photon (thermal quenching), both the lifetime and intensity of the emission are decreased with increasing temperature. Measurement of either lifetime or intensity can thus be used to infer the temperature. Desirable characteristics of such molecular probes include high emission intensity and large change in emission with temperature, stability with time, insensitivity to other factors such as oxygen concentration (oxygen quenching), and non-toxicity. The density of the probe across the substrate should be as uniform as possible.

Two techniques are typically used to measure emission lifetime (Figure 11.1). In the phase lag technique, the intensity of the excitation light is varied sinusoidally and the emission is measured—the phase difference ϕ between the two can be used to determine the lifetime. The decay time technique involves pulsing the excitation light source using a square wave and measuring the characteristic rise or fall time (τ) of the emission intensity. The frequency of the excitation light in both cases needs to be on the order of the lifetime, so high-frequency light sources and cameras can be required. Measurement at single points can be made at much higher frequencies using high-speed photomultipliers.

The emission intensity is characterized by the quantum yield (QY) defined as the number of emissions per photon absorbed by the molecule. Since the intensity of the emission can depend on numerous factors such as

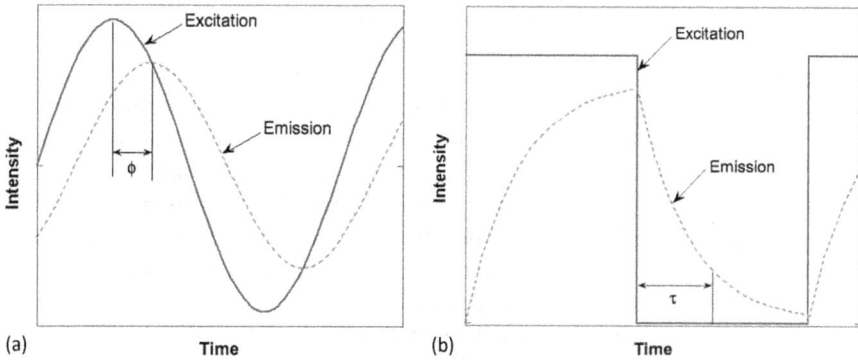

FIGURE 11.1
Phase lag technique (a) and decay time technique (b) for measuring emission lifetime.

illumination intensity, viewing angle, QY, reflections within the substrate, and other sources of emission (e.g., from the substrate itself), a ratiometric technique is often used where emission from the temperature-sensitive probe is normalized on emission from another reference probe with a different (usually lower) sensitivity, thereby accounting for many of the above factors. The change in emission intensity with temperature can be quite large, enabling temperatures to be measured with high accuracy. Many probes are available and are characterized by working temperature range, excitation and emission wavelengths, toxicity, stability, and quantum efficiency. Photobleaching, in which the molecular probe is damaged by the excitation light over time, can affect the rate at which each probe degrades and should be accounted for. The reader is referred to numerous books (e.g., Liu and Sullivan, 2005), review papers (e.g., Wang et al., 2013; Allison and Gilles, 1997), and theses (e.g., Kurits, 2008; Bhandari, 2012) that detail the available probes and their applications. The use of fluorescence techniques to measure local temperature and heat flux distributions during pool and flow boiling is discussed in the remainder of this chapter. All three cases are based on measuring changes in emission intensity with temperature.

11.1.2.1 Fluorescence-Based Temperature Mapping

While almost every molecule or atom can absorb a photon of UV or visible light, thereby forcing the molecule or atom into an excited state, only few of them relax from such an excited state with the delayed emission of a photon. In any case, several subsequent fast vibrational transitions are involved in any kind of relaxation, releasing at least part of the absorbed energy into atom/molecule vibrations, thus resulting in the energy of the emitted photon being lower than the energy of the absorbed photon. This phenomenon is known as *fluorescence*, and the energy decrease is known as the *Stokes shift*.

The energy of the aforementioned relaxation from the excited state into a lower (usually ground) electron state depends on the occupancy of the many vibrational states associated with the lower (ground) electron state. Since the latter depends on temperature (the Boltzmann distribution), the probability and sometimes the energy of the emitted fluorescence photon depend on temperature as well. In practice, this can be detected as temperature-dependent fluorescence emission spectra (Figure 11.2). It is seen that various transitions (involved in the emissions at different wavelengths) can have different, even opposite, temperature coefficients: While some of them increase with temperature (e.g., emission peak at 525 nm), others can decrease with temperature (e.g., emission peak at 545 nm). By measuring the peak intensity of a peak, the local temperature at the location of the emitting atom or molecule can be measured. It should be noted that this temperature dependence often relies on the excitation wavelengths as well.

If the widely spread broadband detection (of 50 nm or a broader emission filter width) of fluorescence with a temperature dependence shown in Figure 11.2 is used, the temperature sensitivity would decrease to almost zero, because the increased fluorescence at lower wavelengths would compensate for those at higher wavelengths. On the other hand, if the emission spectrum can be resolved and integrated separately over the lower and higher emission peaks, or band-pass filters are used to selectively measure emission over an emission peak, the ratio between the two integrals would result in a much higher sensitivity to the temperature and the accuracy of the temperature measurement would increase. Hyperspectral cameras which have different

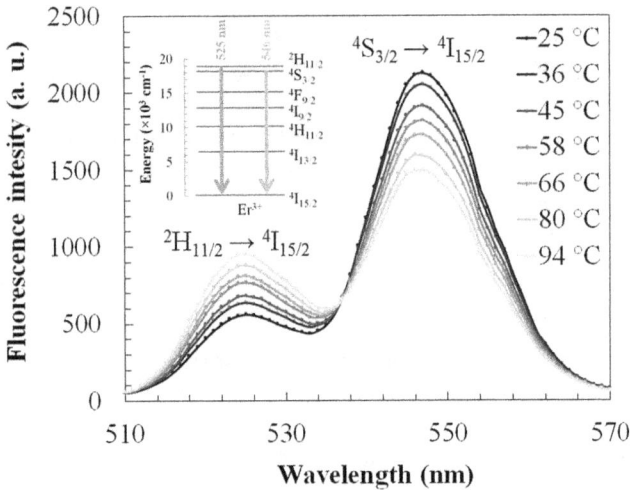

FIGURE 11.2
Fluorescence emission spectra versus temperature for the Er:GPF1Yb0.5Er (Sedmak et al., 2015).

filters installed in front of different pixels are becoming available and can be used to obtain the spectrum at each pixel with sufficient sensitivity. The shift in the peak wavelength can also be used to measure temperature and has the advantage that the measurement is insensitive to the illumination light intensity. Since hyperspectral cameras are not, as of yet, sensitive enough to resolve the small shift in the peak wavelength, only point measurements are currently practical.

11.1.2.2 Temperature-Sensitive Paints

Fluorescent paints that are sensitive to oxygen concentration (pressure-sensitive paints, PSP) and temperature (TSP) have been used to measure shear stress, surface pressure, and wall temperature distributions in aerodynamic and fluid mechanics applications since the 1980s (e.g., Campbell et al., 1994; Ozawa et al., 2015). It has seen very limited use for phase change applications (e.g., Shibuya et al, 2016), however, perhaps because alternate techniques such as IR have been available. If only temperature is to be measured, the paint is encapsulated within a binder or between impermeable barriers to eliminate sensitivity to oxygen. The molecules are typically stimulated using blue-UV light, and emission at a longer wavelength occurs. The intensity of the emitted light decreases with temperature and oxygen concentration, and a filter is used to remove the excitation light before the emitted light is measured by a camera. The emission lifetime is typically on the order of microseconds, much faster than the boiling processes being measured, so the system frequency response is usually limited by the thermal response of the materials onto which the paint is bonded.

The advantages of TSPs over infrared thermography and other techniques such as micro-heater arrays or micro-bolometers to measure temperature are numerous. Readily available CCD or CMOS cameras can be used to measure the intensity distribution instead of expensive IR cameras. Since the TSPs themselves respond very quickly (typically on the order of MHz or greater), the transient response of TSP measurements is usually limited by the speed of video cameras, which are typically much less expensive than IR cameras. Also, since visible light is used for illumination and sensing, glass, clear plastics, PDMS, and other commonly used materials can be used as substrates in place of expensive and difficult-to-machine IR-transparent materials (e.g., silicon, sapphire, CaF_2). Calibration can be performed without having to consider emission from the transparent substrate they are placed on, unlike for IR measurements. Finally, the wavelength of the light used (~0.5 μm) is much smaller than the 3–5-μm or 8–12-μm light that IR cameras are sensitive to, allowing a much higher spatial resolution. The diffraction limit, given by $d = \dfrac{\lambda}{2NA}$, for the 0.6-μm light in air is ~0.3 μm. Since the emission intensity is dependent on the illumination intensity, stable light sources are required

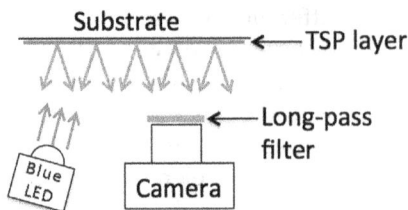

FIGURE 11.3
Schematic of TSP measurement system.

and surfaces need to be opaque to light when performing two-phase measurements since the reflection of excitation light from passing bubbles can change the local excitation intensity. TSPs are inexpensive (~$100/g from Sigma-Aldrich), and formulations that are sensitive to wide temperature ranges are available.

In order to measure the unsteady temperature at many points over a large area, the system shown schematically in Figure 11.3 is used. Limitations on the measurement accuracy are dependent on fluctuations in the excitation light intensity, TSP concentration and brightness, and camera noise. Low-cost UV LEDs have become widely available and can be driven by stable power sources to provide constant illumination. A photomultiplier (not shown) can be used to monitor the illumination intensity and correct for any fluctuations that do occur as the LEDs warm up or age. The LEDs should be positioned to illuminate the TSP as uniformly as possible, and the TSP concentration must be high enough to provide sufficient brightness. The inherent thermal noise within cameras is not of concern when steady-state measurements are desired since the noise can be decreased to acceptable levels by time-averaging over many frames. For transient measurements, noise can be decreased by spatially averaging pixels together at the cost of decreased spatial resolution, and/or sampling at a frequency higher than the desired frequency and time-averaging the noise. Noise decreases as $1/\sqrt{N}$, where N is the factor increase in the number of samples.

11.1.2.3 Phosphor Thermography

Phosphors are inorganic powders that luminesce when excited, and are used in fluorescent lights, cathode ray tubes, and white LEDs, among other applications. Phosphor thermography relies on techniques similar to those for TSP. An excellent introduction to the principles of the phosphor thermography as well as instrumentation-related aspects is given by Allison and Gilles (1997). Phosphors have been used as temperature sensors in many applications, including measurements in aerosol thermometry, combustion, HVAC systems, gas turbine hot section components, cryogenics, and manufacturing. A significant advantage of using phosphors is that they are much less sensitive to photobleaching.

11.2 Pool and Flow Boiling Measurements Using Temperature-Sensitive Paints

The descriptions of two test sections to measure heat transfer during pool and flow boiling of a refrigerant using temperature-sensitive paints are given below. Typical performance and system limitations are described.

11.2.1 TSP Formulation and Characterization

A TSP based on ruthenium tris(1, 10-phenanthroline) dichloride obtained in crystal form from Sigma-Aldrich was used. This fluorophore has a high temperature sensitivity (on the order of -2%/°C), has a short decay time (<1 µs) (e.g., Mills, 1997), and can be excited using UV LEDs. It tends to exhibit a large shift in the spectral peak, so it can also be used as a temperature indicator if desired. The TSP solution consisted of 40 mg Ru compound and 100 mg polyacrylic acid dissolved in 0.5 mL of a solvent (ethanol, isopropanol, etc.). Ozawa et al. (2015) recommended the Ru compound be dissolved in ethanol with a molarity of 0.16 mol/L, while Huang (2006) suggested a ratio of Ru to polyacrylic acid of 1:2. The TSP solution can be spread into a thin film using various techniques (Mayer rods, spinning, aerosol spray, etc.) in a dust-free enclosure. The final TSP thickness for our formulation was 1–2 µm.

11.2.1.1 TSP Characterization

The emission from the TSP measured by a camera (I) is a function of the local TSP concentration (c), temperature (T), LED excitation intensity (I_{LED}), and camera capabilities:

$$I = f(c, T, I_{LED}) \tag{11.1}$$

Calibration of the TSP in the current tests was performed over the temperature range of interest at each pixel at a minimum of two UV LED intensities, so any variation in excitation intensity due to aging or self-heating could be corrected for. A photomultiplier (PMT) was used to monitor the changes in the UV excitation intensity.

11.2.1.2 Temperature and Pressure Dependence

The spectral characteristics of the LED emission and TSP emission wavelengths along with the cutoff spectrum for the long-pass filter are shown in Figure 11.4. A typical calibration curve for two pixels in a CMOS camera (Phantom Miro eX4) is also shown. The intensity averaged over 100 frames decreases with temperature. The temperature resolution was quite high

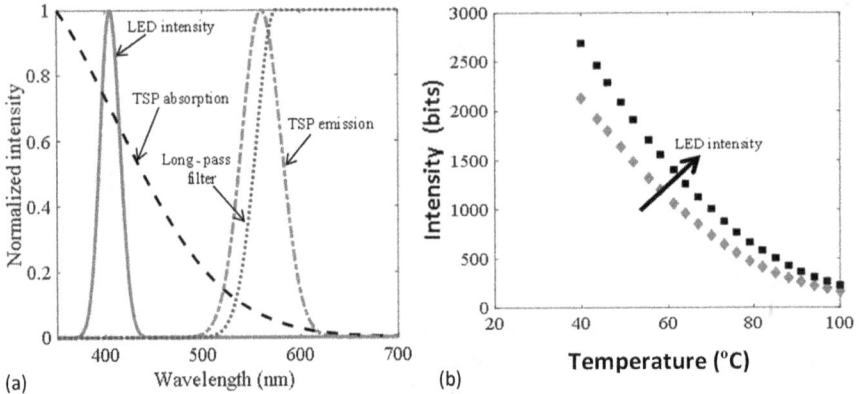

FIGURE 11.4
Spectral and emission intensity characterization. (a) Spectral characteristics of the emission and excitation sources. (b) TSP intensity characteristics with temperature and various LED intensities for two representative pixels.

even where the sensitivity is lowest (~16 bits/°C at 97°C), indicating accurate steady-state measurements can be obtained.

Camera noise can be a significant source of uncertainty for unsteady measurements. Based on the noise and the calibration, the uncertainty in temperature (95% confidence interval) at a given pixel for a mean intensity of 4,000 bits for various uncooled cameras was ±2°C for Sentech (STC-MBCM200U3V), ±0.5°C for Point Grey (FL3-U3–13S2M-CS), and ±0.7°C for Phantom (Miro eX4). Cooled cameras would, of course, have significantly lower noise. The Phantom camera was chosen for the TSP measurements due to its ability to acquire data at higher speeds and spatial resolution with relatively little noise. The variation in TSP intensity was characterized over a range of pressures between vacuum and 2 bar. The measured intensity variation was within the range of the camera noise, indicating pressure has a negligible effect on emission intensity.

11.2.1.3 Aging of TSPs

TSPs can be damaged by exposure to UV and result in lower emission in a phenomenon referred to as photobleaching. For the TSP and conditions used below, photobleaching resulted in a 2%–3% reduction in TSP intensity over 2h, indicating that exposure to UV should be minimized during both calibration and testing.

11.2.2 Pool Boiling Test Apparatus and Demonstration

The TSP was used to measure the temperature variation during pool boiling of HFE-7000 ($C_3F_7OCH_3$, T_{sat} = 34°C at 1 atm). Comparisons of the heat flux derived from the temperature with CHF correlations and energy balances were used to validate the data.

11.2.2.1 Test Section Construction

The test section is shown schematically in Figure 11.5. Germanium dots 200 nm in thickness and 400 μm in diameter were deposited using a liftoff technique in a 3 mm × 3 mm pattern onto one side of a 25-μm-thick adhesive (3M 8146) to serve as an opaque layer. TSP was brushed onto the germanium dots, then the adhesive was attached to the sapphire substrate, so the temperature distribution at this surface could be measured. A TSP layer was then deposited onto one side of a 5-μm double-sided tape (3M 82600) using a Mayer rod (R.D. Specialties, #3) and then attached to the 25-μm adhesive to encapsulate the TSP. To form an opaque heating layer, a sheet of copper (18 μm) onto which a submicron-thick layer of nichrome (NiCr) was deposited (Ticer Technologies TCR foil) was attached to the top of the 5-μm adhesive, which was then etched using hydrochloric acid free from ferric chloride to selectively remove the copper, leaving a NiCr film with an electrical resistance of ~25 Ω/square. The NiCr film was capable of high heat fluxes and blocked any UV light reflected from bubbles, thereby ensuring the steady illumination of the TSP. The 5-μm adhesive between the TSP and NiCr minimized the temperature drop between the temperature measurement and heat generation regions, while the 25-μm adhesive served to amplify the temperature variation across the film produced by boiling so that it could be measured. A DC power supply was used to heat the NiCr heater, which was measured using a four-wire technique. A long-pass filter (Midopt LP590-58) was placed in front of the camera to block any extraneous light from the UV LED (Cree Semi-LED; 400 nm ultraviolet). A PMT (Hamamatsu R3788) was used to monitor and correct for variations in the LED output. A short-pass filter (Alluxa 430 OD6 SP) placed in front of the PMT filtered out the TSP

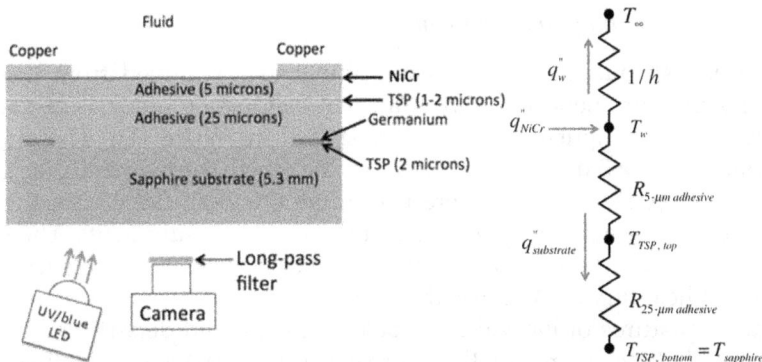

FIGURE 11.5
Thermal resistance circuit representation.

emission. The test section was placed within a black enclosure to minimize external light influence.

11.2.2.2 Data Reduction

The local surface heat flux was computed as a 1D inverse heat conduction problem since the in-plane temperature gradients within the adhesive were much smaller than the normal gradient. The substrate was represented using the thermal circuit shown in Figure 11.5. The temperature variation within the TSP layer was neglected since it was so thin. An energy balance was used to obtain the wall-to-fluid heat transfer. The heat generated in the NiCr layer (q''_{NiCr}) was obtained from the voltage and current through the heater and the heater area. Part of this heat was lost through substrate conduction ($q''_{substrate}$), which could be computed from the measured values of $T_{TSP,top}$ and $T_{TSP,bottom}$ using an unsteady heat conduction code. The substrate conduction is typically small compared to the heat transferred into the fluid due to the low thermal conductivity of the adhesive (~0.15 W/m K) and the large heat transfer coefficients during phase change. The temperature at the adhesive–sapphire interface was obtained by interpolation between the temperatures measured using the dots, spaced 3 mm×3 mm apart. This temperature was quite uniform relative to the wall–fluid temperature difference due to the high thermal conductivity of the sapphire. The transient temperature profile within the 25-μm adhesive at each pixel was computed, which was used to obtain the heat conducted into the substrate. This heat loss was subtracted from the heat generated within the NiCr film in order to obtain the wall heat flux, which corresponds to the heat liberated into the fluid. The heat flux into the fluid (q''_w) is then $q''_w = q''_{NiCr} - q''_{substrate}$. The wall temperature ($T_w$) was computed from $T_{TSP,top}$ and correcting for the temperature drop across the 5-μm adhesive. A description of the calculations can be found in a previous work by Moaveni and Kim (2017).

11.2.2.3 Pool Boiling Measurements

Heat flux distributions during pool boiling were obtained from the TSP intensity measurement and a 1D transient conduction analysis at various values of heater input power, and are shown in Figure 11.6. The circular spots spaced on a 3 mm × 3 mm grid indicate where the temperatures of the adhesive–sapphire substrate were measured. Small cracks within the TSP are seen, but these had negligible effect upon the measurements. The small circular dots are where bubbles actively nucleate and depart the surface. At the lowest heat flux (1.5 W/cm²), the many small bright spots of higher heat flux are signatures of individual bubbles growing and departing from the surface. Bubbles cover nearly the entire surface by 9.0 W/cm². The first signs of bubble coalescence resulting in larger dry spots (darker, lower heat flux regions) occur at 14.7 W/cm², which corresponds to the change in slope in the

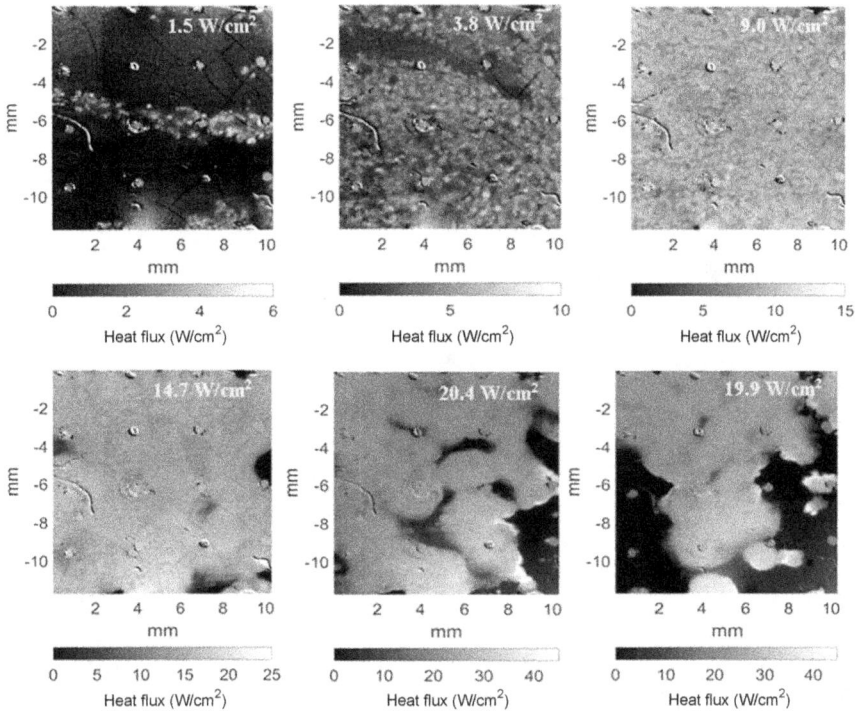

FIGURE 11.6
Surface heat flux distribution at various heat loads.

boiling curve. The surface becomes increasingly covered by dry patches as CHF is reached at ~20.4 W/cm^2.

11.2.3 Flow Boiling Test Apparatus and Demonstration

11.2.3.1 Test Section Construction

A test section that enables the measurement of heat flux and temperature as well as flow visualization during flow boiling is shown in Figure 11.7. The fluid flowed through a 6-mm-ID/8-mm-OD sapphire tube. The test section was oriented vertically, and the inlet stream and outlet stream could be reversed to achieve both upward flow and downward flow. A polyethylene terephthalate film with Carbon NanoBud conductive layer (Canatu Carbon NanoBud (CNB) Flex Film, 100 W/sq) was attached to the outside of the sapphire tube, so it could be heated. One half of the inner surface was covered with a tape (3M 8911) onto which TSP and a NiCr layer were attached, so heat transfer could be measured. The NiCr was used in this application as an opaque layer only. Germanium/TSP dots were used to measure the temperature at the 3M tape–sapphire interface. The other half of the tube was

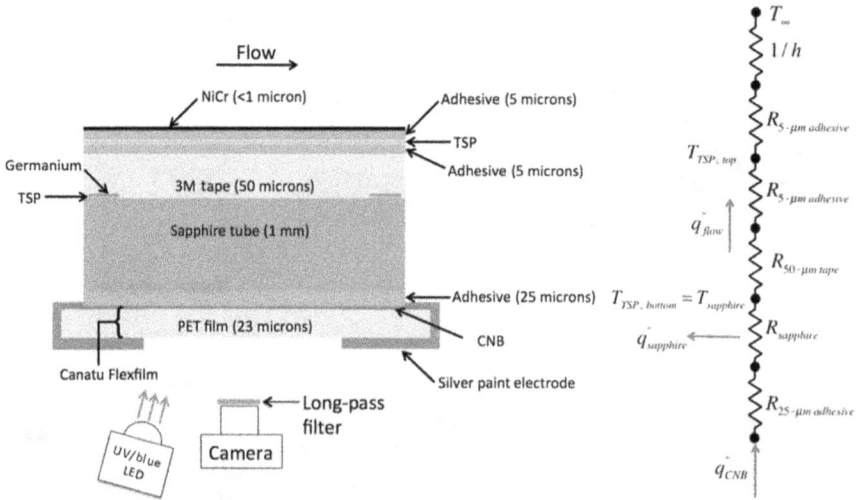

FIGURE 11.7
Schematic of flow boiling heat transfer measurement section and thermal circuit.

covered using a similar laminate, but without the TSP/adhesive/NiCr and allowed for flow visualization. Silver paint was used to make electrical connections to copper electrodes on either end of the heaters. The Phantom camera was used to measure the TSP intensity, and a second camera was used to simultaneously visualize the flow.

UV LEDs were used to excite the TSP, and the temperature was measured using a camera and a long-pass filter. A photomultiplier was used to monitor and correct for any variations in UV light intensity, and the UV was only turned on for 3 s during each run to minimize long-term drift due to photobleaching. Green LEDs were chosen to illuminate the visual side of the tube of the test section since any light leaking through the NiCr layer would not excite the TSP layer and cause extraneous emission, and since the long-pass filter could effectively remove any light that did leak through. It was verified that the measured temperature was independent of whether or not the green LEDs were on.

11.2.3.2 Data Reduction

The thermal circuit shown in Figure 11.7 was used to compute the local heat transfer distribution. Heat generated in the CNB (q''_{CNB}) was conducted into the sapphire through a 25-μm adhesive. Some of this heat could be lost by axial conduction ($q''_{sapphire}$), so the heat into the flow (q''_{flow}) was computed from the temperature difference across a known thermal resistance (the tape) using $T_{TSP,top}$ and $T_{TSP,bottom}$ and an unsteady heat conduction code. 1D heat conduction could be assumed since the spatial resolution of the temperature measurements was 160 μm, about three times as large as the thickness of the 50-μm tape, and since the in-plane temperature gradients within the tape were much smaller than the normal gradient. The measurement technique was validated by comparing single-phase, steady-state data with correlations.

11.2.3.3 Flow Boiling Measurements

Heat transfer distributions for flow boiling during downward flow are shown in Figure 11.8. For the conditions shown, bubbles generated at the wall occasionally merged to form a Taylor bubble that slowly grew as it moved upward against the flow. The heat transfer steadily increased above the bubble as the bubble population increased (upper portion of the tube for

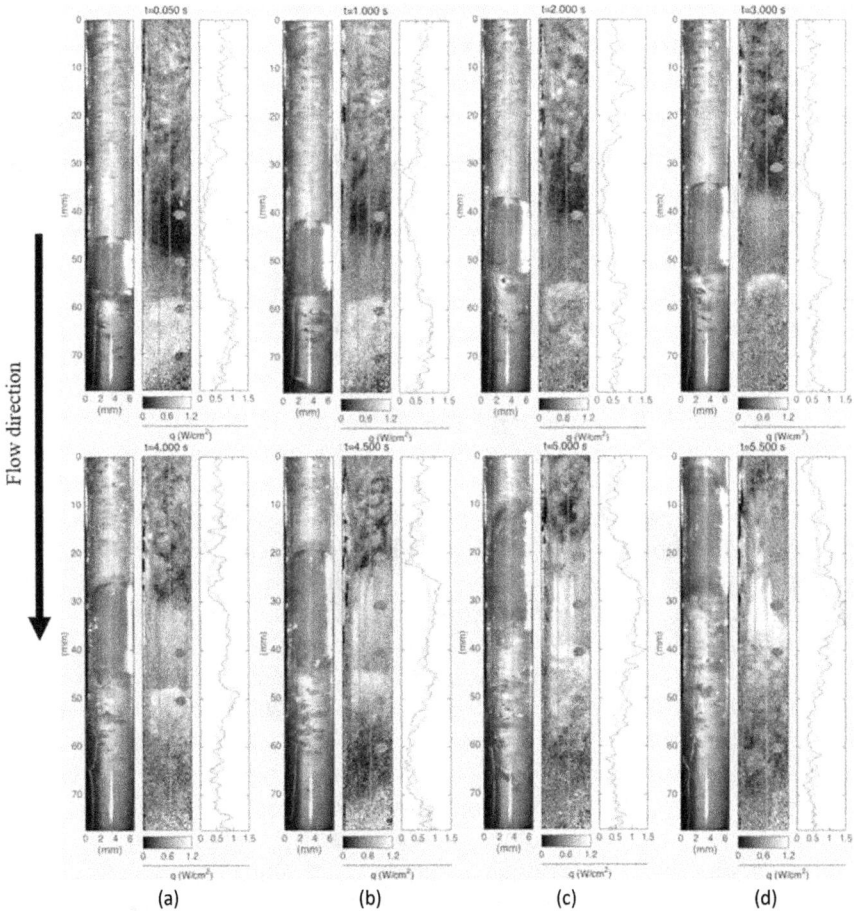

FIGURE 11.8
Visualization (a) and heat flux (b)–(d) vs. time during flow boiling of HFE-7000 during downward flow at 100 mL/min, 7.1°C subcooling, 0.74 W/cm², and $P = 0.89$ bar.

$0.05 < t < 4.0\,\text{s}$). The stagnation region just above the bubble resulted in a low heat transfer region, but the heat transfer increased as the liquid accelerated around the bubble and fell along the tube wall. This film then plunged into the liquid behind the bubble, resulting in turbulent mixing and a spike in heat transfer. Nucleation within the film occurred when the Taylor bubble passed over nucleation sites ($4.5 < t < 5.5\,\text{s}$), and the wake behind the nucleating bubbles resulted in streaks of high heat transfer within the film. The heat flux of $0.74\,\text{W/cm}^2$ for this case was much lower than for the pool boiling data, resulting in much more noise since the temperature variations are correspondingly smaller. Quantitative data could easily be obtained, however.

11.3 Pool Boiling Measurements Using Fluorescence Microscopy

The use of fluorescence to measure wall heat transfer during pool boiling on the submicron-scale is demonstrated below. An erbium/ytterbium-co-doped transparent glass–ceramic sample (Er:GPF1Yb0.5Er) was used as the substrate, and the fluorescence in a plane very close to the surface–fluid interface was measured using confocal microscopy. The substrate itself served as a 2D temperature probe since it was sensitive to heat transfer coupled to the boiling process occurring at the surface of the glass. The emission spectra for the glass–ceramic are shown in Figure 11.2. A typical setup for spectrally resolved microscopy with a liquid crystal tunable filter (LCTF) in front of a detector, called fluorescence micro-spectroscopy is shown in Figure 11.9. With the ability to resolve the emission spectra in every pixel of an image, the peak intensities (or even better, their ratio) could be calibrated against the temperature. To demonstrate the concept of transient temperature mapping during localized heating, an electrically heated tip was brought into contact with the surface. The effect of the tip's heating on the transparent glass–ceramic substrate is shown in Figure 11.10. Temperature gradients of up to $20\,\text{K/}\mu\text{m}$ could be detected by spatially analyzing the fluorescence emission spectra pixel-wise. The technique can be used to map local temperature variations, e.g., during the nucleate boiling process, where rapid temperature drops during bubble growth are observed. This technique can be used to advantage since processes on the submicron-scale affect the overall boiling performance and the current observation and measurement methods lack sufficient spatial resolution to fully characterize the boiling process.

While temperature mapping with submicron and few-K resolution can open up new opportunities in understanding the nucleate boiling process on the submicron-scale, a number of challenges existed with regard to the experimental setup. First, to avoid a large heat flux reaching the objective of the microscope, a considerable working distance was needed reducing

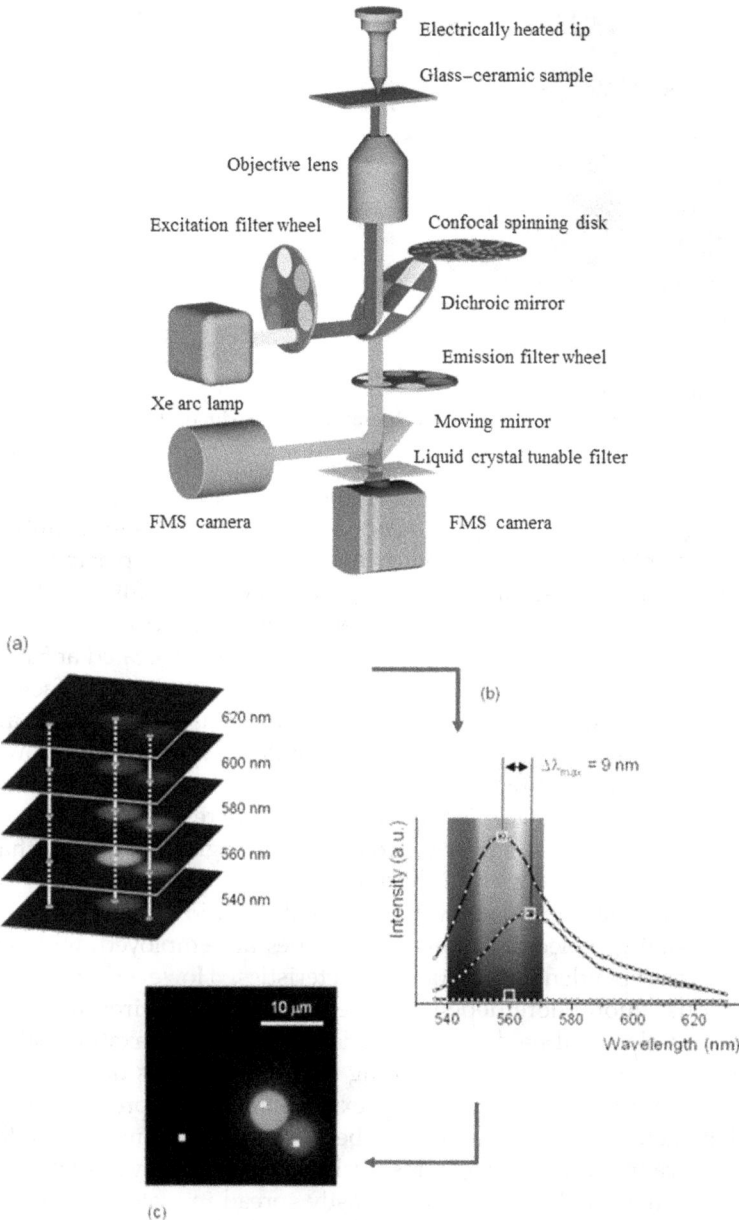

FIGURE 11.9
Spectrally resolved fluorescence imaging, published as a system for fluorescence microspectroscopy, implements a liquid crystal tunable filter (LCTF) as a tunable narrowband-pass filter device enabling the detection of the fluorescence emission spectra in each pixel of an image by acquiring a wavelength stack and image stack post-processing (Sedmak et al., 2015; Arsov et al., 2011; Urbančič et al., 2013). (a) Wavelength (lambda) stack, (b) fluorescence emission spectra, and (c) spectrally contrasted image.

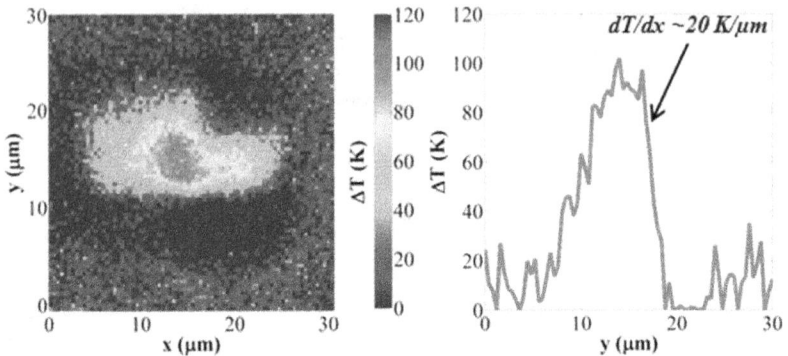

FIGURE 11.10
T-mapping of the transient micron-sized temperature field (Sedmak et al., 2015).

the spatial resolution. Second, to match the speed of the bubble motion during boiling, the acquisition speed needed to be increased significantly. This was implemented partially by removing the LCTF and replacing it with a single narrowband-pass filter (keeping the same 10-nm filter width), thus detecting the fluorescence of only that emission peak, which has the highest coefficient of temperature dependence. (The peak is located at 545 nm in the case shown.) This reduced the temperature sensitivity by a factor of 2, but enabled high-speed acquisition by the camera using a reduced area of interest. Combining the aforementioned solutions, Figure 11.11 shows that it is possible to conduct temperature mapping at 400–500 fps with a spatial resolution of around 0.5–0.74 μm and a temperature resolution of around 1–2 K. The spatial resolution is nearly two orders of magnitude greater than the resolution of commonly used IR thermography.

Temperature measurements within the boiling fluid are much more challenging. Usually, molecular fluorescent probes are employed, with similar temperature-dependent emission characteristics. However, the emission of individual fluorophore depends more on the local environment, including polarity, electrical fields, concentration, and other relaxation pathways, than on the temperature itself, making molecular probes questionable for local temperature mapping. Similarly, extreme pH or the presence of strong oxidative agents or radicals can also be destructive for molecular fluorophores. Entrapment of fluorophores such as Eu in the crystal lattice of an inorganic nanoparticle which can be easily spread in water can be used to address some of these issues. Eu atoms within hydrophilic TiO_2 exhibiting temperature-dependent fluorescence emission are shown in Figure 11.12.

Temperature mapping can be brought to another level with nano-sized temperature probing in the form of nanoparticles with the emission being insensitive to the local environment and with the possibility of redistribution within the entire system, including complex systems such as a living cell. While high spatial, temporal, and temperature resolutions have previously

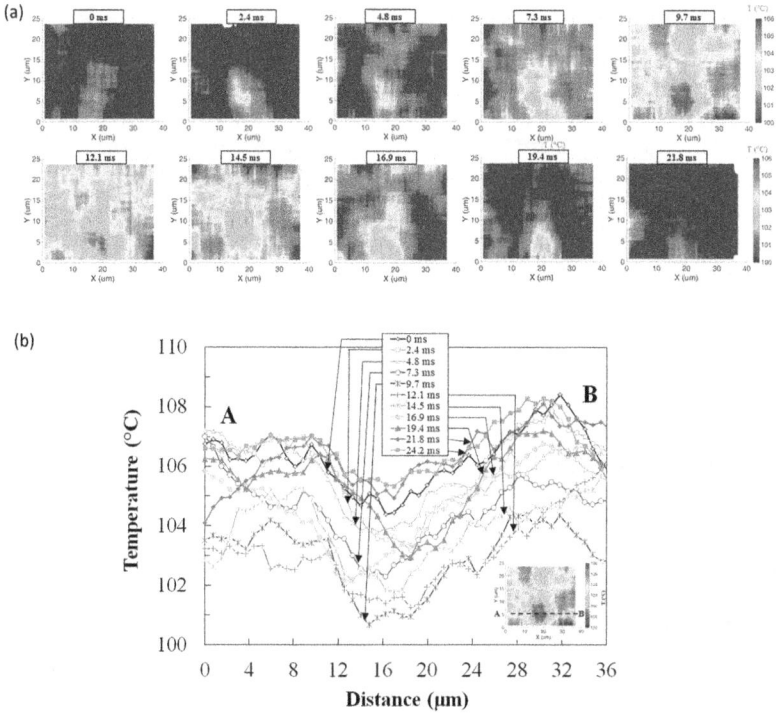

FIGURE 11.11
(a) T-mapping of an individual bubble nucleation event at ~400 fps. (b) Horizontal temperature profile across the observed temperature fields of a growing micro-bubble (Sedmak et al., 2016).

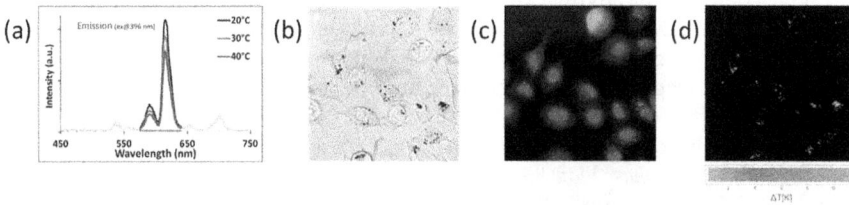

FIGURE 11.12
(a) Emission spectrum of Eu-doped TiO_2 nanotubes and its part used in temperature mapping (d) of a fibroblast L929 cell sample, which is shown in bright field (b) and as a two-channel false-color fluorescence image (c). The nanoparticle location (with Eu signal) is shown with a red color in (c), its associated temperature (relative to the average in the sample) is shown in (d).

been achieved within a substrate just below the interface, the latest application suggests the same can also be achieved within the entire system under observation. Using optical sectioning of a confocal fluorescence microscope, even 3D mapping is theoretically possible with, of course, a much lower temporal resolution.

11.4 Conclusions

The application of optical techniques based on temperature-sensitive paints to phase change heat transfer measurements has been described. The TSP technique uses low-cost fluorophores, low-cost cameras, provides a high spatial resolution, and can be used with many transparent substrates. The frequency response is, in practice, only limited by the speed of the camera. TSP measurement technique was validated for pool boiling heat transfer measurements, including CHF. It was also validated and demonstrated for flow boiling measurements and used to measure local temperature and heat flux in various flow regimes (bubbly, slug, and churn flows). A fluorescence-based technique that exploits the unique thermal properties of an erbium-doped transparent glass to characterize the transient thermal field underneath a growing bubble with submicron spatial resolution was also demonstrated.

References

Allison, S.W. and G.T. Gillies. 1997. Remote thermography with thermographic phosphors: Instrumentation and applications. *Review of Scientific Instruments* 68: 2615–2650.

Arsov, Z., I. Urbancic, M. Gravas, M.D. Biglino, A. Ljubetic, T. Koklic, and J. Strancar. 2011. Fluorescence microspectroscopy as a tool to study mechanism of nanoparticles delivery into living cancer cells. *Biomedical Optics Express* 2: 2083–2095.

Bhandari, P. 2012. Evaluation and improvement of temperature sensitive paint data reduction process through analysis of tunnel data. MS Thesis, University of Maryland, Department of Aerospace Engineering.

Bucci, M., A. Richenderfer, G.Y. Su, T. McKrell, and J. Buongiorno. 2016. A mechanistic IR calibration technique for boiling heat transfer investigations. *International Journal of Multiphase Flow* 83: 115–127.

Campbell, B.T., T. Liu, and J.P. Sullivan. 1994. Temperature sensitive fluorescent paint systems. *Proceedings of the 18th AIAA Aerospace Ground Testing Conference,* Colorado Spring, CO.

Fischer, S., S. Herbert, A. Sielaff, E.M. Slomski, P. Stephan, and M. Oechsner. 2012. Experimental investigation of nucleate boiling on a thermal capacitive heater under variable gravity conditions. *Microgravity Science and Technology* 24: 139–146.

Gerardi, C., J. Buongiorno, L.W. Hu, and T. McKrell. 2010. Study of bubble growth in water pool boiling through synchronized, infrared thermometry and high-speed video. *International Journal of Heat and Mass Transfer* 53: 4185–4192.

Golobic, I., J. Petkovsek, M. Baselj, A. Papez, and D.B.R. Kenning. 2009. Experimental determination of transient wall temperature distributions close to growing vapor bubbles. *Heat and Mass Transfer* 45: 857–866.

Huang, C. 2006. Molecular sensors for MEMS. PhD. Thesis, Purdue University, West Lafayette, IN.

Jung, S. and H. Kim. 2014. An experimental study on heat transfer mechanisms in the microlayer using integrated total reflection, laser interferometry and infrared thermomemetry technique. *Heat Transfer Engineering* 36: 1002–1012.

Kim, T.H., E. Kommer, S. Dessiatoun, and J. Kim. 2012. Measurement of two-phase flow and heat transfer parameters using infrared thermography. *International Journal of Multiphase Flow* 40: 56–67.

Krebs, D., V. Narayanan, J. Liburdy, and D. Pence. 2010. Spatially resolved wall temperature measurements during flow boiling in microchannels. *Experimental Thermal and Fluid Science* 34: 434–445.

Kurits, I. 2008. Quantitative global heat-transfer measurements using temperature-sensitive paint on a blunt body in hypersonic flows. MS Thesis, University of Maryland, Department of Aerospace Engineering.

Liu, T. and J.P. Sullivan. 2005. *Pressure and Temperature Sensitive Paints*. Springer, Berlin/Heidelberg.

Mani, P., R. Cardenas, and V. Narayanan. 2011. Submerged jet impingement boiling on a polished silicon surface. *ASME Pacific Rim Technical Conference and Exhibition on Packaging and Integration of Electronic and Photonic Systems*, Portland, OR, pp. 81–94.

Mills, A. 1997. Optical oxygen sensors utilizing the luminescence of platinum metal complexes. *Platinum Metals Review* 41: 115–127.

Moaveni, S., and J. Kim. 2017. An inverse solution for reconstruction of the heat transfer coefficient from the knowledge of two temperature values in a solid substrate. *Inverse Problems in Science and Engineering* 25: 129–153.

Ozawa, H., S.J. Laurence, J. Martinez Schramm, A. Wagner, and K. Hannemann. 2015. Fast-response temperature-sensitive-paint measurements on a hypersonic transition cone. *Experiments in Fluids* 56: 1853.

Schweizer, N. and P. Stephan. 2009. Experimental study of bubble behavior and local heat flux in pool boiling under variable gravitational conditions. *Multiphase Science and Technology* 21: 329–350.

Sedmak I., I. Urbancic, J. Strancar, M. Mortier, and I. Golobic. 2015. Transient submicron temperature imaging based on the fluorescence emission in an Er/Yb co-doped glass-ceramic. *Sensors Actuators A* 230: 102–110.

Sedmak, I., I. Urbancic, R. Podlipec, J. Strancar, M. Mortier, and I. Golobic. 2016. Submicron thermal imaging of a nucleate boiling process using fluorescence microscopy. *Energy* 109: 436–445.

Sefiane, K., J.R. Moffat, O.K. Matar, and R.V. Craster. 2008. Self-excited hydrothermal waves in evaporating sessile drops. *Applied Physics Letters* 93: 074103.

Shen, J., C. Graber, J. Liburdy, D. Pence, and V. Narayanan. 2010. Simultaneous droplet impingement dynamics and heat transfer on nano-structured surfaces. *Experimental Thermal and Fluid Science* 34: 496–503.

Shibuya, A., Ueki, R., Suzuki, Y., and Tange, M. 2016. Temporal temperature distribution measurement of a heat transfer surface of a flow boiling heat sink with a micro-gap using temperature sensitive paint. *Proceedings of the First Pacific Rim Thermal Engineering Conference*, Hawaii, Paper no. PRTEC–14900.

Solotych, V., D. Lee, J. Kim, R.L. Amalfi, and J. Thome. 2016. Boiling heat transfer and two-phase pressure drops within compact plate heat exchangers: Experiments and flow visualizations. *International Journal of Heat and Mass Transfer* 94: 239–253.

Solotych, V., J. Kim, and S. Dessiatoun. 2014. Local heat transfer measurements within a representative plate heat exchanger geometry using infrared (IR) thermography. *Journal of Enhanced Heat Transfer* 21: 353–372.

Theofanous, T.G., J.P. Tu, A.T. Dinh, and T.N. Dinh. 2002a. The boiling crisis phenomenon Part I: Nucleation and nucleate boiling heat transfer. *Experimental Thermal and Fluid Science* 26: 775–792.

Theofanous, T.G., J.P. Tu, A.T. Dinh, and T.N. Dinh. 2002b. The boiling crisis phenomenon Part I: Critical heat flux and burnout. *Experimental Thermal and Fluid Science* 26: 793–810.

Urbancic, I., Z. Arsov, A. Ljubetic, D. Biglino, and J. Stancar. 2013. Bleaching-corrected fluorescence microspectroscopy with nanometer peak position resolution. *Optics Express* 21: 25291–25306.

Wang, X., O.S. Wolfbeis, and R. Meier. 2013. Luminescent probes and sensors for temperature. *Chemical Society Reviews* 42: 7834–7869.

12

Practical Heat Flux Measurement

T. E. Diller

Virginia Tech

CONTENTS

12.1 Introduction

Temperature is a well-acknowledged measure of the thermal energy of a substance. Heat is often erroneously considered to be the amount of this thermal energy that a substance contains. Thermodynamics, however, clearly defines heat as the movement of that energy from a heat source to a heat sink and temperature as the potential for that energy to move. This is usually referred to as the heat transfer or heat transfer rate, or when taken per area as the heat flux in units of W/m^2. Consequently, heat flux and temperature are intimately related. One cannot really talk about one without the other. The first chapter of most heat transfer textbooks [1] establishes these basic concepts.

Temperature is one of the fundamental properties of a substance. It is well recognized and appreciated by everyone, particularly because it is associated with the comfort of the human body. Actually, however, the temperature of the surroundings is often less important than the heat transfer in dictating the level of thermal comfort. Conversely, heat flux is not a property, but a transport function of the system like mechanical work or power.

How and where thermal energy is transferred is often equally important as or more important than the temperature. For example, a person usually feels warm in 27°C air, but cool when jumping into 27°C water because of the much higher heat transfer to water than to air. Wind chill factor is another common example of the importance of convection heat transfer in addition to air temperature. Controlling this thermal energy transfer is often crucial to the performance of the system. Consequently, sensors that can be used to directly sense heat flux can be extremely important. It is rather amazing that heat flux sensors are not more widely used today in modern technology.

Over the years, researchers have developed many innovative methods for measuring heat flux to investigate heat transfer problems. They are typically challenging to build and operate for accurate measurements. Consequently, they are usually only used for those special applications and do not find their way into widespread or industrial use. This article is not focused on these specialized measurement systems which have been covered in previous reviews geared toward academic readers [2–4]. An example is transient infrared measurement of surface temperatures. This method uses data processing programs for specific geometries and material properties to infer the surface heat flux from the measured temperature.

The present review is meant to be practical in nature for users who are not necessarily experts of heat transfer, but recognize its importance in the real world. Consequently, the most useful sensors and manufacturers have been updated from previous reviews [5]. In addition, there are a number of new heat flux sensors that are now readily available with many new applications open. Practical details of how to make cost-effective heat flux measurements with these sensors are given.

12.2 Heat Flux Sensing

Most of the methods for measuring heat flux are based on temperature measurements on the surface or in the solid material. Usually, this involves insertion of a device either onto or into the surface, which has the potential to cause both a physical disruption and a thermal disruption of the surface. As with any good sensor design, the goal for good measurements is to minimize the disruption caused by the presence of the sensor. It is particularly important to understand the thermal disruption caused by the sensor because it cannot be readily visualized and because all heat flux sensors have a temperature change associated with the measurement. Consequently, a wise selection of the sensor type and operating range is important for good heat flux measurements [3,4].

The following sections describe the basic design and operation of the available heat flux sensors. Heat flux measurement is particularly challenging because of the wide range of values to be measured. Values range from below 1 W/m² for geothermal and insulation systems to over 1 MW/m² for high-speed aerodynamics and combustion systems. In addition, the desired time response can vary from the microsecond timescale for investigating high-speed events to many hours or even days when measuring thermal building response or geothermal events. Consequently, different sensors are needed for each specific application. The most common commercially available gages are listed in Table 12.1 with the general characteristics of the system. Prices have not been included because there are usually many specific variations of gages with different costs. Contact information for the manufacturers is given in Table 12.2. Many also have dealers and representatives located around the world.

TABLE 12.1

Available Heat Flux Instrumentation

Sensor	Description	Mounting Method	Manufacturer
Micro-foil	Foil thermopile	Surface	RdF
Heat flow sensor	Wire-wound thermopile	Surface	Concept
Heat flux transducer	Direct-Write Thermal Spray	Surface	MesoScribe
Thermal flux meter	Thermopile	Surface	ITI
HFM	Microsensor thermopile	Insert	Vatell
Gardon gage	Circular foil design	Insert	Medtherm and Vatell
Schmidt–Boelter	Wire-wound thermopile	Insert	Medtherm
FHF	Foil thermopile	Insert	Hukseflux
gSKIN	Semiconductor thermopile	Surface	greenTEG
PHFS	Printed thermopile	Surface	FluxTeq
HTHFS	Welded thermopile	Surface, insert	FluxTeq

TABLE 12.2

Manufacturers of Heat Flux Sensors.

Concept Engineering Old Saybrook, CT 06475 (860) 388-5566 http://www.conceptheatsensors .com/	RdF Corporation Hudson, NH 03051-9981 (603) 882-5195 http://www.rdfcorp .com/	Hukseflux Thermal Sensors Delft, The Netherlands 31-15-214-2669 http://www.hukseflux.com/
International Thermal Instrument Del Mar, CA 92014 (858) 755-4436 http://www.iticompany.com/	Vatell Corporation Christiansburg, VA 24073 (540) 961-3576 http://www.vatell.com/	MesoScribe Technologies, Inc. St. James, NY 11780 (631) 981-7081 http://www.mesoscribe.com/
Medtherm Corporation Huntsville, AL 35804 (205) 837-2000 http://medtherm.com/	greenTEG AG Zürich, Switzerland 41-44-632-0420 https://www.greenteg. com/	FluxTeq Blacksburg, VA 24060 (540) 257-3735 http://www.fluxteq.com/

12.2.1 Basic Fundamentals of Temperature and Heat Flux Measurement

Heat transfer is driven by a spatial difference in temperature from a source to a sink. Figure 12.1 illustrates the potential heat transfer to and from a surface exposed to a fluid at a different temperature. Conduction occurs in the solid (q_{cond}). Convection is shown from the fluid to the surface (q_{conv}), but can be in either direction. Radiation is emitted from the surface ($q_{emit,r}$) to the environment and from any external sources absorbed on the surface ($q_{abs,r}$). The resulting surface energy balance is

$$q_{cond} = q_{abs,\,r} - q_{emit,\,r} + q_{conv} \tag{12.1}$$

where the conduction heat flux is equal to the net heat flux at the surface, q_{net}. Using the gray body surface emissivity as ε and expressions for the convection and emitted radiation gives Eq. (12.2) in terms of the incident radiation, $q_{inc,r}$

$$q_{cond} = q_{net} = \varepsilon q_{inc,\,r} - \varepsilon \sigma T_s^4 + h\left(T_{gas} - T_s\right) \tag{12.2}$$

This expresses the balance of the incident and emitted radiation with the convection. The net heat flux to the surface must then be equal to the conduction from the surface into the material. The goal of a heat flux sensor at the surface is to measure this conduction heat flux.

Measuring the surface conduction has been attempted in many different ways, including surface temperature as a function of time, temperature distributions as a function of time, electric power into the material, lateral temperatures, and normal temperatures. The most successful, however, has been temperature difference measurements, particularly using differential thermocouples.

Why are differential temperature sensors the best way to measure heat flux in most situations? The answer lies in the difficulty of accurately measuring temperature at a specific location, particularly with simultaneous heat transfer. Any physical sensor for temperature has location inaccuracies

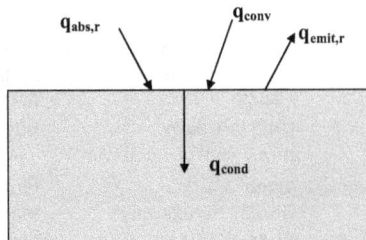

FIGURE 12.1
Surface energy balance.

and thermal contact resistance. Depositing temperature sensors directly on the material is complex and inconvenient and will only give one temperature at the surface. Inferring the heat flux from the transient temperature requires knowledge of the material properties, which must be independently measured. This can be appropriate for research environments, but not for general use. It is more reasonable to measure the temperature difference across the material of the sensor itself. The properties can then be calibrated by the manufacturer and incorporated directly into the sensitivity of the gage. One common example is shown in Figure 12.2. Here, the temperature difference across a thermal resistance layer is measured in the direction of the conduction heat flux through the material, which is related to the surface flux as expressed in Eq. (12.2). For steady-state conduction, the heat flux is expressed in terms of the thermal conductivity, k, the thickness of the material, δ, and the temperature difference across the thermal resistance layer.

$$q = \frac{k}{\delta} (T_1 - T_2) \qquad (12.3)$$

The crucial issue is how to accurately measure the temperature difference, T_1–T_2. To illustrate this, imagine an example with $T_1 = 50°C$ and $T_2 = 49°C$ measured by a system at room temperature of 25°C. A one percent measurement error would be 0.25°C for each individual temperature [0.01 times (50°C–25°C)], which would give a combined error for the difference of the two measurements of 0.35°C. This would be a 35% error in the 1°C temperature difference for the heat flux measurement. Conversely, if the temperature difference was measured directly, even a 5% error in the 1°C temperature difference would be 0.05°C, which is a factor of seven smaller. Consequently, it is desirable to measure the temperature difference directly, if possible. This advantage of measuring the temperature difference versus taking the difference of two temperature measurements is not appreciated even by many awarded heat flux sensor patents.

Unfortunately, most temperature measurement systems cannot be arranged as a differential. This is true for infrared cameras, liquid crystals,

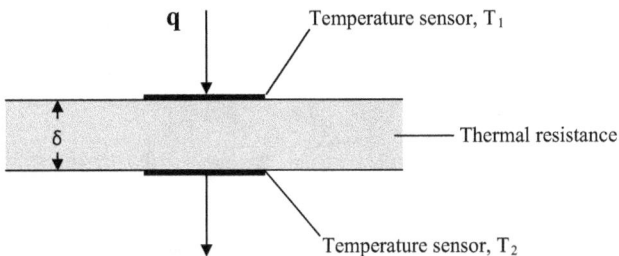

FIGURE 12.2
Basic heat flux sensor concept [6].

thermographic phosphors, and electrical resistance devices (RTDs). Consequently, these all have limited use for heat flux sensors. Thermocouples, however, inherently only measure temperature difference between the two junctions of dissimilar materials, as illustrated in Figure 12.3. The corresponding voltage output E is proportional to the temperature difference and the Seebeck coefficient S_T of the two materials.

$$E = S_T \left(T_1 - T_{\text{ref}} \right) \tag{12.4}$$

Although the values are often slightly nonlinear, this effect is included as a functional dependence of the Seebeck coefficient with temperature, as described in the section on calibration.

Thermocouple readout devices normally measure the temperature difference between the measurement junction and the reference junction, T_{ref}, as illustrated in Figure 12.3. The electronic chips for measuring thermocouple temperatures often use a diode with a calibrated temperature coefficient that is mounted in the circuitry. The lower leg of the circuit from the reference junction is usually internal to the circuit board. The accuracy of the reference temperature mostly depends on how closely this diode follows the temperature of the terminal block where the thermocouple wires are attached. Although it is assumed that the device is at isothermal conditions, this is rarely the case and this adds greatly to the absolute temperature uncertainty. To observe this effect, simply blow hot air on the thermocouple terminals of any digital thermocouple readout and observe the response of the recorded temperature. Note that the voltage output will change because the actual reference temperature at the terminal junction changes, while the measured reference temperature from the chip diode will remain the same.

For measuring heat flux, however, there is not any reference junction needed because the system is used to directly measure the temperature

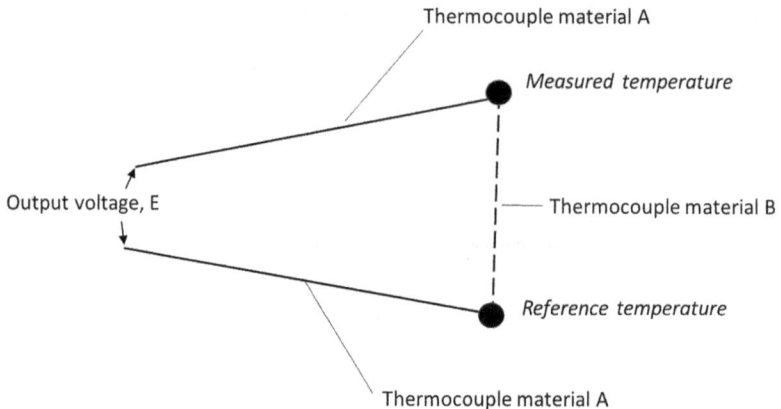

FIGURE 12.3
Basic thermocouple measurement.

difference, T_1–T_2, as shown in Figure 12.4. Consequently, thermocouples are ideally suited for measuring heat flux. The heat flux is normal to the thermal resistance layer, and the voltage output is measured laterally from the leads at the end of the sensor. Observe the similarity between Figures 12.3 and 12.4. The only difference is that the labels have been changed to reflect the measurement of the temperature difference across the thermal resistance. In essence, this is a simpler measurement than that of an absolute temperature. Moreover, the junctions can be arranged sequentially across a thermal resistance layer as illustrated in Figure 12.5 to form a differential thermopile. The voltage output from each pair of thermocouple junctions adds, with the total output represented by

$$E = N \, S_T \, (T_1 - T_2) \tag{12.5}$$

where N represents the number of junction pairs on the sensor. This would be five for the illustration in Figure 12.5. For T-type thermocouples, the

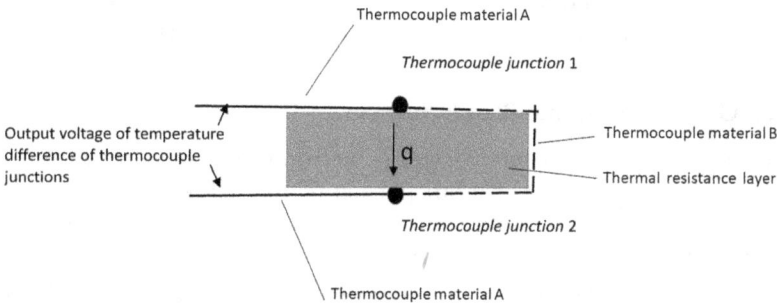

FIGURE 12.4
Measurement of temperature difference for heat flux.

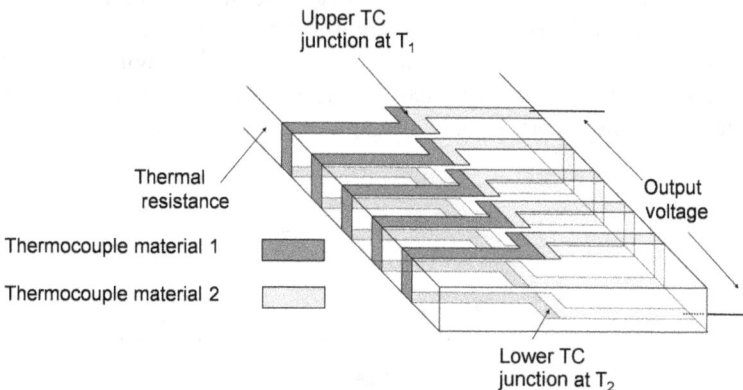

FIGURE 12.5
Thermopile heat flux gage.

approximate value of $S_T = 40 \, \mu V/°C$, which gives a voltage output of $E = 200 \, \mu V$ for a 1°C temperature difference across the sensor. This amplification of the signal by this differential thermopile can greatly increase the sensitivity of the sensor while keeping the temperature disruption minimal. The corresponding heat flux sensitivity is ideally given by

$$S_{cond} = \frac{E}{q} = \frac{N \, S_T \delta}{k} \tag{12.6}$$

although direct calibration is required to account for the two-dimensional features and manufacturing tolerances of actual heat flux gages. Because the values of both k and S_T have a weak dependence on the operating temperature range, the calibrated sensitivity is also a function of the gage temperature, as discussed in Section 12.3. Differential temperature heat flux gages are usually categorized as either planar sensors that mount onto surfaces [6] or insert gages that mount through a hole in the material flush with the surface [7]. These ASTM standards provide the basic information on how the sensors work and how to use them. A review of the characteristics of the commonly available sensors is given in the following sections.

12.2.2 One-Dimensional Planar Sensors

These heat flux sensors are designed for mounting onto a surface with assumed one-dimensional heat flux perpendicular to the surface. Figure 12.6 shows a heat flux gage mounted onto a plate with the surface temperature of the gage of T_s and the surface temperature of the surrounding plate of T_p. The goal is to keep the gage surface temperature as close as possible to the plate temperature to minimize the thermal disruption of the gage. Because of the nonlinear response of the thermal boundary layer when convection occurs, the effect of any temperature disruption is amplified [8]. This requires the thermal resistance of the gage and adhesive to be minimized along the thermal pathway from T_s to T_p. Another method to avoid the surface temperature disruption problem is to cover the entire surface with the heat flux gage material. This effectively ensures that the thermal resistance through the gage is matched with that of the surrounding plate. It is important to

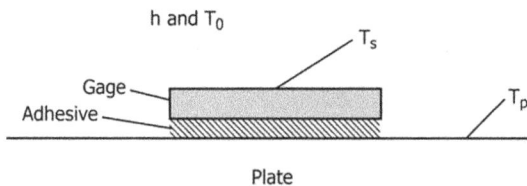

FIGURE 12.6
Mounting of a one-dimensional planar sensor [6].

have independent measures of the substrate surface temperature and the surface temperature of the gage. The gage surface temperature is useful for defining the thermal disruption of the gage and for calculating the value of the convective heat transfer coefficient.

A thin plastic thermopile gage (HFS) is produced by RdF Corp. Thin foils of two types of metal are alternately wrapped around a thin polyimide sheet and butt-welded on either side to form thermocouple junctions, as illustrated in Figure 12.5. A separate thermocouple is included to provide a measure of the sensor temperature. The flexible sensors are 75–400 μm thick and can be attached to a variety of surface shapes. Although they are limited to temperatures below 250°C and heat fluxes less than 100 kW/m², this covers many general-purpose industrial and research applications. The time response can be as fast as 20 ms, but transient signals can be attenuated unless the frequency of the disturbance is less than a few hertz. These sensors act much like a first-order system with an exponential time constant.

A similar thin plastic gage is made by Hukseflux except the metal foils are all placed on the same side of the thermal resistance material. Additional strips of plastic sheet are placed alternately on the top and bottom of the main sheet containing the metal foils. This effectively gives a temperature difference between adjacent foil junctions located at different thickness positions within the gage. Although the output is proportional to heat flux, it is most reliable when "heat spreaders" are used to make the heat transfer more one-dimensional across the gage. Both of these thin plastic sensors have a moderate cost.

A low-cost alternative has been developed by FluxTeq. Their PHFS heat flux gage consists of a differential thermopile made through holes in a thin sheet of polyimide. The resulting sensors are flexible and can be mounted on contoured shapes. They can be made in custom sizes and shapes up to 25 cm by 25 cm, although the price increases with sensor size. They are typically about 150 μm thick with a corresponding time response of 0.6 s. The temperature limit for long-term operation is 120°C. The maximum measurable heat flux is 150 kW/m², while the minimum is less than 1 W/m². A separate thermocouple is mounted integral to the gage for surface temperature measurement along with the heat flux.

The gSKIN heat flux sensor by greenTEG is a differential thermopile made by depositing bismuth telluride semiconductor materials. These thermocouples give a particularly high thermoelectric output. The sensors are typically encapsulated between metal sheets and have a thickness of 0.5 mm. The resulting time response of the gage is about one second. They are offered at a moderate cost.

A high-temperature heat flux gage has been developed by MesoScribe Technologies that can be deposited directly onto test surfaces using a Direct-Write Thermal Spray process [9]. Up to eighty thermocouple pairs (type K or N) are scribed around a layer of dielectric serving as the thermal resistance for the heat flux with a separate thermocouple giving the surface temperature. A layer of ceramic dielectric is first deposited on a metal surface to provide electrical isolation and can also be deposited over the gage to encapsulate it for better durability. All of the layers together are less than 0.4 mm thick, which gives the sensor a low profile. The sensor is sufficiently thick, however, for good survivability at high temperatures, and it gives a good output signal. The sensor can also be embedded in existing barrier coatings used for thermal, environmental, or erosion resistance. Because it can operate at temperatures of up to 850°C, it has a higher cost.

A different approach for a high temperature heat flux gage uses chromel and alumel strips that are welded together [10] as illustrated in Figure 12.7. Instead of a separate thermal resistance between thermocouple junctions, the thermal resistance is provided by the thermocouple elements themselves. Ceramic (ZTA) strips are included, but only to provide electrical insulation between elements. An additional thermocouple wire welded to the top of the gage provides the surface temperature, which is useful for interpreting the heat flux signal and calculating surface heat transfer coefficients. The gage can be operated and cycled to over 1000°C, which means it has a higher cost. The thickness is about 3 mm with a thermal time constant less than one second. Because of the all-metal construction, however, the thermal resistance is small. The measurement chip can also be mounted in an insert-type housing which can be air-cooled or water-cooled for gage temperature control. The HTHFS is currently made by FluxTeq.

A heat flux sensor design using welded wire to form the thermopile across the thermal resistance layer is made by International Thermal Instrument Co. The sensors are about 3 mm thick with thermal time constants in the order of 1 s and an upper temperature limit of 300°C. Sensors with higher sensitivity are made with semiconductor thermocouple materials for geothermal applications. Sensors are also available for operating temperatures up to 1000°C with higher cost.

A different type of differential thermopile can be created by wrapping a wire around the thermal resistance layer and plating one side of it with a different metal. A common combination is constantan wire with copper

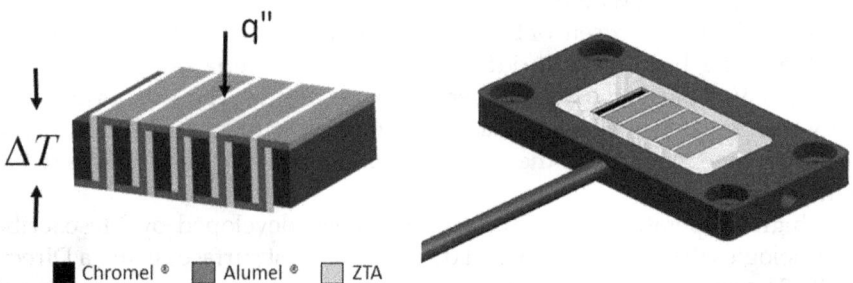

FIGURE 12.7
HTHFS; schematic of the measuring element on the left, and complete sensor on the right [10].

plating. The resulting wire-wound sensor looks similar to the sensor shown in Figure 12.5, except that the constantan wire is continuous all around the sensor and the copper plating covers the side to form the other thermocouple [11,12]. It does not produce as much voltage output as two separate thermocouples, but it is generally easier to manufacture. Concept Engineering offers a range of these types of sensors at moderate cost. Because of the many wire windings around 2-mm-thick plastic strips, the sensitivity to heat flux is high, but the corresponding thermal resistance is also large.

Heat flux gages are normally mounted on a good heat sink to provide a pathway for the thermal energy to flow. Not all real applications provide such an ideal situation, however. Hubble and Diller [13] have shown with a simple modification of the signal how to use heat flux gages on any surface, including thermal insulators. Adding a second term to account for the energy storage in the sensor itself creates what is termed the "hybrid heat flux."

$$q = q_{dif} + \frac{1}{2}\, \rho C\delta \frac{dT}{dt} \tag{12.7}$$

The first term is the usual differential voltage signal from the sensor, while the second term is the transient response of the sensor itself. The best transient temperature to use is the average of the top and bottom temperatures of the sensor if this is available. This gives the correct heat flux whether the gage is mounted on a good heat sink (high-conductivity material) or a good insulator (low-conductivity material). It also decreases the time response of the gage by about an order of magnitude.

12.2.3 Insert Heat Flux Gages

Insert gages are mounted through a hole in the material flush with the surface as shown in Figure 12.8. Their use is covered by an ASTM standard [7]. Because they are usually designed for high heat fluxes, they are often internally water-cooled and mounted in high-conductivity materials. The challenge with water cooling is to match the gage temperature with that of the surrounding plate. If not water-cooled, the plate must act as the heat sink.

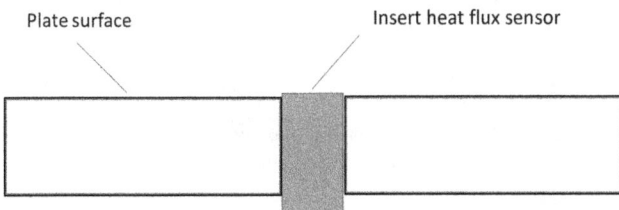

FIGURE 12.8
Mounting of an insert heat flux gage.

Consequently, the thermal contact between the plate and the gage is crucial to keep the gage at the plate temperature. Because insert heat flux gages are made individually by hand, their cost is high and they are made for research applications.

One popular version uses a plated wire wrapped around a small anodized piece of aluminum that is potted into a circular housing as shown in Figure 12.9. The usual material is a constantan wire that is plated with copper. This type is commonly known as a Schmidt–Boelter gage. Kidd and Nelson [14] have analyzed these gages to determine the lower voltage output of the plated wire relative to a true differential thermopile. The sensors are commercially available from Medtherm in sizes as small as 1.5 mm diameter. There is also some ability to contour the surface of the sensor to match a curved model surface for complex test article shapes. Because of the low conductivity of the potting material, the temperature of the gage surface is higher than the surroundings. This is particularly important when convection is present [15].

A thin thermopile sensor called the Heat Flux Microsensor (HFM) is manufactured by Vatell Corp. Because it is made with thin-film sputtering techniques, the entire sensor is less than 2 μm thick. The thermal resistance layer of silicon monoxide and the metal layers are sputtered directly onto the high-conductivity aluminum nitride substrate of the gage. The resulting physical and thermal disruptions of the surface due to the presence of the sensor are extremely small. They are best suited for heat flux values above 1 kW/m², with no practical upper limit. Because of the oxidation of the thin layers, however, it is not recommended for long-duration measurements at

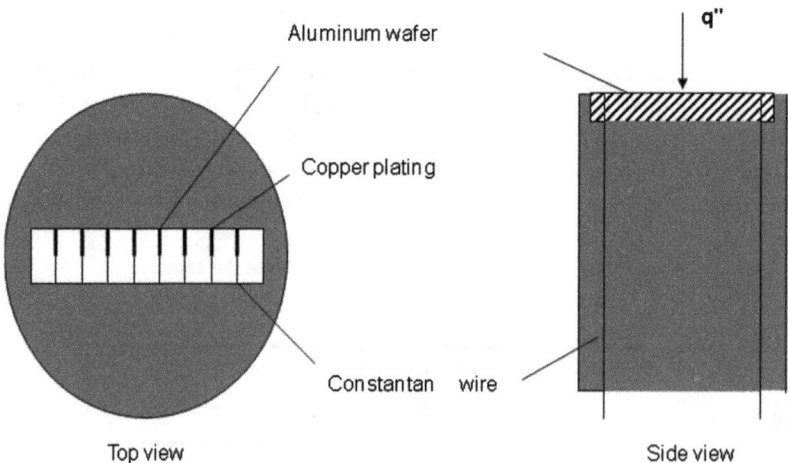

FIGURE 12.9
Schmidt–Boelter gage design [15].

high temperature. They are ideal for fast response measurements. The thermal response time is less than 10 μs [16], giving a good frequency response well above 1 kHz. A temperature measurement that is integrated into the sensor surface is very useful for determining heat transfer coefficients.

The circular foil or Gardon gage consists of a hollow cylinder of one thermocouple material with a thin foil of a second thermocouple material attached to one end. A wire of the first material is attached to the center of the foil to complete a differential thermocouple pair between the center and edge of the foil as illustrated in Figure 12.10. The body and wire are commonly made of copper with the foil made of constantan. Heat flux to the gage causes a radial temperature distribution along the foil as measured by this single thermocouple pair. These sensors are covered by an ASTM standard [17]. They are manufactured by two companies at moderate to high cost (Medtherm and Vatell) and are often used as secondary standards for the measurement of radiation. The biggest problems with the circular foil gages arise when they are used with any type of convection heat transfer. It has been shown analytically and experimentally that the output is altered for convective heat transfer because of the distortion of the temperature profile in the foil from the parabolic profile for incoming radiation [18]. For the usual 10 millivolt output, the center-to-edge temperature difference is approximately 200°C, which is a very large temperature non-uniformity on the surface. Consequently, for convection measurements, the range of the sensor should be much larger than the expected heat flux (low heat flux sensitivity) to keep the temperature disruption small and then amplify the small signal. Generally, however, it is much better to use a different sensor for convection.

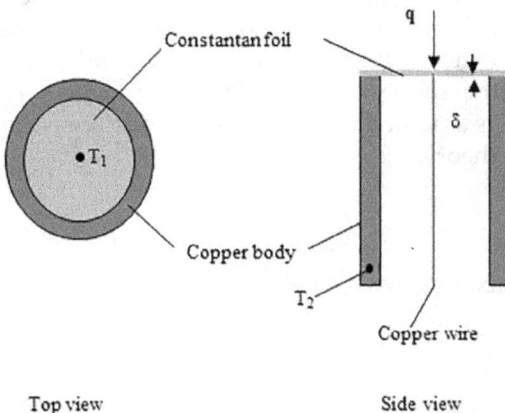

Top view   Side view

FIGURE 12.10
Schematic of a circular foil heat flux gage.

12.3 Calibration

Calibration of heat flux sensors is a complicated issue because some heat flux sensors respond to different modes and conditions of heat flux differently. For example, a sensor calibrated by radiation can have a substantially different response to the same amount of heat flux in convection. The manufacturers generally use conduction calibration for the flat planar sensors and radiation for the insert gages. The conduction calibrations are usually done at low heat flux levels. The radiation calibrations can be done at a wide range of heat flux values, but the gage coating is critical to how the wavelength and angle of the radiation interact with the coating. It is important to know how the calibrations are performed because they may not apply to other scenarios. Unfortunately, no one does convection calibrations, and usually, the sensors have a different response in convection than in either radiation or conduction. Some manufacturers, however, have tested their sensors in all three modes of transfer and can provide guidance on how they perform. It is reassuring if the sensor has the same calibration/response in all three heat transfer modes [10].

For the low heat fluxes seen in building applications, the guarded hot-plate method for calibration has been well established [19,20]. The sensor to be calibrated is placed between a heated plate and a cooled plate. The electrical power into the heated plate surrounded by guard plates is measured at steady state to provide the known heat flux. The National Institute of Standards and Technology (NIST) provides calibrated insulation samples to provide traceability against their standards.

Radiation calibrations are generally done by comparison with a secondary standard that has an absolute calibration in reference to a primary standard [21]. The gage is cooled to keep it at room temperature when exposed to a radiation source. The secondary standards are used to calibrate the manufactured heat flux gages by direct comparison either simultaneously or sequentially with matching radiation fluxes. The calibration is given in terms of either the incident radiation $q_{inc,r}$ or the absorbed radiation $q_{abs,r}$. The difference is whether the absorptivity of the gage surface coating is included. The absorptivity is assumed to be equal to the surface emissivity, ε, according to gray body theory.

$$q_{abs.r} = \varepsilon\, q_{inc.r} \qquad (12.8)$$

When the heat flux sensitivity is defined in terms of the absorbed radiation,

$$S_{rad} = \frac{E}{q_{abs,\, r}} \qquad (12.9)$$

where the voltage output is E and the sensitivity is S in units of voltage per heat flux. Here, it is assumed that the radiation emitted from the gage is negligible compared to the incident because the gage is kept at near-ambient temperature by water cooling and the surroundings are at the same temperature. This also minimizes any convection effects. The measured conduction into the gage at steady state is found by

$$q_{cond} = \frac{E}{S_{rad}}$$ (12.10)

To find the absorbed radiation or the convection heat flux, use this heat flux with Eqs. (12.1) and (12.2). To include transient effects, use Eq. (12.7) where $q_{dif} = q_{cond}$. Although the same can be done for convection calibration [15], this is rarely, if ever, done.

As expressed in Eqs. (12.8)–(12.10), differential temperature heat flux sensors are assumed to be linear with heat flux when the gage temperature is maintained constant. Consequently, manufacturer calibrations are usually performed at a single value of heat flux at room temperature. At zero heat flux, they should output zero voltage. Consequently, the value of sensitivity is all that is needed when the gages are used at the temperature of the calibration. However, the properties of the sensor are almost always a function of the sensor temperature. As seen in Eq. (12.6), the sensitivity is a direct function of Seebeck coefficient and the thermal conductivity, both of which are a function of temperature. This means that although the sensitivity is not a function of the heat flux, it is a function of the sensor temperature. Whenever the sensor is used at a temperature outside of the room temperature range of calibration, a correction of the sensitivity is needed. The manufacturer should supply this information with the sensor as an equation to correct the sensitivity. This is an additional reason to have a temperature measurement included on the sensor. This then becomes part of the data reduction to incorporate the sensor temperature with the sensitivity correction to calculate the correct measured heat flux. A high-temperature radiation calibration system that can operate with the heat flux sensors at up to 900°C has been used with the high-temperature sensors of Section 12.2.2 [9,22]. The temperature effect is usually expressed as a correction of the room-temperature sensitivity. Conduction calibrations can also be done at more modest elevated temperatures.

12.4 Thermal Measurement Details (Errors to Avoid)

As with many measurement systems, one major issue to address with heat flux measurement is the error caused by the disruption of the thermal field by the sensor itself. This is a design problem that starts with the thermal resistance of the sensor, R'', relative to the other resistances present in the system. It also includes the thermal contact resistance between the sensor and the surface material. A general example is shown in Figure 12.11 with the gage and plate temperatures labeled, where the fluid is the heat source at temperature T_∞ and the plate is the heat sink at temperature T_p. A simple model [15] of the ratio of steady-state heat flux of the gage q_g to that of the undisturbed plate q_p is

$$\frac{q_g}{q_p} = \frac{1}{\dfrac{h_p}{h_g} + h_p R''}$$

(12.11)

The effective heat transfer coefficient from the fluid to the gage h_g may be different than for the plate h_p because of the thermal disturbance of the gage. Consequently, it is important to minimize the temperature difference between the gage and the plate.

$$\frac{T_g - T_p}{T_\infty - T_p} \ll 1$$

(12.12)

This means that the gage sensitivity must be matched with the application. A high-sensitivity sensor should not be used if the convective heat transfer coefficient is high because this would give a large temperature drop in the gage when the overall temperature difference is small. It is also important to provide a good method of attachment of the sensor to minimize the thermal contact resistance with the plate. For an insert gage, this means that either a press-fit in the hole, or the use of a flange that can tighten the gage to the plate is preferred [7]. For a sensor mounted onto the surface, a thin layer of adhesive is best for relatively low temperatures. If thermal grease is used, it should be a thin layer. The thermal conductivity of thermal grease is much less than that of most metals, so it should be used carefully. For conduction, high-conductivity flexible pads can often be used.

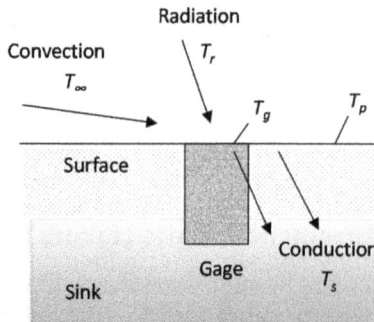

FIGURE 12.11
Schematic of temperature and heat flux on a surface with a gage [15].

To minimize the temperature drop across the sensor, the voltage signal should generally be kept at about one millivolt or less. To measure these microvolt signals requires a good voltmeter or a modern data acquisition system with a 24-bit analog-to-digital conversion circuit. Ideally, an auto-zero amplifier should be used to minimize the noise and zero-drift of the circuit. These low-noise data acquisition systems can usually be purchased with the heat flux sensors and can also be programmed to read thermocouples for the gage, surface, and fluid temperatures. It is helpful to twist the wires and shield them, if necessary, to decrease the signal noise.

Surface temperatures are challenging to measure accurately when significant heat flux is occurring, particularly if bead thermocouples or RTDs are used. The contact area of a sphere is so small on a hard, solid surface that many times it will not give a representative value of the surface temperature, particularly when substantial heat flux is occurring. Consequently, surface temperatures should be measured either with thin-film thermocouples on the surface or with thermocouples directly embedded into the material. Because it is difficult to solder directly to large pieces of high-conductivity metals, one method is to solder the thermocouple inside a small-diameter copper tube and then press-fit this into a hole in the metal piece. The surface can then be sanded smooth. Simply inserting a thermocouple into a hole does not guarantee a good thermal connection to the material. Even when a thermal paste is used (which does not have high thermal conductivity), the thermocouple wire acts as a fin, which will alter the measured temperature if the thermocouple is not attached very well to the metal piece. Even with thin-film thermocouples mounted directly to the surface, it is important to remember to include thermal contact resistance in models of surface temperature and heat transfer. Thermal contact resistance has the same direct effect on the measured surface temperature as the heat flux. For example, a typical thermal resistance of $0.001 \, \text{m}^2\text{-K/W}$ will give a 1°C error in the measured temperature for a modest heat flux of $1,000 \, \text{W/m}^2$.

12.5 Example Uses

The value of measuring heat flux is often not appreciated, even by engineers. One example is in teaching heat transfer courses to undergraduate engineering students. After passing a heat transfer course, many students still have misconceptions about temperature and heat transfer [23]. This should not be a surprise when the method of teaching these courses is considered. Not only are they normally taught as passive lecture courses, but the focus is on the solutions of the temperature field. Students have no physical feel of what heat flux is or even typical values. Installing hands-on workshops has been shown to help [23]. A series of fourteen educational workshops using heat flux sensors is available at http://www.me.vt.edu/heat-transfer-mobile-lab-3/. One example is to have students observe the transient temperature and heat flux when placing their hand on carpet versus concrete. Carpet acts like an insulator, while concrete is a relatively good conductor. One of the flat plastic heat flux sensors with a thermocouple (such as those made by RdF, Hukseflux, and FluxTeq) can be used between their hand and the material to make the measurements. An example of the results is shown in Figure 12.12. There is very little difference in the temperature response, but the heat flux is much higher for the concrete than for the carpet. The major conceptual lesson is that the heat flux is what dictates why the concrete feels much colder than the carpet. Students also plot the thermal resistance ($R = (T_s-T_{initial})/q$) as a function of time and compare it with the theory of semi-infinite materials. Because heat flux is measured directly, R is easy to calculate from Figure 12.12 and is seen to increase in proportion to the square root of time.

An example that is more research oriented is the measurement of heat flux in mixed mode and difficult environments, such as combustion. One very challenging measurement in a fire is the separation of the radiation

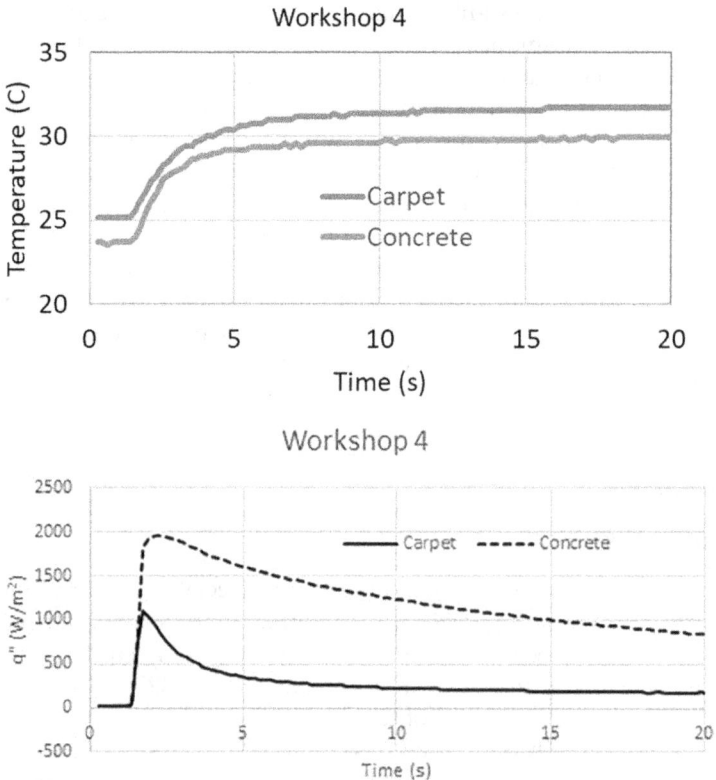

FIGURE 12.12
Measurements of temperature and heat flux on two surfaces from an individual hand.

and convection components of the heat flux. Vega et al. [24] give a method of doing this using a combination of heat flux and surface temperature measurements for an uncooled heat flux gage. This requires a high-temperature gage (such as those made by MesoScribe and FluxTeq) mounted to the surface in question. The system was initially tested under controlled conditions at low temperatures ($< 60°C$) to demonstrate the validity. A two-minute test is shown in Figure 12.13 with the gas and surface temperatures along with the heat flux measured by the gage. As the surface temperature increases with time, the heat flux decreases. An exposure heat flux is defined as the total flux absorbed at the surface by both convection and radiation. Combining Eqs. (12.2) and (12.9) shows how the exposure heat flux can be calculated from the voltage produced by the gage, E, and the emitted heat flux from the surface at the surface temperature and emissivity.

$$q_{exp} \equiv \varepsilon \ q_{inc, r} + h\left(T_{gas} - T_s\right) = \frac{E}{S_{rad}} + \varepsilon\sigma T_s^4 \tag{12.13}$$

As the surface temperature changes in time, the value of the exposure heat flux changes because the convection changes. This can be expressed in terms of the derivative to give the corresponding convection heat transfer coefficient, h.

$$\frac{dq_{exp}}{dT_s} = -h \tag{12.14}$$

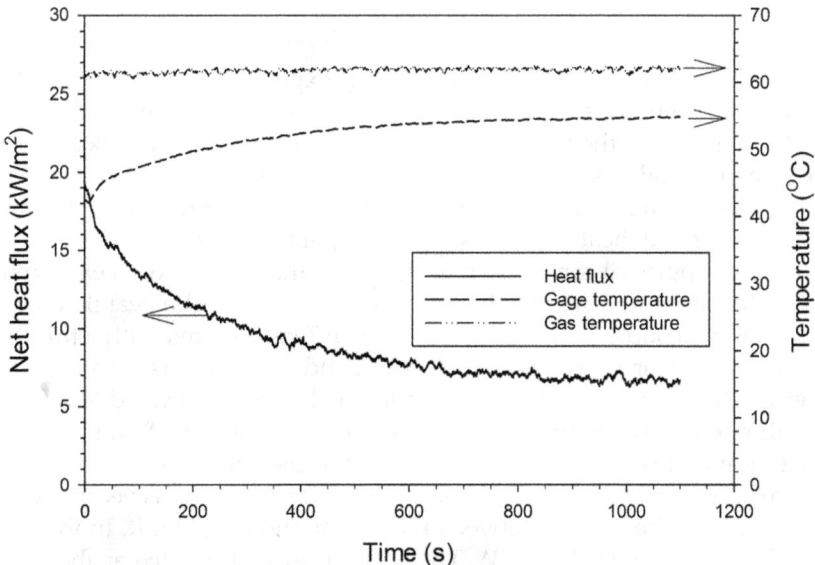

FIGURE 12.13
Measurements of temperature and heat flux on a surface with radiation and convection [24].

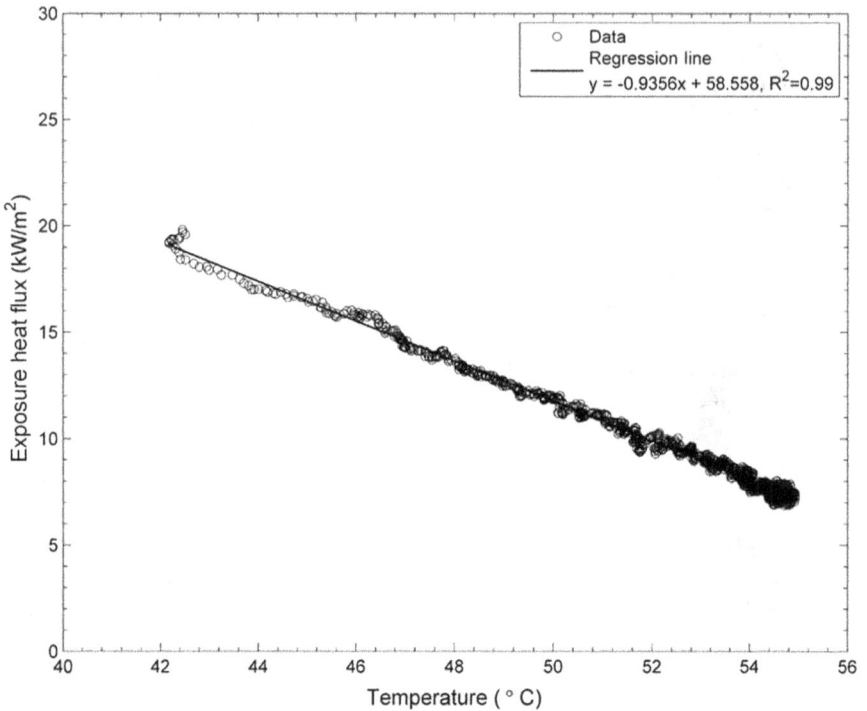

FIGURE 12.14
Illustration of a method to determine the convection heat transfer coefficient in mixed mode using the exposure heat flux [24].

The results are shown in Figure 12.14 for the experimental measurements of Figure 12.13. The slope of the line of data gives a heat transfer coefficient of $h = 936 \, \mathrm{W/m^2 \, K}$ for the high-velocity impinging jet used in the experiment. The same method has been used in simulated fires [25] and in natural convection flames and forced convection jet flames. This emphasizes the value of measuring both heat flux and surface temperature in real time.

A similar approach can be used to thermally interrogate complex systems, such as blood perfusion in the human body [26]. A thin-film heat flux sensor with a thermocouple (RdF, Hukseflux, or FluxTeq) is paired with a thin-film heater. The sensor is attached to the skin, and the heater is placed on top. After the sensor reaches the steady state, the heater is activated to produce a small rise in temperature with an appropriate heat flux. A sample of the resulting measurement is shown by the solid lines in Figure 12.15. The difference between the tissue surface and sensor temperatures is because of the thermal contact resistance between the sensor and the skin, R. In this case, the value of $R = 0.00037 \, \mathrm{m^2 \, K/W}$. The heat flux event is started at about nine seconds and continues for fifty seconds. The temperature data are used to

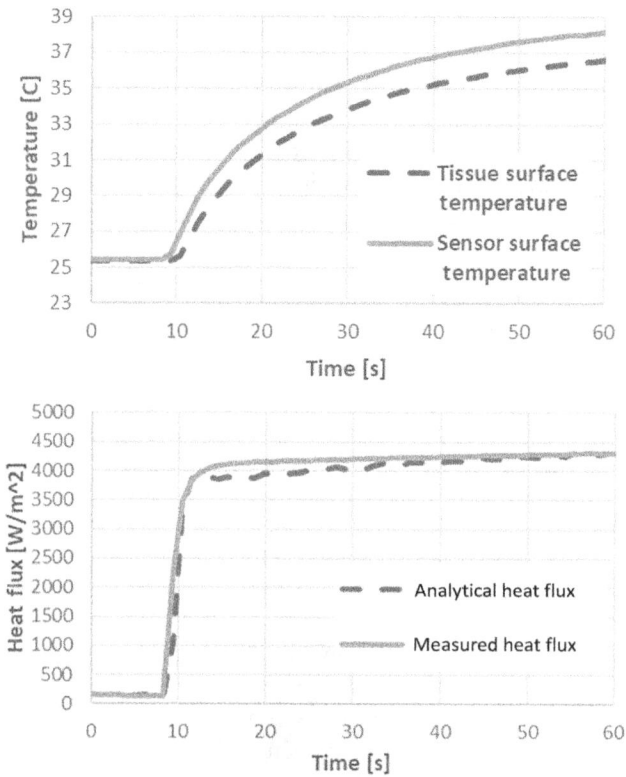

FIGURE 12.15
Measured and analytical values of temperature (a) and heat flux (b) for determining in vivo blood perfusion.

establish the boundary condition for a solution of the conduction equation including a perfusion term, w_b.

$$\rho C \frac{\partial \theta}{\partial t} = k \frac{\partial^2 \theta}{\partial x^2} - \rho C w_b \theta \qquad (12.15)$$

The density is ρ, the specific heat is C, the thermal conductivity is k, and θ represents the temperature difference between the tissue temperature and the core body temperature. Each temperature data point is used as a step function input, which when added together gives the complete response. The resulting surface heat flux from this solution is matched with that measured, as shown in Figure 12.15. The optimal values of the blood perfusion (w_b), thermal contact resistance, and core body temperature are found when the best fit between the analytical solution and experimental data is achieved according to a least-squares criterion. Estimating these parameters is made possible by having the transient measurements of the actual heat flux that enters the tissue at the surface. In essence, one is using the difference in response of the surface temperature to the applied heat flux to find the system properties. The shape of the measured output curve is a function of these system parameters. This is but one example of many possible applications of using surface measurements of heat flux and temperature to investigate the interior of material systems.

12.6 Summary

Currently, there are a number of heat flux sensors available for use in a wide variety of applications. The most common commercially available sensors are listed in Table 12.1, and the information for contacting the manufacturers is given in Table 12.2. These differential temperature devices provide a direct readout of the heat flux over the surface of the sensor. With the proper choice of sensor for the application and care in measurement method, the heat flux results are easily obtained. Because of the multiple modes of heat transfer that occur simultaneously and the effects of transients, however, interpretation can often be challenging. This is where understanding the sensor and the system model can be very helpful. Some of the common issues with calibration and measurement error have been addressed for the different types of sensors. The most important criterion is to properly match the sensor with the type and range of measurement desired. There is no one heat flux sensor that is right for every measurement. Three example novel applications have been discussed. One is to teach students the basic concepts of heat transfer and temperature by making measurements of both of these quantities simultaneously. The second is to show the value of measuring both heat flux and temperature to directly determine heat transfer coefficients and separate convection and radiation effects. The third is to use heat flux and temperature at the surface of solids to interrogate the material noninvasively. More details on all aspects of heat flux measurement can be obtained from the manufacturers and the references listed.

References

1. Incropera, F., D. DeWitt, T. Bergman, and A. Lavine, *Fundamentals of Heat and Mass Transfer*, 7th Ed., New York: John Wiley & Sons, 2011.
2. Diller, T. E., "Heat flux measurement," in *Mechanical Engineers' Handbook*, Vol. 4, Energy and Power, 4th Ed., Ed. M. Kutz, New York: John Wiley & Sons, 2015, pp. 285–312.

3. Diller, T. E., "Advances in heat flux measurement," in *Advances in Heat Transfer,* J. P. Hartnett et al., Vol. 23, Boston, MA: Academic Press, 1993, pp. 279–368.

4. Keltner, N. R., "Heat flux measurements: theory and applications," Ch. 8, in *Thermal Measurements in Electronics Cooling,* K. Azar, Boca Raton, FL: CRC Press, 1997, pp. 273–320.

5. Diller, T. E., "Heat flux," Ch. 67, in *Measurement, Instrumentation and Sensors Handbook,* J. G. Webster and H. Eren, Boca Raton, FL: CRC Press, 2014, pp. 67.1–15.

6. ASTM E2684-17, Standard Test Method for Measuring Heat Flux Using Surface-Mounted One-Dimensional Flat Gages. Ann. Book ASTM Standards, 15.03, 2017.

7. ASTM E2683-17, Standard Test Method for Measuring Heat Flux Using Flush-Mounted Insert Temperature-Gradient Gages. Ann. Book ASTM Standards, 15.03, 2017.

8. Moffat, R. J., J. K. Eaton, and D. Mukerji, "A general method for calculating the heat island correction and uncertainties for button gages," *Measurement Science and Technology,* Vol. 11, 2000, pp. 920–932.

9. Trelewicz, J. R., J. P. Longtin, D. O. Hubble, and R. J. Greenlaw, "High-temperature calibration of direct write heat flux sensors from 25°C to 860°C using the in-cavity radiation method," *IEEE Sensors Journal,* Vol. 15, 2015, pp. 358–364.

10. Gifford, A. R., D. O. Hubble, C. A. Pullins, S. T. Huxtable, and T. E. Diller, "A durable heat flux sensor for extreme temperature and heat flux environments," *AIAA Journal of Thermophysics and Heat Transfer,* Vol. 24, 2010, pp. 69–76.

11. Hauser, R. L., "Construction and performance of in situ heat flux transducers", in *Building Applications of Heat Flux Transducers,* Vol. 885, E. Bales et al., West Coshocton, PA: ASTM STP, 1985, pp. 172–183.

12. Van der Graaf, F., "Heat flux sensors", in *Sensors,* Vol. 4, W. Gopel et al., New York: VCH, 1989, pp. 295–322.

13. Hubble, C. O. and T. E. Diller, "A hybrid method for measuring heat flux," *ASME Journal of Heat Transfer,* Vol. 132, 2010, 031602, 8 pages.

14. Kidd, C. T. and C. G. Nelson, How the Schmidt-Boelter gage really works. *Proc. 41st Int. Instrum. Symp.,* Research Triangle Park, NC: ISA, 1995, 347–368.

15. Gifford, A., A. Hoffie, T. Diller and S. Huxtable, "Convection calibration of schmidt-boelter heat flux gages in stagnation and shear air flow", *ASME Journal of Heat Transfer,* Vol. 132, 2010, 031601, 9 pages.

16. Holmberg, D. G. and T. E. Diller, "High-frequency heat flux sensor calibration and modeling," *ASME Journal of Fluids Engineering,* Vol. 117, 1995, pp. 659–664.

17. ASTM E511-07, Measurement of Heat Flux Using a Copper-Constantan Circular Foil Heat-Flux Gage. Ann. Book ASTM Standards, 15.03, 2009.

18. Kuo, C H. and A. K. Kulkarni, "Analysis of heat flux measurement by circular foil gages in a mixed convection/radiation environment," *ASME Journal of Heat Transfer,* Vol 113, 1991, pp. 1037–1040.

19. Bomberg, M., "A workshop on measurement errors and methods of calibration of a heat flow meter apparatus," *Journal of Thermal Insulation and Building Environments,* Vol. 18, 1994, pp. 100–114.

20. ASTM C1130-17, Standard Practice for Calibrating Thin Film Heat Flux Transducers, Ann. Book ASTM Standards, 4.06, 2017.

21. Murthy, A. V., B. K. Tsai and R. D. Saunders, "Radiative calibration of heat-flux sensors at NIST: Facilities and techniques," *Journal of Research of the National Institute of Standards and Technology,* Vol. 105, 2000, pp. 293–305.

22. Pullins, C. A. and T. E. Diller, "In situ high temperature heat flux sensor calibration," *International Journal of Heat and Mass Transfer*, Vol. 53, 2010, pp. 3429–3438.
23. Cirenza, C. F., T. E. Diller, and C. B. Williams, "Hands-on workshops to assist in students' conceptual understanding of heat transfer," *ASME Journal of Heat Transfer*, Vol. 140, 2018, 092001, 10 pages.
24. Vega, T., R. A. Wasson, B. Y. Lattimer, and T. E. Diller, "Partitioning radiative and convective heat flux," *International Journal of Heat and Mass Transfer*, Vol. 84, 2015, pp. 827–838.
25. Vega, T., B. Y. Lattimer, and T. E. Diller, "Fire thermal boundary condition measurement using a hybrid heat flux," *Fire Safety Journal*, Vol. 61, 2013, pp. 127–137.
26. O'Brien, T. J., A. Roghanizad, P. Jones, C. Aardema, J. Robertson, and T. Diller, "The development of a thin-filmed, non-invasive tissue perfusion sensor to quantify capillary pressure occlusion of explanted organs," *IEEE Transactions on Biomedical Engineering*, Vol. 64, No. 7, 2017, pp. 1631–37.

13

Heated Meter Bar Techniques: What You Should Know and Why

Roger Kempers

York University

Anthony Robinson

Trinity College Dublin

CONTENTS

13.1 Introduction and Principle

13.1.1 Characterizing Heat Transfer Properties

Steady-state heat transfer is generally modeled using the Ohm's law analogy [1]. As illustrated in Figure 13.1, the potential nodes are temperature, the energy flow is heat, and the resistance is the thermal resistance. This technique for analyzing heat transfer is suitable for analyzing the simplest to the most complex thermal networks and has been used for as long as heat transfer has been an engineering discipline in its own right.

In order to model heat transfer systems using the Ohm's law analogy, one must estimate the thermal resistance. Depending on the scenario, the thermal resistance could involve conduction, convection, radiation, or a combination of these acting together. There exists, of course, an extremely diverse range of materials, fluids, flow configurations, surface finishes, architectures,

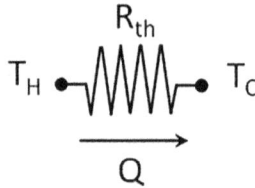

FIGURE 13.1
Ohm's law analogy of heat transfer.

and geometries. Further, in many scenarios the thermal resistance is coupled in a nonlinear way with the temperature. This makes analytic estimations of the thermal resistance possible for only the very simplest of heat transfer situations. For most other situations, empirical techniques must be used.

Considering Ohm's law analogy, the thermal resistance can be estimated directly, provided that the driving temperature difference as well as the heat flow can be measured since

$$R_{th} = \frac{\Delta T}{Q} \qquad (13.1)$$

where R_{th} is the thermal resistance (in K/W), $\Delta T = T_H - T_C$ is the driving temperature difference (in K), and Q is the rate of heat transfer (in W). This becomes more application-specific when the effective conductance, σ (in W/m²K), is defined such that

$$R_{th} = \frac{1}{\sigma A} \qquad (13.2)$$

where A is the overall heat transfer surface area. For conduction heat transfer across a medium of thermal conductivity, k, and thickness, t,

$$\sigma = \frac{k}{t} \qquad (13.3)$$

For convection and/or radiation heat transfer from an exposed surface to/from its surroundings,

$$\sigma = h \qquad (13.4)$$

where h is generally referred to as the heat transfer coefficient.

This illustrates that the empirical estimation of thermal resistance can be very useful in characterizing conduction, convection, radiation, and mixed heat transfer problems because of its capacity to predict more specific properties and parameters such as thermal conductivity and heat transfer coefficients. However, these characterizations require that suitable experiments be performed and that nodal temperatures and heat transfer be measured with adequate accuracy.

13.1.2 The Heated Meter Bar Techniques

One measurement method that has been extensively used by the heat transfer community for characterizing thermal resistance and related properties is the heated meter bar technique (HMBT). Figure 13.2 shows a generic embodiment of the HMBT for a scenario where the heated surface is exposed to an ambient heat sink (e.g., convection, boiling, radiation). As shown, the measurement system is comprised of a heat source embedded in the meter bar located at some distance from the exposed heat transfer surface. As the heat travels to the surface, parasitic heat loss occurs. These heat losses are difficult to quantify accurately, making the use of the heater power, $Q_{in} = IV$, inappropriate for characterizing the heat transfer at the surface.

To circumvent this issue, the "neck" of the meter bar is made long enough and far enough away from the heat source that one-dimensional heat transfer occurs in the near vicinity of the exposed surface. In this way, measuring

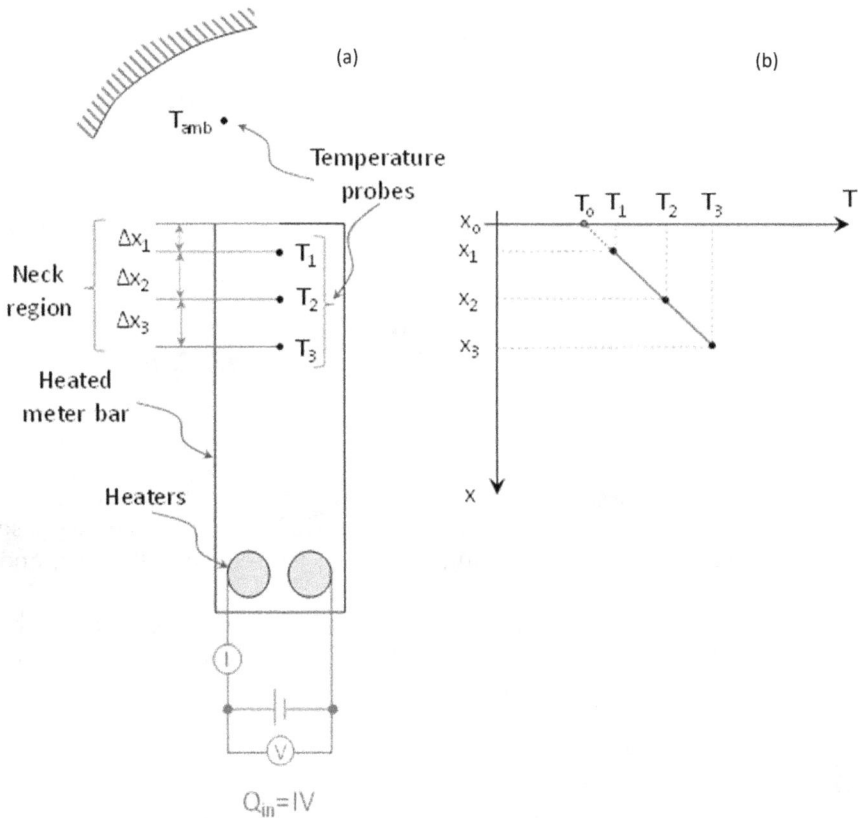

FIGURE 13.2
(a) Illustration of heated meter bar, showing embedded heaters and temperature probes; (b) schematic of the temperature vs. distance plot.

the temperature gradient, dT/dx, with simple embedded temperature probes along with accurate knowledge of the thermal conductivity of the meter bar, k_{MB}, is sufficient to measure the heat flux to the exposed surface, given that for one-dimensional heat flow

$$q'' = -k_{MB} \frac{dT}{dx} \tag{13.5}$$

As illustrated in Figure 13.2, the temperature gradient is simply determined by the least squares linear regression fit to the measured data on the temperature vs. distance plot. From the same plot, it is then possible to indirectly measure the surface temperature by simply extrapolating the T vs. x curve to the surface, such that

$$T_0 = T_1 - \left|\frac{dT}{dx}\right|(x_0 - x_1) \tag{13.6}$$

Importantly, this is a non-contact estimation of the surface temperature.

To estimate the overall thermal resistance between the exposed surface and the ambient, one simply requires the exposed surface area, A, as well as a measurement of the ambient temperature, T_{amb}, such that

$$R_{th} = \frac{(T_0 - T_{amb})}{q''A} \tag{13.7}$$

or, in the context of the effective thermal conductance,

$$\sigma = \frac{q''}{(T_0 - T_{amb})} \tag{13.8}$$

The example illustrated above has extensively been used to characterize a broad range of convective heat transfer scenarios, including single-phase convection [2], nucleate pool boiling [3], forced convective boiling [4], and condensation heat transfer [5].

A simple adaptation of the method described above can be used for the measurement of thermal contact resistance, thermal interface material (TIM) technology, and the bulk thermal conductivity of materials. As illustrated in Figure 13.3a, thermal contact resistance between two surfaces pressed together can be measured by simply using two meter bars—one heated and one cooled. This scenario requires that the upper and lower surface temperatures be estimated using the extrapolation technique described above (Figure 13.3b), whereas the heat flux can be measured using Eq. (13.5) for one of the meter bars or the average of the two. Importantly, the use of two meter bars allows for an energy balance to be performed across the thermal joint.

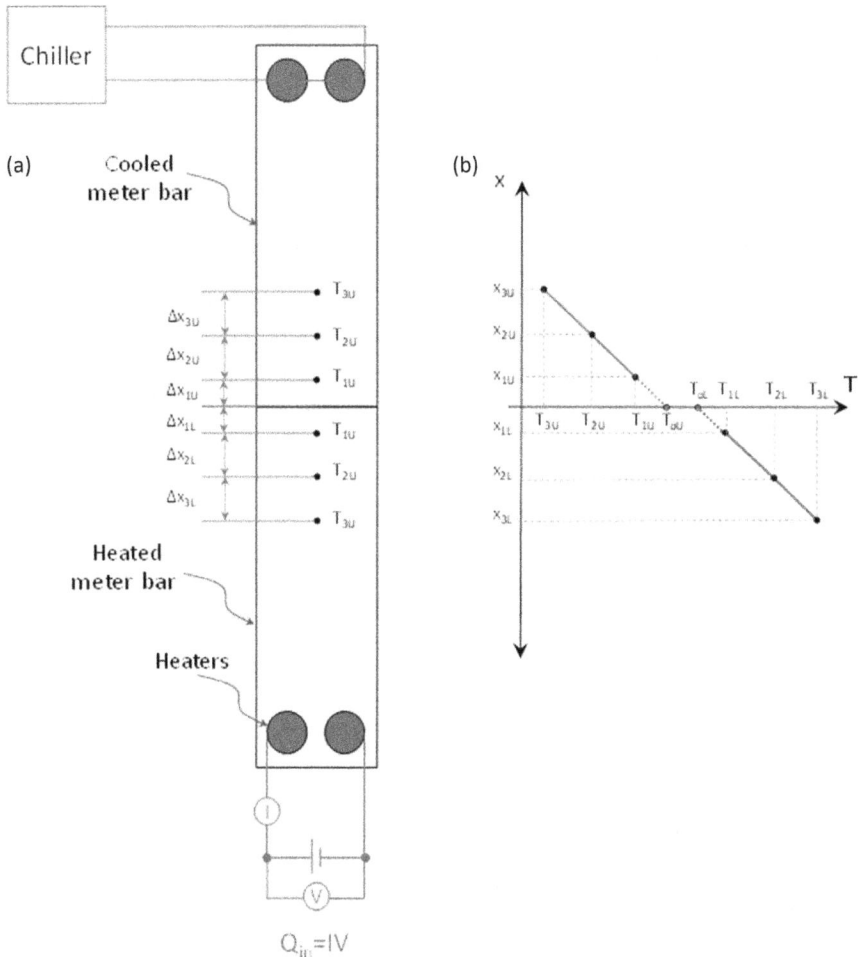

FIGURE 13.3
(a) The HMBT setup for contact thermal resistance measurement, and (b) the associated temperature vs. distance plot.

The specific thermal resistance, RA, of the joint or thermal interface material can then be characterized as

$$\mathrm{RA} = \frac{(T_{0L} - T_{0U})}{q''} \tag{13.9}$$

where T_{0L} and T_{0U} are the extrapolated surface temperatures of the meter bars.

Examples of the use of the HMBT for dry contact thermal resistance can be found in [6,7]. Of course, there are many situations, such as electronics cooling, where the thermal contact resistance must be mitigated. To do so, TIMs are deployed between the mating surfaces [8]. Estimating the effective resistance of TIMs between two surfaces is straightforward in that the technique described above for dry contact resistance can be used with a TIM sandwiched between the two opposing meter bars. However, this results in an effective measurement of the thermal resistance across the joint, which includes the contact resistances between the meter bars and the TIM, along with the bulk resistance of the TIM itself. In order to measure the bulk thermal conductivity of a TIM—or any material or structure for that matter—the net contact resistances must be effectively subtracted from the overall resistance. This can be accomplished by testing the same bulk material at two or more thicknesses, as depicted in Figure 13.4.

Figure 13.4b shows that for the same bulk material the overall resistance increases linearly with test material thickness and the intercept with the vertical axis gives a measurement of the net contact resistance (one per side). The bulk material thermal conductivity can then be approximated from the slope of the linear relationship as

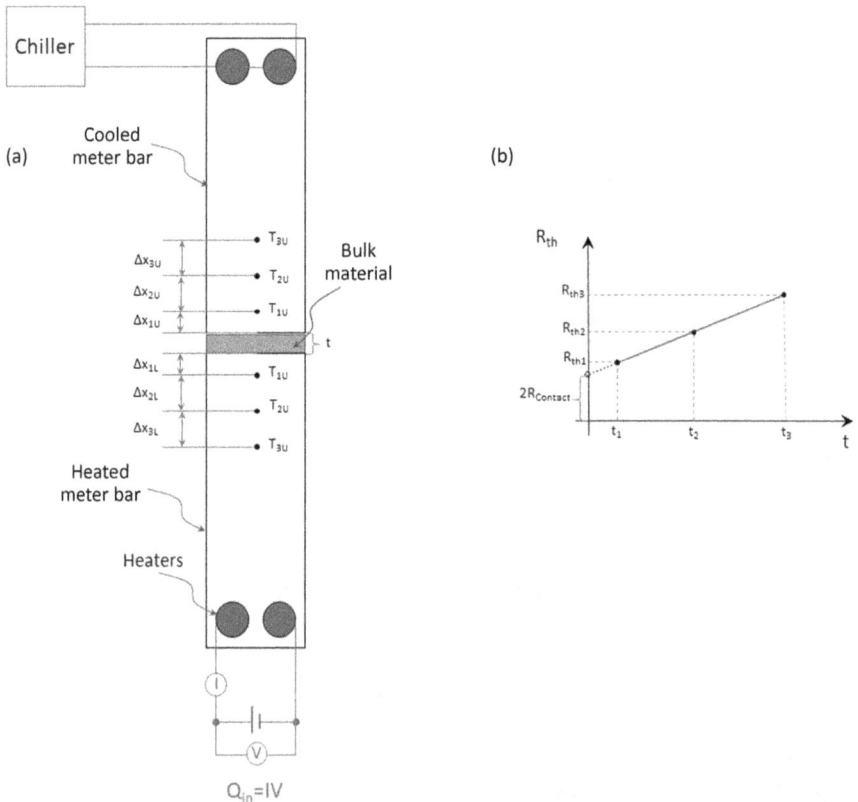

FIGURE 13.4
(a) The HMBT setup for bulk thermal conductivity measurement, and (b) schematic plot of thermal resistance vs. bulk material thickness.

$$\frac{dR_{net}}{dt} = \frac{1}{k_{bulk}A} \tag{13.10}$$

such that

$$k_{bulk} = \frac{1}{\dfrac{dR_{net}}{dt}A} \tag{13.11}$$

Examples of this method for bulk material thermal conductivity measurement can be found in [9,10].

13.2 Assessing Uncertainty in the HMBT

The HMBT is a function of several measured variables, each with its own level of uncertainty. In order to estimate the uncertainty of the thermal resistance, the error associated with the extrapolated temperature (Eq. 13.6) and the heat flux (Eq. 13.5) must be determined. This is not trivial since each data point used to estimate the heat flux has errors associated with both the spatial location of the temperature probe and the individual temperature measurements themselves. Thus, the compounding of the error must be considered for each measurement point during the least squares regression analysis which is further complicated by its influence on the extrapolated surface temperature error. Until recently, the methods used to characterize the uncertainty of the thermal resistance and related properties associated with its use have not adequately addressed this.

In the sections below, a method for the accurate quantification of heat flux and surface temperature using the heated meter bar technique based on the work of Kempers et al. [11] is discussed. The foundation of the method is a Monte Carlo approach for combining the spatial and temperature uncertainties of multiple measurement points in order to achieve a robust uncertainty of the linear temperature gradient, which is then extended to the estimation of uncertainty of the heat flux and surface temperature measurements. Importantly, the technique is straightforward and can be implemented at the initial design stage in such a way that the experimental layout, instrumentation, and calibrations can be considered, and target uncertainties can be achieved for the heat transfer scenario and range of parameters to be tested.

13.2.1 Calculating Uncertainty Using the HMBT

As described in the previous section, the extrapolated surface temperature, T_o, and heat flux, q'', for each meter bar are obtained by performing a least squares regression of the axial temperature distribution to a straight line and computing the resulting y-intercept and slope at the contact surfaces. As a result, the uncertainty of T_0 and q'' depends on both the temperature and spatial uncertainties of each temperature probe. This is depicted in Figure 13.5.

There are several approaches to calculating how the uncertainty propagates through a least squares regression. Wald [12] and Bartlett [13] outlined methods for fitting a straight line when both variables are subject to error. These methods are mathematically involved and are restricted to the assumption that the uncertainties are uniform and normally distributed. Press et al. [14] described an analytical method for calculating the uncertainties of the slope and y-intercept of a straight line model based on the assumption of a normal distribution and the standard deviations in both the x- and y-data. The resulting expressions are nonlinear and unwieldy and best solved by numerical methods. Kedzierski and Worthington [15] presented low-order and relatively simple expressions for estimates of the uncertainties in wall temperature and gradient as originally obtained by Ku [16]. These expressions demonstrate that the lowest uncertainties are obtained by using a meter bar of high thermal conductivity with a large number of well-spaced,

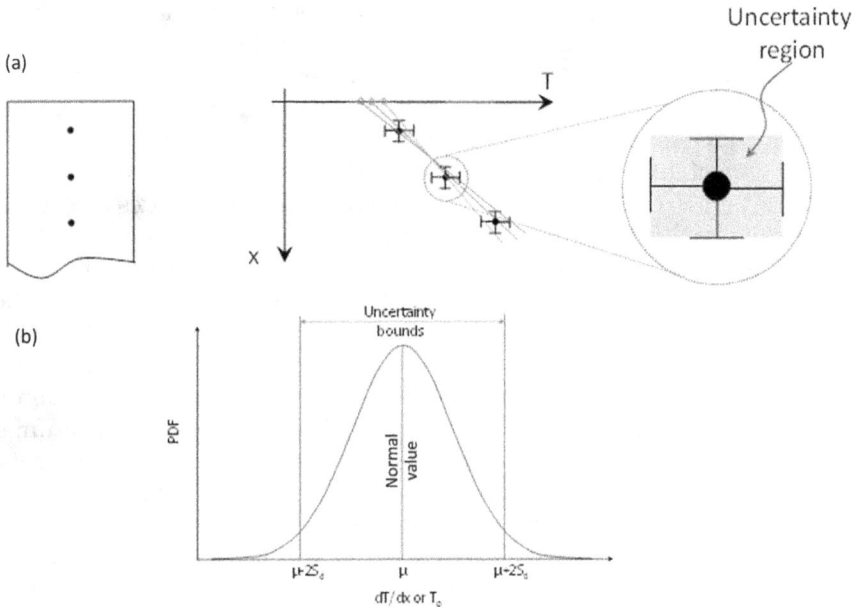

FIGURE 13.5
(a) Schematic possible linear regression fits to measured temperature vs. displacement data with uncertainties associated with both temperature and probe location, and (b) diagrammatic representation of a normal probability distribution curve for dT/dx or T_0 based on Monte Carlo simulation method.

small-diameter holes. They further demonstrate that the calculated surface temperature can have a greater precision than the individual temperature measurements.

Regardless, the techniques described in the earlier literature are based on relatively classical approaches and do not adequately address the fundamental issue with this measurement approach, which is that each measurement point has a region, contained within an area defined by the magnitudes of the spatial and temperature measurement error, within which the measurement may confidently lie, as depicted in Figure 13.5a. Thus, albeit bounded, there are an infinite number of least squares linear regression possibilities (three shown in Figure 13.5), each giving a unique estimation of the temperature gradient and the extrapolated wall temperature. Thus, the key challenge is to accurately and rigorously quantify the resulting uncertainty of dT/dx and T_o which includes multiple measurements, each with multiple uncertainties.

13.2.2 Monte Carlo Simulation of Uncertainties

To address the shortcomings of other approaches to the problem of propagation of both thermal and spatial uncertainties through least squares regression through meter bars, we suggest the straightforward and robust approach of computing the uncertainties of the surface temperature (intercept) and slope (temperature gradient) numerically using a Monte Carlo simulation. The technique is straightforward to implement, robust, and accurate. It is also advantageous in its ability to deal with non-uniform uncertainties among the temperature and position measurements and differing uncertainty distributions.

For the methodology outlined here, the temperature uncertainty is assumed to have a normal distribution, where the overall uncertainty depends on the factory specifications or the temperature calibration of the sensor and is equal to two standard deviations. Previous studies have suggested that the sensor location uncertainty also has a normal distribution; however, this implies that it is statistically possible for the probe to be located outside the hole, which is unphysical [15]. Additionally, there is no physical justification for the location uncertainty to have a normal distribution. Other authors have argued that, once the temperature sensors are fixed in place, this error is systematic and thus does not have as pronounced an effect. Here, we suggest the conservative yet realistic approach of modeling this uncertainty as a flat distribution bounded by difference in radius between the temperature probes and the holes.

The standard deviations for the slope (temperature gradient) and intercept (surface temperature) are then calculated by performing numerous randomized linear regression fits to the data constrained by the temperature and displacement uncertainty distribution at each point, as depicted in Figure 13.5a. The uncertainties of the slope and *y*-intercept are then taken as two standard deviations of this data set, as shown in Figure 13.5b. The general procedure is outlined below:

1. Define the measured temperatures and positions of each probe.
2. Define the measurement and positional uncertainty of each probe.
3. For each probe, use the Monte Carlo method to randomly generate a T–x data point within the defined uncertainty region.
4. Perform least squares linear regression on the data set.
5. Calculate the slope, dT/dx, and the intercept, T_o, for the data set.
6. Repeat steps 3–5 for numerous data sets (generally thousands to tens of thousands of data points) until the cumulative average of both dT/dx and T_o converges.
7. Calculate the nominal dT/dx and T_o based on the converged cumulative averages.
8. Estimate the uncertainty of dT/dx and T_o based on two standard deviations of the resulting normally distributed data sets.
9. Estimate the uncertainty of the heat flux, q'', based on the calculated uncertainty of the temperature gradient, dT/dx, and the thermal conductivity of the meter bar, k_{MB}.

The above is a generic description of the Monte Carlo simulation method for the estimation of HMBT uncertainty. However, it is important to recognize, as noted earlier, that this technique is used in a broad range of applications. For example, when used in convective boiling applications as in [4], the heat fluxes can be in the region of hundreds of W/cm^2, meaning that the temperature gradients are very steep. Thus, even with copper used as the meter bar, the calibration error and spacing of the temperature probes can be moderate while still achieving adequate measurement accuracy of the convective heat transfer coefficients. In contrast, [11] used the dual-meter-bar technique for characterizing high-performance TIMs. Here, a delicate balance was required between probe spacing, heat flux, and probe type and calibration in order to achieve the precision required to confidently characterize very low thermal resistance thermal joints. Albeit the two relatively extreme examples, together these illustrate that the meter bar design, probe type, and the associated calibrated accuracy depend on the situation(s) within which the HMBT is to be applied. It is thus imperative that the above method for error estimation is implemented during the initial design stage of the experiment. In this way, the desired uncertainty and/or uncertainty range can be defined at the outset of the experimental design, and the technique is used to inform all aspects of the experiment, including but not limited to the meter bar material, meter bar size, temperature probe type, size and calibrated accuracy, and probe spacing. This is the essence of the *"design for uncertainty"* approach to experimental design, where robust error analysis is the cornerstone around which the experiment is designed. In the following two sections, the examples mentioned above will be discussed in order to illustrate the HMBT in two very different applications.

13.3 An Example: High-Precision Thermal Interface Material Testing

The mitigation of thermal contact resistance in high-heat-flux systems is especially important in electronics industries where a poor contact between high-power components and their cooling systems can represent a major bottleneck for effective cooling, which results in reduced reliability or failures. Increased power densities and improved cooling architectures (particularly liquid and two-phase cooling) have resulted in further attention being given to reducing contact resistance. This has spawned the development of a new generation of high-performance thermal interface materials (TIMs) [8,17]. The specific thermal resistance of a given TIM depends on its thickness, L, and its effective thermal conductivity, k_{eff}, and is given by

$$RA = \frac{L}{k_{eff}} \qquad (13.12)$$

As such, the thin bond lines and increasingly thermally conductive properties of these materials can make accurate quantification of their thermal characteristics very challenging.

As outlined in the previous section, the thermal characterization of TIMs represents one of the classical applications of the HMBT, and this approach is perhaps the most common implementation of ASTM D5470 "Standard test method for thermal transmission properties of thermally conductive electrical insulation materials" which has emerged as the *de facto* standard for the steady-state characterization of TIMs by both researchers and manufacturers worldwide.

The specific thermal resistance of a TIM is measured by the HMBT apparatus using Eq. (13.9). Thus, the precision and accuracy of these measurements depend on a variety of factors, including heat flux, contact area, meter bar length, meter bar thermal conductivity, sensor locations, sensor size, and sensor accuracy and precision.

The primary challenge for these types of apparatuses is to quantify the very small temperature difference across the TIM samples. One strategy to measure extremely low thermal resistances is to increase the applied heat flux in order to drive up the temperature difference across the interface. However, doing so can vastly increase the temperature excursions of the meter bars above the ambient, resulting in increased heat loss to the surroundings, and this contributes to nonlinearities in the meter bar temperature distribution, thereby negating the principle of the HMBT. The alternative approach detailed in [11] is to use low heat fluxes and extremely carefully calibrated sensors to maintain good meter bar linearity and low measured and calculated uncertainties.

Ultimately, for given meter bar design and sensor positions, the uncertainty of the thermal resistance measurements obtained using the HMBT for TIM characterization relies primarily on the magnitude of two elemental uncertainties—specifically, the thermal uncertainties of the sensors and the spatial uncertainties in their locations. To examine the influence of only the thermal accuracy on the uncertainty of the measured specific thermal resistance, Eq. (13.9) can be expressed as

$$U_{RA} = \sqrt{\left(\frac{A}{Q}\right)^2 (U_{\Delta T})^2 + \left(\frac{-A\Delta T}{Q^2}\right)^2 (U_Q)^2} \tag{13.13}$$

where

$$U_Q = \sqrt{\left(\frac{KA}{\Delta x}\right)^2 (U_{\Delta T})^2} \tag{13.14}$$

and

$$U_{\Delta T} = \sqrt{U_{Ta}^2 + U_{Tb}^2}. \tag{13.15}$$

Here, it is assumed that the temperature uncertainty at the contacting surfaces is equal to the elemental sensor uncertainty and that all other uncertainties are zero. Thus, for given values of k_{mb}, A, Δx, and Q, the variation of the uncertainty U_{RA} in the specific thermal resistance Eq. (13.13) can be plotted as a function of sensor thermal uncertainty, U_T.

The U_{RA} evaluated for an apparatus with meter bars with a thermal conductivity of $k_{mb} = 215$ W/mK, a contact area of $A = 1600$ mm², a sensor spacing of $\Delta x = 20$ mm, and a heat flux of 2500 W/m² (or $Q = 4$ W) is shown in Figure 13.6.

Here, the values of the thermal uncertainty, U_T, correspond to the uncertainties reported in the literature from studies which employ the HBMT. These largely depend on the type of sensor used in the meter bars, and their calibration and instrumentation. Examples include embodiments which use thermocouples [18–20] where typically $U_T \approx 0.2$ K, RTDs [21] where $U_T \approx 0.05$ K and calibrated RTDs [22–25] where $U_T \approx 0.01$ K.

The horizontal line in Figure 13.6 represents 10% of the theoretical specific thermal resistance of a TIM with a thickness of 50 μm and a thermal conductivity of 5 W/mK, calculated using Eq. (13.12).

This demonstrates that, based on the assumptions of this simplified analysis and all other parameters being equal, the uncertainty of the

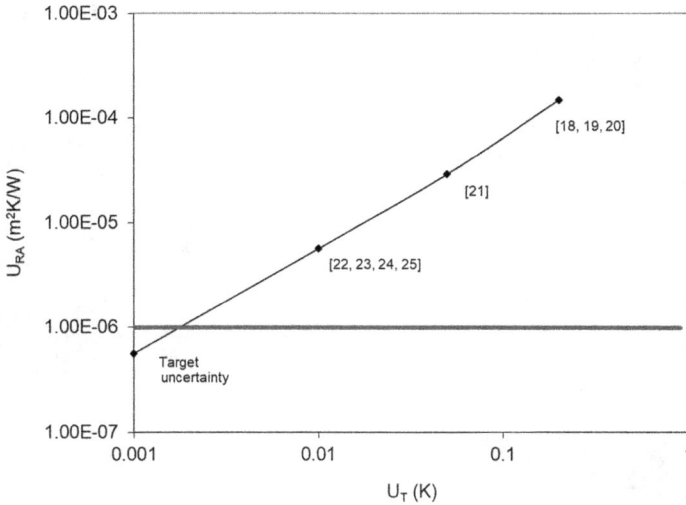

FIGURE 13.6
Variation of specific thermal resistance uncertainty, U_{RA}, with thermal sensor uncertainty, U_T, for a common set of experimental parameters.

thermal measurements must be on the order of 0.001 K in order to accurately characterize the performance of a TIM with a specific thermal resistance of $1e^{-5} m^2 K/W$.

This thermal uncertainty target was achieved using the embodiment of the HMBT described in [11] by using small (0.38 mm) and very sensitive thermistors. Thermistors tend to be very nonlinear, making them of limited use over large temperature ranges; however, due to the low heat fluxes required, temperature excursions were kept to a minimum and the thermistors could be carefully calibrated over a small temperature range (15°C–40°C). Additionally, by keeping the temperature differences low, this helps to minimize heat leakage to the ambient, which ensures a linear temperature distribution in the meter bars. Due to their nominally high electrical resistance, thermistors can be prone to self-heating errors during resistance measurement. This issue was overcome by employing a Lake Shore Model 370 AC resistance bridge, which uses AC excitation and low excitation current (3.16 μA), to perform the 4-wire resistance measurements. Considering any temperature instabilities in the calibration environment and curve fitting errors, the relative temperature uncertainty measurement between the thermistors was reduced to $U_T = 0.001$ K.

A photograph and sketch of the apparatus is shown in Figure 13.7, which illustrates the ancillary instrumentation for force, sample thickness, and the large stiff die set frame which was used to align the meter bars. Additional details regarding the meter bars, instrumentation, and calibration techniques can be found in [11].

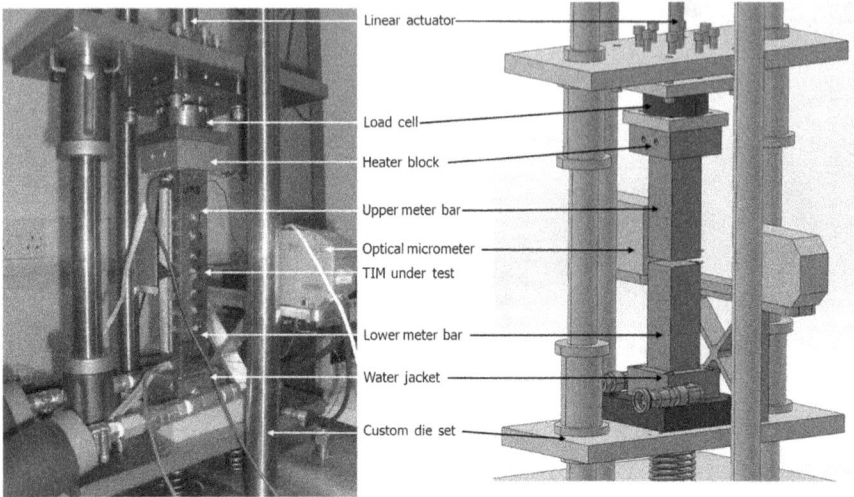

FIGURE 13.7
Photograph and sketch of HMB TIM test rig developed in [11].

A sketch of the upper and lower meter bars is shown in Figure 13.8 with the location of the thermistor holes dimensioned in the x-direction with respect to the surface of the meter bar in contact with the TIM. The location of the sensor closest to the contacting surface is crucial to the design of the meter bars: too far from the surface necessitates greater extrapolation; too close results in the perturbation of the ideal one-dimensional heat flow through the meter bar to be compromised by three-dimensional heat flow around the hole. Finite element simulations for a range of expected operating powers minimize the distance of this hole location from the contacting surface while maintaining a temperature uniformity of 0.001 K at the contact surface.

The locations of the remaining sensors were then tuned in to balance the effect of sensor positional uncertainty on the Monte-Carlo-simulated surface temperature and temperature gradient with the overall length and thermal conductivity of the meter bars (which dictates the maximum temperature excursion at the hot and cold ends of the meter bars and, therefore, the degree of heat loss). Note that the actual positions of the holes were measured using a microscope and micrometer stage after machining to ensure accuracy.

The sensitivity and precision of the meter bars were verified by performing self-contact interface tests to characterize meter bar performance for low thermal resistance scenarios. The temperature distribution in the meter bars during the self-contact test is shown in Figure 13.9. This extremely low power case ($Q = 0.262 \pm 0.006$ W) can be used to demonstrate the thermistor temperature sensitivity and linearity of the meter bars measurements with respect to the thermal uncertainty indicated by the error bars ($U_T = 0.001$K). The previously detailed Monte Carlo simulation was used to estimate the

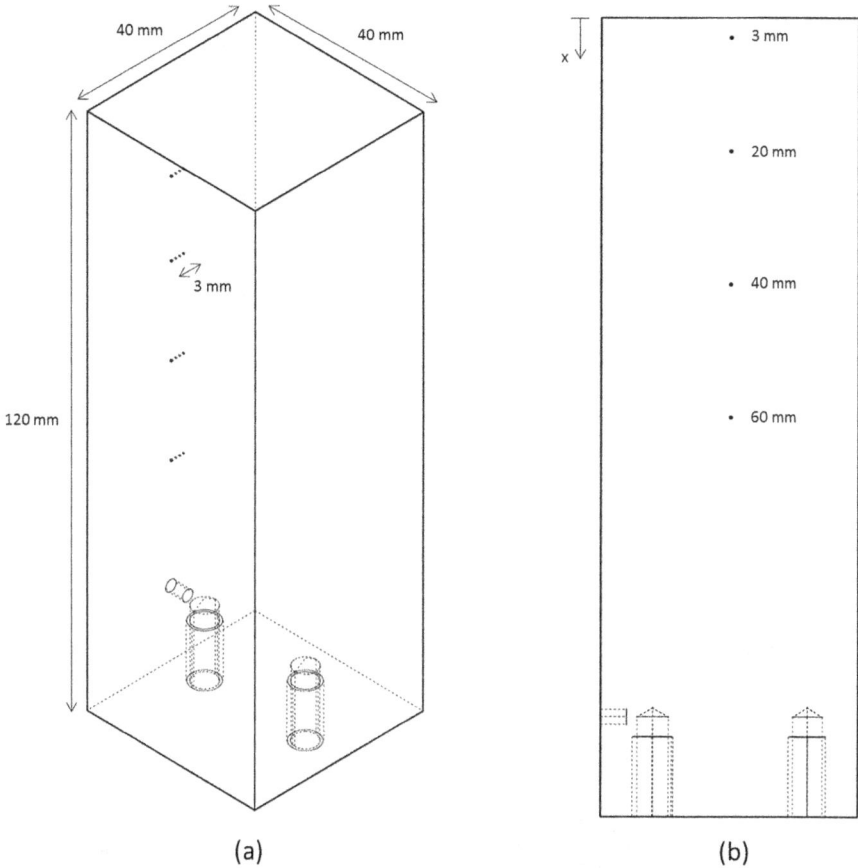

FIGURE 13.8
Heated meter bars used in [11] showing key dimensions and hole spacing.

uncertainty of the temperature gradient of each of the meter bars. The uncertainty of the temperature gradient was then propagated through Eq. (13.14) along with associated uncertainties in meter bar thermal conductivity, k_{mb}, and area, A, to calculate $U_Q = 0.006$ W.

Similarly, the Monte Carlo regression fits to the data indicate a temperature difference between the contacting surfaces of 0.0089 ± 0.0012 K. The heat transfer in the meter bars agrees to within 0.4%, demonstrating accurate energy balance. As the heater power is increased, these uncertainties drop significantly. Indeed, over a large range of input powers, the calculated heat fluxes between the two meter bars balance within their computed uncertainties, as illustrated in Figure 13.10.

A minimum specific thermal resistance of $4.68\,e^{-6}\,m^2\,K/W$ was measured with an uncertainty of 2.7% using a heat transfer rate of 16.8 W for this embodiment of the HMBT [11], which was below the target uncertainty for this apparatus.

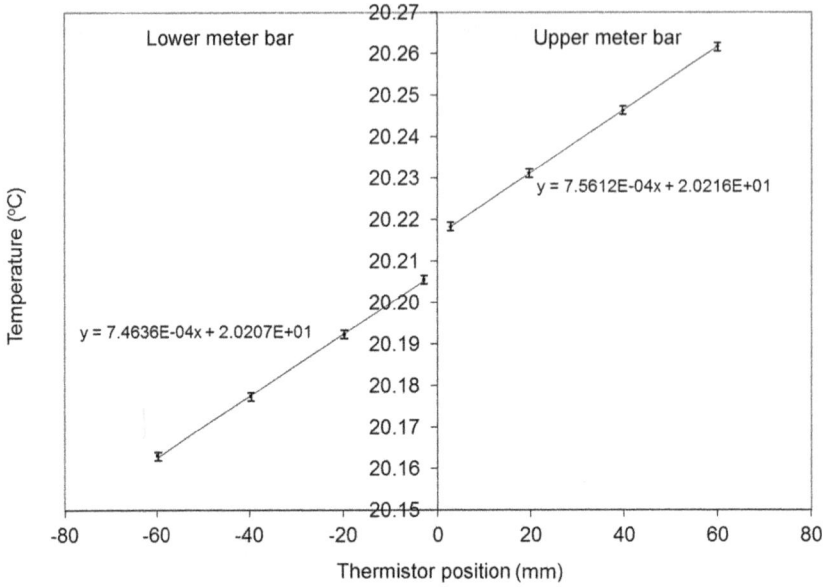

FIGURE 13.9
Low-heat-flux meter bar temperature distribution during self-contact, $Q = 0.262 \pm 0.006$ W, $P \approx 0.2\,\text{MPa}$ [11].

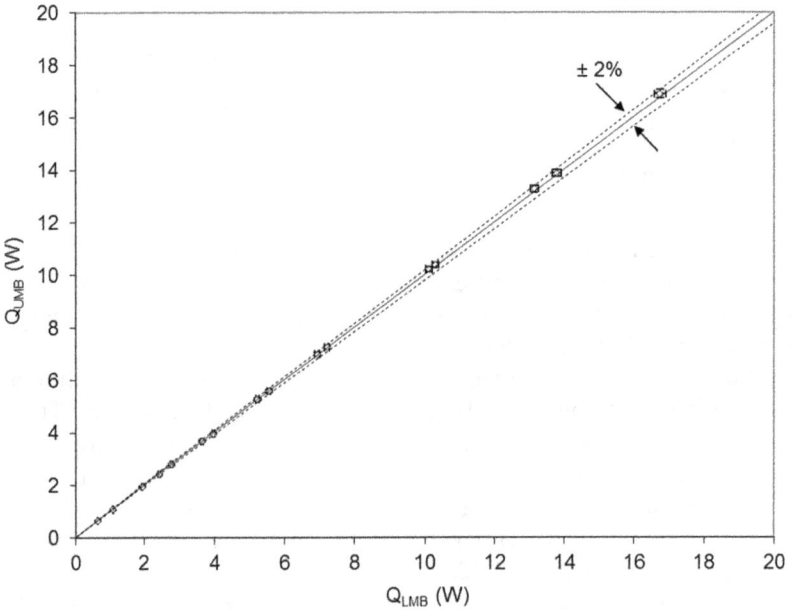

FIGURE 13.10
Energy balance between upper and lower meter bars for a range of powers [11].

13.4 An Example: High-Heat-Flux Forced Convective Boiling

The heat flux of some electronic components, such as power amplifiers and microprocessors, has escalated to the point that not only is air cooling no longer feasible, but conventional liquid cooling solutions are also no longer of adequate performance. The 100 W/cm² threshold has been passed, and components are now edging toward the 1000 W/cm² mark. This not only requires that very aggressive new heat transfer processes be explored to meet cooling requirements, but it also creates the experimental challenge of characterizing the heat transfer coefficient at heat fluxes in the 100–1000 W/cm² range.

One potential heat transfer process that can achieve extremely high heat transfer performance is jet impingement boiling. In a recent study by de Brún et al. [4], the HMBT was used to characterize the forced convective heat transfer coefficients and critical heat fluxes (CHFs) of confined boiling jet arrays. In this work, an experiment was designed for target heat fluxes of 500 W/cm² from a 1.5×1.5 cm heat transfer surface and, as will be discussed below, required careful coordination of the Monte-Carlo-based uncertainty analysis and finite element analysis (FEA) in order to implement a feasible HMBT.

Figure 13.11a shows a schematic of the jet impingement boiling test section. As shown, a jet manifold (heat sink) is positioned atop a heated meter bar block (heat source) in such a way that forced convective boiling occurs on the exposed surface. This is better illustrated in Figure 13.11b, which shows a close-up of the manifold–HMB interface. The figure shows the neck of the HMB with three thermocouple wells onto which a copper chip is situated. The detachable copper chip is used in this experiment such that different boiling surface structures can be tested without having to manufacture a heater system for each surface. This to some degree differentiates the use of HMBT from its traditional configuration since the thermal contact resistance and associated temperature drop between the HMB neck and the copper chip negate the direct extrapolation of the surface temperature. Here, a thermocouple is

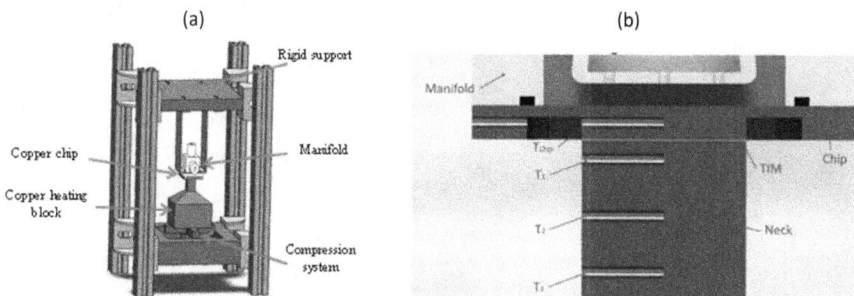

FIGURE 13.11
(a) Schematic of jet impingement boiling test section showing the jet manifold atop the heated meter bar system, and (b) a close-up diagram of the heated meter bar and jet manifold [4].

embedded in the copper chip in such a way that the heat flux is determined in the classical way via least squares linear regression of the temperature distribution in the HMB neck, which is then used in conjunction with the copper chip temperature, T_{chip}, to extrapolate the surface temperature.

It is noted in Figure 13.11 that the heater block is comprised of a large lower section and a proportionately small neck region. This is better depicted in Figure 13.12a and b. In Figure 13.12a, the assembled heater is shown with an exposed upper surface of a 1.5×1.5 cm neck. Figure 13.12b shows an assembly drawing of the heater block and copper chip system. This configuration, which differs considerably from that described in Section 13.3 and others like it, is a direct challenge to providing sufficient thermal power through the neck to achieve the desired heat flux level of circa 500 W/cm^2 without exceeding the temperature limit of the embedded cartridge heaters. At the same time, the HMBT demands that the neck be of a length such that the three thermal probes can be inserted with adequate distance between them and in a region where one-dimensional heat flow is assured. However, since the heat sink of the system is the small exposed boiling surface, the source-to-sink thermal resistance is very sensitive to the length of the heater bar neck. Thus, it must be as short as possible while still facilitating the HMBT. At the same time, thermal power in the region of 1500 W must be supplied by the embedded heater, which requires very careful positioning in the lower block.

It is clear from the above that there are tensions between reaching the desired thermal power supply through the neck, constraints on the maximum operating temperature of the heater elements, and a requirement of multiple adequately spaced probes in the neck and in the region of one-dimensional heat flow to facilitate the use of the HMBT at target accuracy. In order to achieve the required design requirements, an iterative design approach was implemented, whereby the Monte Carlo simulation and FEA were coordinated in order to achieve a suitable test heater block design, with the final design illustrated in Figure 13.12b. Figure 13.12c shows the FEA simulation results of the design configuration. As illustrated, five cartridge heaters, rated at 500 W each, are inserted in a U-shaped staggered interdigitated fashion in order to provide sufficient thermal power while

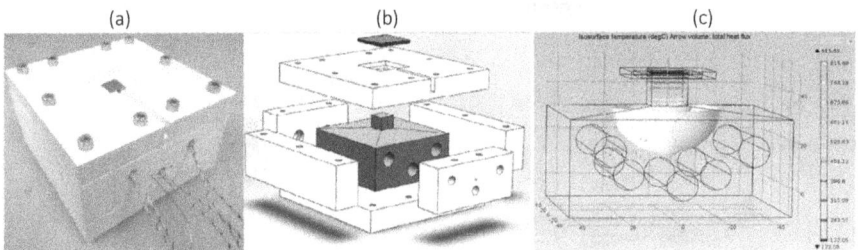

FIGURE 13.12
(a) Assembled heater block, (b) assembly drawing of heater block design, and (c) FEA of heater block design [4].

allowing a sufficiently low thermal resistance between each heater and the neck region—i.e., toward the heat sink—such that their temperature limits are not reached at the required power. The neck region is notably short since, as mentioned, its small cross-sectional area creates a significant heat flow restriction. However, as the FEA shows, 12mm was deemed sufficient to receive three 1.0-mm-diameter drilled wells, 3.0mm apart, into which the thermocouples were inserted and carried out in a region of one-dimensional heat flow. In contrast to the TIM apparatus described in Section 13.3, the high temperature gradients associated with the range of heat fluxes in these experiments were such that considerably smaller probe-to-probe spacing and standard, yet calibrated to ±0.1 K, thermocouples could be used. Also, in contrast to the TIM apparatus in Section 13.3, where extremely small temperature differences were required to be measured, the differences between the surface temperature and the ambient temperature (i.e., T_o-T_{amb} in Eq. (13.7)) are of the order of 1–40 K in convective boiling applications with water. This also relieves the accuracy requirement of the temperature probes. Finally, it should be noted that ultra-pure OFE copper was used in order to reduce to negligible levels the uncertainty associated with the thermal conductivity of the meter bar material when calculating the heat flux error.

Figure 13.13 shows an example of the measured convective boiling curves for different jet array configurations and jet Reynolds numbers. The graph shows that the driving temperature for heat transfer, here the wall superheat, ranges from 1 to 40 K, as mentioned earlier. The heat flux ranges from 20 to 450W/cm². However, since the temperature gradient in the neck of the heater block and the wall temperature increase with heat flux, the relative

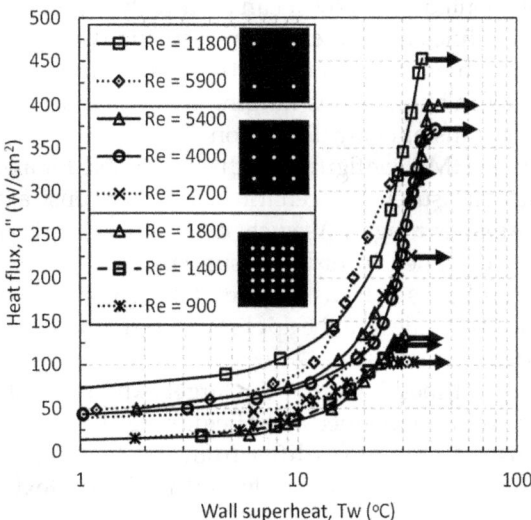

FIGURE 13.13
Flow boiling curves for different jet array configurations and jet Reynolds numbers.

TABLE 13.1

Selected Uncertainty Estimates Using Measured Data with the Monte Carlo
Error Estimation Method for the HMBT

	$q'' = 49.8\,\text{W/cm}^2$	$q'' = 386.6\,\text{W/cm}^2$
Error on q''	19.3%	2.5%
Error on ΔT	20.1%	3.0%

uncertainty of the heat flux and wall superheat decrease with increased heat
flux. In the experimental design phase, the target uncertainty range for both
heat flux and superheat was selected so as not to exceed 20% at a heat flux of
$50\,\text{W/cm}^2$ and a superheat of 1 K, as well as not to exceed 5% at a heat flux of
$500\,\text{W/cm}^2$ and a superheat of 40 K. Table 13.1 shows an example of the estimated experimental uncertainties using the technique described in Section
13.2 for the first and second last data points of the 3×3 jet configuration and
Re = 5400. (The last data point was unsteady due to partial dry-out.) As the
table shows, the estimated uncertainties are in adequate agreement with the
design targets for this experimental design.

13.5 Summary

The heated meter bar technique is a cornerstone measurement method in the
thermal-fluid science community. Its pervasive use is largely based on the
fact that, if implemented correctly, it can accurately measure the heat flux to
or from a surface while simultaneously measuring the surface temperature
in a non-contact fashion. It is an elegantly simple measurement technique
and has been used in applications ranging from thermal contact resistance
to convective heat transfer characterization.

There is no single HMB configuration that is suitable for all applications in
the sense that each measurement scenario has its own constraints, geometries,
and target performance ranges. As such, each new scenario demands that a
bespoke HMB be designed in order to suit specific needs and constraints.
At present, there still does not exist a general design guideline for the HMBT
that can be applied regardless of the application scenario, and this is what this
chapter is intended to address.

Specifically, it is outlined here that the target experimental uncertainty or
uncertainty ranges must be defined at the beginning of the initial experimental
design stage. This will, of course, differ from application to application and
thus must be considered very carefully and in the context of the application environment. The target uncertainties are then used as the fixed center around which the HMB is designed and will inform all key aspects of
the measurement system, including but not limited to the choice of sensors,

the level of required calibration, the layout of the sensors, and the geometry, layout, and material of the HMB. In this work, a straightforward and robust Monte Carlo method is described in order to resolve the issues associated with the estimation of uncertainty of least squares linear regression of three or more measurements with two or more error modes. This method, along with analytical and/or simulation techniques, is the key design tool required to construct HMBT experiments which achieve target uncertainties.

Since the HMBT can vary considerably between applications, two very different case studies were discussed in order to outline the proposed design approach. The first case study described an apparatus that used the dual-meter-bar technique with unprecedented accuracy and precision for TIM characterization. Here, very low heat fluxes were targeted; these required extremely accurate temperature probes, meter bar material of moderate thermal conductivity, and relatively large spacing between probes. In the second case study, the HMBT was described for the experimental characterization of very high heat flux convective boiling. Compared with the precision TIM application, this application resulted in a very different HMB because the design had to find a balance between performance range, temperature constraints, and target uncertainty ranges. This required the coordination of FEA and the Monte Carlo method to iteratively converge to a feasible design solution.

In summary, the HMBT is an ideal measurement technology for characterizing a broad range of heat transfer scenarios. However, it is not without pitfalls. It is hoped that this chapter provides a solid foundation and pragmatic approach to help navigate these pitfalls and aid practitioners in the design and application of the HMBT for their unique engineering or scientific applications.

References

1. Incopera FP, De Witt DP. *Introduction to Heat Transfer*. 2nd ed. John Wiley and Sons Inc., 1990.
2. Robinson AJ, Schnitzler E. "An experimental investigation of free and submerged liquid miniature jet array impingement heat transfer," *Experimental Thermal & Fluid Science*, 32, 2007, pp. 1–13.
3. Petrovic S, Robinson AJ, Judd RL. "Marangoni heat transfer in subcooled nucleate pool boiling," *International Journal of Heat and Mass Transfer*, 47 (23), 2004, pp. 5115–5128.
4. de Brún C, Jenkins R, Lupton TL, Lupoi R, Kempers R, Robinson AJ. "Convective boiling of confined impinging jet arrays," *Experimental Thermal Fluid Sciences*, 86, 2017 pp. 224–234.
5. Kim K, Jeong JH. "Condensation mode transition and condensation heat transfer performance variations of nitrogen ion-implanted aluminum surfaces," *International Journal of Heat and Mass Transfer*, 125, 2018, pp. 983–993.

6. McWaid T, Marschal E. "Thermal contact resistance across pressed metal contacts in a vacuum," *International Journal of Heat and Mass Transfer*, 35 (2), 1992, pp. 2911–2920.

7. Zhang P, Xuan Y, Li Q. "A high-precision instrumentation of measuring thermal contact resistance using reversible heat flux," *Experimental Thermal and Fluid Science* 54, 2014, pp. 204–211.

8. Razeeb KM, Dalton E, Cross G, Robinson AJ. "Present and future thermal interface materials for electronic devices," *International Materials Reviews*, 63, 2018, pp. 1–21.

9. Kempers R, Lyons AM, Robinson AJ. "Modelling and experimental characterization of metal micro-textured thermal interface materials," *ASME Journal of Heat Transfer*, 136, 2013, p. 011301.

10. Karimi G, Li X, Teertstra P. "Measurement of through-plane effective thermal conductivity and contact resistance in PEM fuel cell diffusion media," *Electrochimica Acta*, 55, 2010, pp. 1619–1625.

11. Kempers R, Kolodner P, Lyons A, and Robinson AJ. "A high-precision apparatus for the characterization of thermal interface materials," *Review of Scientific Instruments*, 80, 2009, p. 095111.

12. Wald A. "The fitting of straight lines if both variables are subject to error," *Annals of Mathematical Statistics*, 11, 1940, p. 284.

13. Bartlett MS. "Fitting a straight line when both variables are subject to error," *Biometrics* 5, 1949, pp. 207–212.

14. Press WH, Teukosky SA, Vetterling WT, Flannery BP. "*Numerical Recipes in FORTRAN - The Art of Scientific Computing*," 2nd ed. Cambridge University Press, Cambridge, chapter 15, 1992.

15. Kedzierski MA, Worthington JL. "Design and machining of copper specimens with micro holes for accurate heat transfer measurements," *Experimental Heat Transfer* 6, 1993, pp. 329–344.

16. Ku HH. "An introduction to the statistical treatment of measurement data". NBS Report 8677, 1965.

17. Hansson J, Nilsson TMJ., Ye L, Liu J. "Novel nanostructured thermal interface materials: A review," *International Materials Reviews*, 63(1), 2018, pp. 22–45, DOI: 10.1080/09506608.2017.1301014.

18. Rao VV, Bapurao K, Nagaraju J, Krishna Murthy MV. "Instrumentation to measure thermal contact resistance," *Measurement Science and Technology*, 15, 2004, pp. 275–278.

19. Misra P, Nagaraju J. "Test facility for simultaneous measurement of electrical and thermal contact resistance", *Review of Scientific Instruments* 75, 2004, pp. 2625–2630.

20. Singhal V, Litke PJ, Black AF, Garimella SV. "An experimentally validated thermo-mechanical model for the prediction of thermal contact conductance", *International Journal of Heat and Mass Transfer* 48, 2005, pp. 5446–5459.

21. Gwinn JP, Saini M, Webb RL. "Apparatus for accurate measurement of interface resistance of high performance thermal interface materials". *Proceedings of the 2002 IEEE Inter-Society Conference on Thermal Phenomena*, 2002, pp. 644–650.

22. Kearns D. "Improving accuracy and flexibility of ASTM D5470 for high performance thermal interface materials". *Proceedings of the 19th IEEE Semi-Therm Symposium*, 2003, pp. 129–133.

23. Culham JR, Teertstra P, Savija I, Yovanovich MM. "Design, assembly and commissioning of a test apparatus for characterizing thermal interface materials", *Proceedings of the 2002 IEEE Inter-Society Conference on Thermal Phenomena*, 2002.
24. Savija I, Culham JR, Yovanovich MM. "Effective thermophysical properties of thermal interface materials: Part 2 experiments and data", *Proceedings of InterPACK* 2003, Maui, HI, July 6–11, 2003.
25. Teertstra P. "Thermal conductivity and contact resistance measurements for adhesives", *Proceedings of InterPACK 2007*, Vancouver, BC, July 8–12, 2007.

14

Inverse Problems in Heat Conduction: Accurate Sensor System Calibration

Stefan Loehle and Fabian Hufgard

University of Stuttgart

CONTENTS

14.1 Introduction

Inverse heat conduction problems (IHCPs) are problems related to the determination of unknown surface quantities from in-depth sensor data. One prominent example of an IHCP is the determination of surface heat flux during the atmospheric entry maneuver of a space capsule from in-depth temperature data measured using thermocouples. The aerodynamic heating during reentry is so high that a direct measurement of surface quantities is impossible [1].

The calculation of a solution for a particular IHCP is difficult, because inherently occurring inaccuracies during the measurement at the sensor location may result in unstable solutions of the IHCP.

There are many ill-posed problems and a wide variety of solutions in order to determine a reasonable quantity from noisy data sets. For an overview on the various methods, the reader is referred to the excellent books of Alifanov,

Beck, Özisik, and Thikonov [2–5]. In the present text, the focus is set to one of the fundamental problems in high-temperature thermal engineering: the measurement of surface heat flux from in-depth measurement data. This chapter introduces us to the accurate determination of surface heat flux by the calibration of the sensor system and the application of modern system identification methods. Figure 14.1 shows a heat shield material sample during testing in a high-enthalpy flow filed comparable to the situation during atmospheric entry. The gas temperatures are of the order of 10,000 K. The material surface reaches 3,000 K [6,7].

For the illustration of the method within this chapter, a simple problem is considered. A block made of copper is heated on one side, and the temperature is measured inside the material using a conventional K-type thermocouple. Figure 14.2 illustrates this problem. On the left side, a computer model is

FIGURE 14.1
Heat shield material test in high-enthalpy air plasma flow.

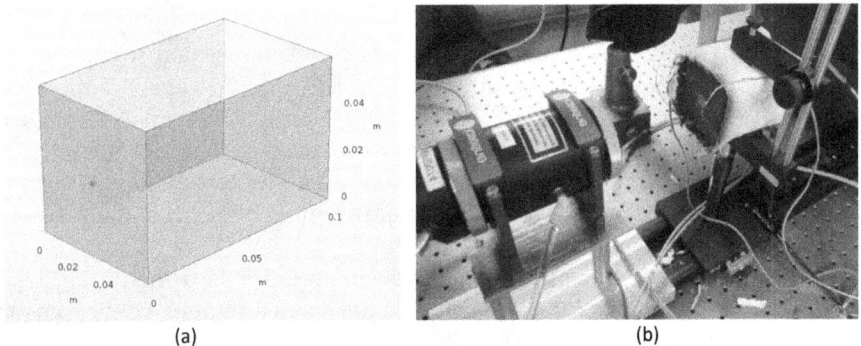

(a)   (b)

FIGURE 14.2
The *inverse heat conduction problem* as a copper block with in-depth thermocouple: (a) model; (b) experiment.

shown with the heated surface showing to the left and the temperature sensor just below the surface marked as a dot. On the right side, the corresponding experimental setup is shown with a laser as the heating source in front of the heated surface. An aluminum oxide layer was wrapped around the copper block aiming at an improved adiabatic boundary.

The variable of interest is the unknown transient heat flux at the surface $q''(0, t)$. The double prime stands for the *area-specific* heat flux. The measured data are the in-depth transient temperature $T(x = d, t)$. Note that throughout this chapter, the surface heat flux is considered as the heat flux into the surface; that is, all effects at the surface such as the interaction with the ambient are not considered with the surface heat flux. This heat flux is commonly named *net heat flux*.

14.2 Mathematical Model

This section provides the mathematical background for the system identification using the example of a one-dimensional, semi-infinite heat conduction problem. Using the analytical solution of this problem, the identification procedure is described, which uses the *non-integer system identification* (NISI) approach [8,9].

14.2.1 One-Dimensional, Semi-Infinite Heat Conduction Problem

The principle of the system identification procedure is best described by the analytical mathematical description of the one-dimensional, semi-infinite heat conduction problem as it is sketched in Figure 14.3. The heat conduction problem can be described by the heat equation as, for example, given in [4] with boundary and initial conditions:

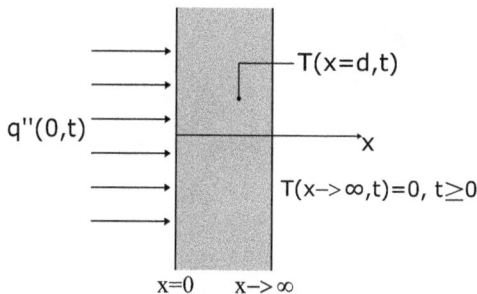

$q''(0,t)$

$T(x=d,t)$

x

$T(x->\infty,t)=0,\ t\geq 0$

$x=0$ $\quad x->\infty$

FIGURE 14.3
One-dimensional, semi-infinite heat conduction problem.

$$\rho c_p \frac{\partial T(x,t)}{\partial t} = \frac{\partial}{\partial x} \lambda \frac{\partial T(x,t)}{\partial x} \quad x \in [0,w], t \geq 0,\tag{14.1a}$$

$$-\lambda \frac{\partial T(0,t)}{\partial x} = q''(0,t),\tag{14.1b}$$

$$T(x \to \infty, t) = 0, \quad t > 0,\tag{14.1c}$$

$$T(x, t = 0) = T_0, \quad x \geq 0.\tag{14.1d}$$

Equation (14.1a) is a partial differential equation, and Eq. (14.1b) is the so-called flux or Neumann boundary condition [3]. The temperature in this formulation is the temperature difference to an isothermal initial condition $T(x, 0) = T - T_0 = 0$. In this work, the problem described by Eqs. (14.1a)–(14.1d) is solved by using the Laplace transformation and the thermal diffusivity $\alpha = \lambda/(\rho c_p)$ is introduced as, for example, in [10]:

$$\mathcal{L}\left\{\frac{1}{\alpha}\frac{\partial T(x,t)}{\partial t}\right\} = \mathcal{L}\left\{\frac{\partial^2 T(x,t)}{\partial x^2}\right\} \quad x \in [0,w], Re(s) > 0$$

$$\frac{1}{\alpha}\left(s\hat{T}(x,s) - \hat{T}(x,0)\right) = \frac{d^2\hat{T}}{dx^2} \quad \hat{T}(x,0) = 0\tag{14.2}$$

$$\frac{d^2\hat{T}}{dx^2} - \frac{s}{\alpha}\hat{T}(x,s) = 0.$$

With the Laplace transformation, the time dimension in the original partial differential Eq. (14.1a) is removed. The solution to the resulting ordinary differential Eq. (14.2) is found with the ansatz

$$\hat{T}(x,s) = A(s)\,e^{-kx} + B(s)e^{kx} \quad x \in [0,\infty], Re(s) > 0,\tag{14.3}$$

with $k = \sqrt{\dfrac{s}{\alpha}}$.

The second boundary condition, Eq. (14.1c), in the Laplace domain is

$$\mathcal{L}\{T(x \to \infty, t)\} = \mathcal{L}(0)\tag{14.4}$$

$$\hat{T}(x \to \infty, s) = 0\tag{14.5}$$

Applying Eq. (14.5) and

$$\lim_{x \to \infty} e^{-x} = 0\tag{14.6}$$

to Eq. (14.3) results in

$$\hat{T}(x \to \infty, s) = 0 = \lim_{x \to \infty} \left(0 + Be^{kx}\right), \tag{14.7}$$

which gives $B = 0$.

The second constant A is found with the heat flux boundary condition. In the Laplace domain, Eq. (14.1b) is

$$\hat{q}''(x = 0, s) = -\lambda \frac{\partial T}{\partial x}. \tag{14.8}$$

The equation to solve is

$$\hat{T}(x, s) = Ae^{-kx}. \tag{14.9}$$

Differentiation of Eq. (14.9) with respect to x is

$$\frac{\partial \hat{T}}{\partial x} = -Ake^{-kx}. \tag{14.10}$$

Applying Eq. (14.8) to this equation results in

$$-\frac{\hat{q}''}{\lambda} = -Ake^{-kx} \tag{14.11}$$

and at $x = 0$, this gives

$$-\frac{\hat{q}''}{\lambda} = -Ak \tag{14.12}$$

and hence

$$A = \frac{\hat{q}''}{\lambda k}. \tag{14.13}$$

The solution of the differential equation in the Laplace domain is then

$$\hat{T}(x, s) = \hat{q}'' \frac{1}{\lambda k} e^{-kx}. \tag{14.14}$$

14.2.2 Heat Conduction System

If the heat conduction problem is considered as a *heat conduction system*, the problem can be sketched as shown in Figure 14.4 with the system input $q''(0, s)$ and the system output $T(x, s)$. The transfer function $H(x, s)$ is the system's behavior with respect to the transport of heat into the body resulting in

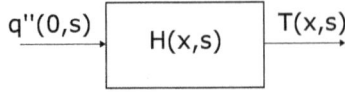

FIGURE 14.4
Heat conduction system with transfer function (*heat kernel*) $H(x, s)$, input heat flux $q''(x, s)$, and output temperature $T(x, s)$.

a temperature rise at a certain location $x = d$. A system is fully characterized by the so-called impulse response [11]; in the present case, this corresponds to the temperature rise following a *Dirac* (or *Heaviside* function) input. From an impulse response, the transient behavior of a system is fully known, for example the delay until the heat applied to the surface reaches the temperature sensor to a measurable extent, or the way the system stores heat due to heat capacity. If a complete transient problem is to be solved, i.e., heat flux of different magnitudes occurred over a certain amount of time, *Duhamel's theorem* can be applied. The transient problem is divided in time steps Δt of impulses of different magnitudes, and these impulses are superposed [3]. From Eq. (14.14), this means

$$H(x,s) = \frac{\hat{T}(x,s)}{\hat{q}''(0,s)} = \frac{1}{\lambda}\frac{1}{k}e^{-kx}, \tag{14.15}$$

or reformulating with $k = \sqrt{\dfrac{s}{\alpha}}$

$$H(x,s) = \frac{1}{\lambda}\frac{1}{\sqrt{\dfrac{s}{\alpha}}}e^{-\sqrt{\frac{s}{\alpha}}x}. \tag{14.16}$$

In the time domain, this results in

$$H(x,t) = \frac{1}{\sqrt{\lambda c_p \rho 4\pi}}\frac{1}{\sqrt{t}}e^{\frac{-x^2}{4\alpha t}}. \tag{14.17}$$

In the classical treatment of differential equations, the heat equation is a linear partial differential equation of the type (see, for example, [12])

$$\frac{\partial}{\partial t}T(x,t) = -\alpha\frac{\partial^2}{\partial x^2}T(x,t). \tag{14.18}$$

The kernel of this system with the boundary conditions as stated in Eqs. (14.1b) and (14.1d) is Eq. (14.17).

14.2.3 System Identification

A system identification approach is a method for the calibration of an actual sensor system [11]. In heat conduction, this means the identification of a system with the application of a known calibration heat flux, for example, provided by a laser source. The general idea is to identify the system's behavior when an input (i.e., a heat flux) occurs. If a system's impulse response can be determined, the system and, in the present case, the heat conduction system are fully characterized [11].

The system identification essentially means to find the correct transfer function $H(x, s)$ for a particular system. Therefore, Eq. (14.16) is modified by applying the series expansion

$$e^{-\sqrt{\frac{s}{\alpha}}x} = \sum_{n=0}^{\infty} \frac{\left(-x\sqrt{\frac{s}{\alpha}}\right)^n}{n!} \tag{14.19}$$

to

$$H(x,s) = \frac{1}{\lambda} \frac{1}{\sqrt{\frac{s}{\alpha}}} \sum_{n=0}^{\infty} \frac{\left(-x\sqrt{\frac{s}{\alpha}}\right)^n}{n!} = \frac{1}{\sqrt{\lambda \rho c_p}\sqrt{s}} \sum_{n=0}^{\infty} \frac{\left(-x\sqrt{\frac{s}{\alpha}}\right)^n}{n!}, \tag{14.20}$$

which can be reformulated to

$$H(x,s) = \frac{\hat{T}(x,s)}{\hat{q}''(0,s)} = \frac{1}{\sqrt{\lambda \rho c_p}} \sum_{n=0}^{\infty} \frac{(-1)^n s^{(n-1)/2} x^n}{\alpha^{n/2} n!}. \tag{14.21}$$

The transformation of Eq. (14.21) into the time domain leads to non-integer derivations. Without going into further mathematics here, one can imagine a full derivative (d/dt) as twice a half-derivative $(d^{0.5}/dt^{0.5})$. Defining D as the operator for derivation, i.e., $D^n = (d^n/dt^n)$ with $n \in \mathbb{R}$, Eq. (14.21) in time is

$$\sum_{n=-M_0}^{M} \alpha_n D^{n/2} T(x,t) = \sum_{n=-L_0}^{L} \beta_n D^{n/2} q''(0,t). \tag{14.22}$$

The parameters α_n and β_n are unknown constants for problems of the type in Eqs. (14.1a)–(14.1d). Knowing the parameters allows the transfer of heat flux occurring at the surface (and into the solid) to a temperature for this particular system. Equation (14.22) was derived on the basis of the one-dimensional, semi-infinite heat conduction problem. In the past, it was shown that the equation holds also for problems with 2D effects or transpiration cooling environments [13]. For an adiabatic back boundary condition, the approach was also analytically derived [9].

The calibration of a system, i.e., the identification of the parameters α_n and β_n in the NISI Eq. (14.22), requires the solution of a linear least-square regression problem [14,15]. The method shown here is a matrix approach as described in more detail in Ljung [11]. During calibration, a signal $Y(t)$ measured at time t differs from the real one $y(t)$ by an error $e(t)$ which is written as

$$Y(t) = y(t) + e(t) \tag{14.23}$$

Equation (14.22) can be written as

$$D^{M_0/2}Y(t) = \sum_{n=L_0}^{L} \beta_n D^{n/2}q''(t) - \sum_{n=M_0+1}^{M} \alpha_n D^{n/2}Y(t) + \sum_{n=M_0}^{M} \alpha_n D^{n/2}e(t), \tag{14.24}$$

where α_{M_0} is set to 1. Equation (14.24) can be written as a regression formula of the form

$$D^{M_0/2}Y(t) = \mathbf{H}(t)\theta + \varepsilon(t) \tag{14.25}$$

where

$$\mathbf{H}(t) = \left[-D^{(M_0+1)/2}Y(t)\ldots - D^{M/2}Y(t) \quad D^{L_0/2}q''(t)\ldots D^{L/2}q''(t) \right] \tag{14.26}$$

and

$$\theta = \left[\alpha_{M_0+1}\ldots\alpha_M\beta_{L_0}\ldots\beta_L \right]^T. \tag{14.27}$$

During data acquisition, K discrete values are recorded for K constant time steps Δt. The linear regression Eq. (14.25) is hence the matrix formulation

$$D^{M_0/2}Y_K(t) = \mathbf{H}_K(t)\theta + E_K \tag{14.28}$$

where the measured temperatures $Y(t)$ become a vector $Y_K(t) = [Y(t)Y(t + \Delta t)\ldots Y(t + K\Delta t)]$ and $\mathbf{H}$ becomes a matrix. The error $e(t)$ converts to the vector for discrete times E_K. The unknowns in Eq. (14.25) are the vector of parameters θ. Equation (14.25) is solved for θ with the minimum error $E_K = 0$ to

$$\theta = \left(\mathbf{H}_K^T\mathbf{H} \right)^{-1} \mathbf{H}_K^T D^{M_0/2}Y_K. \tag{14.29}$$

A system's parameters are thus identified when Eq. (14.29) is solved. The vector θ contains the parameters α_n and β_n. The found parameters can be used to calculate the impulse response. This is realized by solving Eq. (14.22) with the known parameters α_n and β_n for a heat flux vector $q'' = [10,000\ldots K_0]^T$.

From the system point of view, this is calculating a forward solution, known $q''(t)$ into unknown $T(x = d, t)$.

14.2.4 Inverse Heat Conduction Problem Solution Using NISI

When the system is calibrated, this means that the system's behavior to a heat impulse is known. Thus, an unknown transient temperature recording can be converted into a heat flux. This is called *inversion*.

There are some assumptions to be considered. First, the problem treated within this chapter is linear. This means that in Eq. (14.1a), the thermophysical properties are temperature independent. Second, the system identification presented is based on a heat flux input at the surface and into the surface. This means that an inverse heat conduction determination results *only* in the net heat flux into the surface. If chemical reactions close to the surface result in the heating of the solid body and the temperature rises, this is interpreted as a heat flux when NISI is applied. If a sensor is meant to measure the radiative component only, this boundary condition would result in an incorrect heat flux value.

On the other hand, an enormous advantage of this method based on a sensor system calibration is that due to the calibration procedure, the thermophysical properties and details of the temperature sensor have not to be known, since its influence on the data acquisition is taken into account by the sensor calibration.

The temperature signal inside the body for a linear problem, i.e., constant thermophysical properties, can be formulated as a linear combination of scaled and time-shifted impulse responses. For a linear problem, the scaling is linear. In an interval $[0, \Delta t]$, the heat flux $q''(t)$ is

$$T(t) = q''(t) \cdot I(t), \quad t \geq \Delta t, \qquad (14.30)$$

which is a simple inversion of a scalar product. For many time steps, the signal can be calculated by

$$T(t) = \sum_{i=0}^{\frac{t}{h}-1} q''(ih)I(t - ih). \qquad (14.31)$$

Figure 14.5 shows the resulting temperature signal from three heat fluxes at the surface at different times (0, 5, and 15 s). The very first increase in temperature is the actual impulse response since a heat flux of $q'' = 1$ W/m² is applied. For the inverse method, it can be seen that in order to identify the heat flux at later times, the heat flux until that time starting from an isothermal system has to be known or determined. It is therefore required to provide an isothermal starting condition of the system to be used. It is less important when this isothermal state is realized, but it is important that

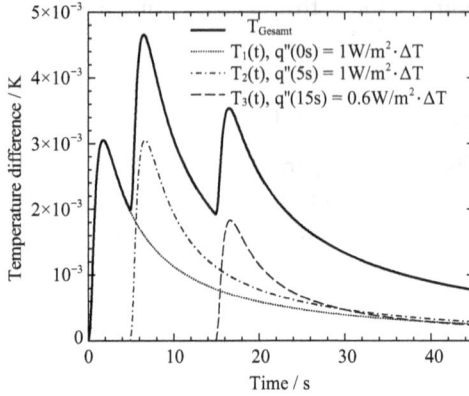

FIGURE 14.5
Calculation of a transient temperature signal based on the impulse heat fluxes of different amplitudes and times. (Reproduced from Ref. [16].)

from this time on, the temperature data are recorded. Based on Eq. (14.31), the rise in heat flux can be determined from the temperature data. For the first step Δt, i.e., the isothermal system, there is no temperature rise to be considered. Every next step induces a certain temperature rise. In order to calculate the heat flux at an inversion step $t > t = 0$, the fluxes ahead have to be removed with

$$T_S(t) = T(t) - \sum_{i=0}^{\frac{t}{h}-2} q''(ih)I(t-ih).$$ (14.32)

The temperature step can then be inverted to

$$q''(t-1) = \frac{T_S(t)}{I(1)}.$$ (14.33)

A potential regularization of the inversion is required to keep the inversion process stable. One method is Beck's sequential function estimation method [3].

14.3 Sensor System Calibration

The sensor system calibration means the collection of data in order to identify the system's behavior, and for the present case of a heat conduction system, this means a setup for the determination of the unknown model parameters

α_n and β_n. From the perspective of system identification theory, a particular sensor system has to be calibrated using a heat flux step (Heaviside) or a heat flux impulse. Differentiating the Heaviside function results in the Dirac (δ) function.

However, for real systems, an infinitely short pulse with infinite amplitude is not possible. Also, a true step function has an infinitely sharp rise, which is not possible in a real experiment. Therefore, the calibration is realized by a chain of short pulses of different (pseudorandom) pulses. This ensures for a certain pulse length variation that the system's temporal behavior is sufficiently covered.

A further problem occurs in the system calibration if there is a time delay between the surface heat flux and the temperature response. The fundamental heat equation implies that in the instant a heat flux affects the surface, an—extremely—small temperature rise occurs in the whole domain. However, in reality, the temperature does not rise until a significant amount of heat reaches the sensor. Kaviany provides a time estimate for this duration called the *penetration time*, defined as [17]

$$t_p = \left(\frac{d}{3.6 \sqrt{\frac{\lambda}{\rho c_p}}} \right)^2 .$$ (14.34)

There are two possibilities to overcome this issue: an increase in the time stepping so that a temperature signal during the first time step becomes measurable, or the delay has to be accounted for in the inversion and identification routine by setting the temperature to $T = 0$ for the time steps $t < t_p$. If this is not considered, the calibration and hence the inversion are unstable.

Since 2006, the system identification methods are applied by the High Enthalpy Flow Diagnostics Group (HEFDiG) for heat conduction problems using a diode laser system as a radiative calibration heat flux. Laser systems provide very short pulses at comparably high radiating powers while keeping comparably short rise times in the order of 0.1 ms. The main drawback of using laser systems is that the surface absorptivity must be provided for the system to be identified. The radiation can be reflected, absorbed, or transmitted. Opaque bodies have zero transmission, so the measurement of the surface reflectivity is also a measure of the surface absorptivity. The surface reflectivity is temperature, wavelength, and angular dependent. In the application of thermal problems, the heating is usually realized normal to the surface and, therefore, literature data may be found. For copper, the reflectivity was measured including wavelength dependency for a polished and an oxidized surface. Figure 14.6 shows these curves (courtesy CEA). In the sense of calibration, it is required that there is no surface state change between calibration and application.

(a)

(b)

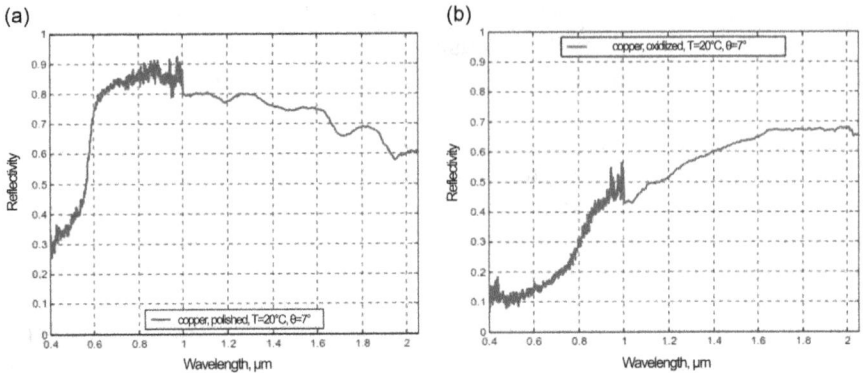

FIGURE 14.6
Surface reflectivity for copper: (a) polished; (b) oxidized. (Courtesy CEA.)

14.4 Application

In this section, the presented approach is applied to an experimental system as sketched in the introduction (Figure 14.2). The dimensions of the copper block are chosen such that a semi-infinite heat conduction problem is designed. The same problem was tried to design using COMSOL. This section now aims at highlighting that even a very detailed and powerful 3D numerical simulation of the problem is not covering all experimental aspects of a particular system. In the present case, the system was chosen to be very simple. But even this simple system cannot be modeled accurately, i.e., in a way that the experimental data are reproduced. The calibration approach, however, allows taking all the unknown susceptibilities into account, resulting in an impulse response of the real system.

Thus, the advantages of the application of system identification are shown here by comparing real experiments with a numerical simulation of the calibration using finite element modeling. The problem shown in Figure 14.2a was modeled in COMSOL as a 3D finite element model. The material properties taken are similar to those of copper ($\rho = 8920 \text{ kg/m}^3$, $c_p = 387 \text{ kJ/kg/K}$, and $\lambda = 394 \text{ W/m/K}$). Moreover, on the front side, the surface is assumed to re-radiate with an emissivity of 0.85. In the experiment, the heating at the front side is realized using a diode laser system which heats a circular surface area with a diameter of 48 mm. A thermocouple is glued in a hole drilled in parallel to the surface to the center of the sample at a depth from the surface of $d = 2 \text{ mm}$. The thermocouple was not modeled in the COMSOL model.

Figure 14.7 shows the heated surface of the numerical model. Figure 14.8 shows the calibration runs for both the numerical model and the experiment. In the experiment, the heat flux is applied as a radiative heat flux to the

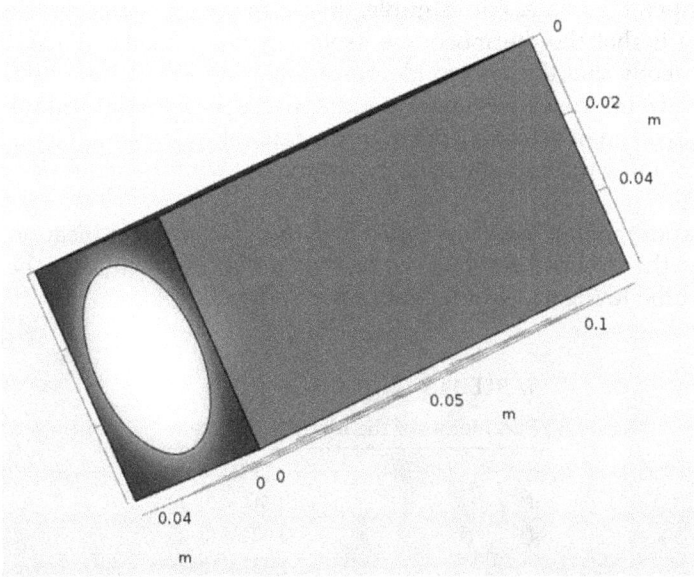

FIGURE 14.7
Heating as modeled in COMSOL.

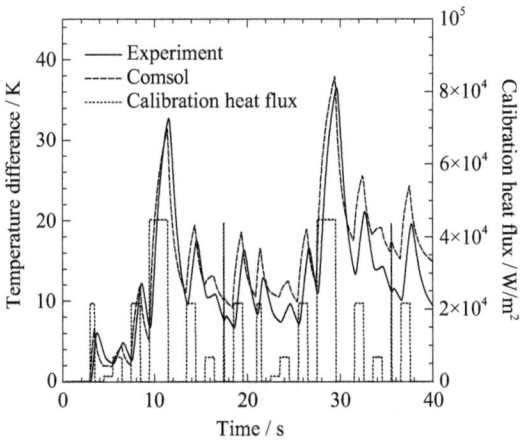

FIGURE 14.8
Comparison of temperature result from calibration in COMSOL and the experiment.

surface and the net heat flux is calculated with the assumption of a surface absorptivity of 0.85. This net heat flux was taken in COMSOL as the surface heat flux. As can be seen, there is a difference between the two tempera-ture data sets. The exact source is hard to find, because many different influ-ences in the experiment as well as in the numerical simulation can affect the local temperature data. For example, one of the biggest inaccuracies in the modeling is that the thermocouple itself was not modeled. Thus, this plot shows directly that the system identification approach of the actual system significantly improves the measurement, because potential unknowns are inherently taken into account through the calibration process. The calibrated system contains three α_i and nine β_i parameters with the values as given in Table 14.1.

The experimental data are taken for the system identification, which results in the impulse response as plotted in Figure 14.9b after the system identification to the measured temperature (Figure 14.9a).

TABLE 14.1

Parameters of the Identified System

α	
d^0	1.611×10–22
$d^{0.5}$	-4.024×10–23
d^1	7.532×10–24
β	
$d^{-0.5}$	4.178×10–26
d^0	4.148×10–26
$d^{0.5}$	5.261×10–26
d^1	1.611×10–22
$d^{1.5}$	7.04×10–27
d^2	1.159×10–27
$d^{2.5}$	-1.147×10–28
d^3	6.294×10–30
$d^{3.5}$	-1.465×10–31

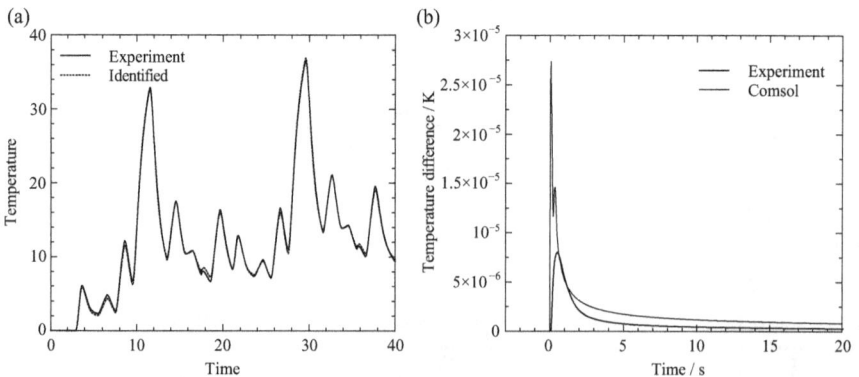

FIGURE 14.9
NISI result (a) and calculated impulse response (b).

14.5 Further Applications

From the present analysis, it can be concluded that a system identification approach with subsequent inversion of a uniquely calibrated system offers advantages particularly for problems with complex materials or unknown temperature sensor positioning. The concept of calibration is also followed by Frankel et al. [18].

Based on the approach presented in this chapter, the NISI method was further developed in the past for the application of the method to more complex problems.

A nonlinear problem was analyzed by a piecewise linearization of the problem through the calibration of the problem at different temperatures [19]. It was shown analytically that the amplitude of the impulse response is mainly driven by the heat capacity and the temporal form results from the conductivity of the problem. Knowing this, impulse responses for different temperatures can be calculated. Experiments with calibration at higher temperatures have been conducted in order to show the temperature-dependent behavior of the nonlinear problem. The resulting impulse responses are shown in Figure 14.10. Applying these impulse responses in the inversion depending on the current temperature allows the analysis of nonlinear problems.

Furthermore, the application of the NISI method was further developed for porous media. It is shown that the impulse response changes with cooling gas mass flow through porous surfaces. Calibrating a system for various transpiration cooling gas mass flows allows the measurement of the net heat flux for a transpiration-cooled surface [20].

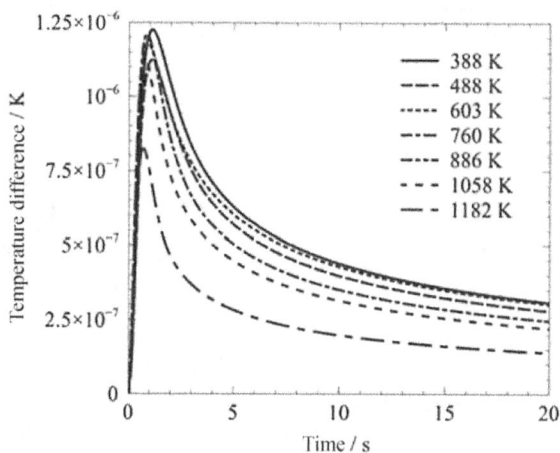

FIGURE 14.10
Impulse responses for varying temperatures.

Very recently, the NISI method has been investigated for the measurement of heat flux in a transpiration-cooled environment from the measured plenum pressure data of the actual system [21,22]. The plenum pressure follows the surface heating similar to the in-depth temperature data, and therefore, this offers the opportunity to measure heat flux non-intrusively.

14.6 Summary

This chapter presented theoretical and experimental details for a useful sensor calibration procedure for heat flux gages. The theoretical approach which is presented here on the basis of a one-dimensional, semi-infinite heat conduction problem shows the basic underlying idea of treating heat conduction problems as heat conduction systems. Based on this theoretical approach, the adaptations to system identification procedures for heat conduction problems are shown. The approach is verified experimentally using a copper block with an in-depth thermocouple.

Acknowledgments

The development of this work was funded through DFG Lo 1772/1-1. This work was partly funded through the ESA NPI Program under contract No. 4000121220/17/NL/MH. Ulf Fuchs provided a lot of fundamental work for the NISI method and the NISI code. His commitment is greatly appreciated.

References

1. J. D. Anderson: *Hypersonic and High-Temperature Gas Dynamics*, 2nd Edition, AIAA Education Series. AIAA, Reston, Auflage, 2006.
2. O. M. Alifanov: *Inverse Heat Transfer Problems*. Springer Science and Business Media, Berlin, Heidelberg, 2012.
3. J. Beck, B. Blackwell and C. R. St. Clair: *Inverse Heat Conduction: Ill-Posed Problems*. John Wiley & Sons, Hoboken, NJ, 1985.
4. M. N. Ozisik: *Heat Conduction*, 2nd Edition. John Wiley & Sons, Hoboken, NJ, Auflage, 1993.
5. A. N. Thikonov: *Solutions of Ill-Posed Problems*. John Wiley & Sons, Hoboken, NJ, 1977.

6. F. Zander, S. Loehle, T. Hermann and H. Fulge: Fabry-Perot spectroscopy for kinetic temperature and velocity measurements of a high enthalpy air plasma flow. *Journal of Applied Physics D*, Vol. 50, No. 33, 2017 doi: 10.1088/1361-6463/aa7b0c.

7. S. Loehle, T. Hermann and F. Zander: Experimental assessment of the performance of ablative heat shield materials from plasma wind tunnel testing. *CEAS Space Journal*, Vol. 10, 203–211, 2017.

8. J.-L. Battaglia, O. Cois, L. Puigsegur and A. Oustaloup: Solving an inverse heat conduction problem using a non-integer identified model. *International Journal of Heat and Mass Transfer*, Vol. 44, 2671–2680, 2001.

9. S. Loehle: Derivation of the non-integer system identification method for the adiabatic boundary condition using laplace transform. *International Journal of Heat and Mass Transfer*, Vol. 115, 1144–1149, 2017.

10. B. Lawton and G. Klingenberg: *Transient Temperature in Engineering and Science*. Oxford University Press, Oxford, 1996.

11. L. Ljung: *System Identification: Theory for the User*. Prentice Hall, Upper Saddle River, NJ, 1987.

12. L. C. Evans: *Partial Differential Equations*. American Mathematical Society, Providence, 2010.

13. S. Loehle and U. Fuchs: Theoretical approach to surface heat flux distribution measurement from in-depth temperature sensors. *AIAA Journal of Thermophysics and Heat Transfer*, Vol. 26, No. 2, 352–356, 2012.

14. J.-L. Battaglia: Multiple heat fluxes estimation using the noninteger system identification approach: application on the milling process. *Inverse Problems in Science and Engineering*, Vol. 13, No. 1, 1–22, 2005.

15. S. Loehle, J.-L. Battaglia, J.-C. Batsale, O. Enouf, J. Dubard and R.-R. Filtz: Characterization of a heat flux sensor using short pulse laser calibration. *Review of Scientific Instruments*, Vol. 78, 53501, 2007.

16. U. Fuchs: Transiente Wärmestromdichtebestimmung durch System Identifikation auf Basis gebrochen rationaler Funktionen und Erweiterung auf dreidimensionale Probleme. Master's thesis, Institut für Raumfahrtsysteme, Universität Stuttgart, 2010. IRS-10-S22.

17. M. Kaviany: *Principles of Heat Transfer*. John Wiley & Sons, Hoboken, NJ, 2001.

18. J. Frankel, M. Keyhani and B. Elkins: Surface heat flux prediction through physics-based calibration: Part 1 theory. *In: 50th AIAA Aerospace Sciences Meeting including the New Horizons Forum and Aerospace Exposition*, 2012 Nashville, TN, USA.

19. S. Loehle, U. Fuchs, P. Digel, T. Hermann and J.-L. Battaglia: Analysing inverse heat conduction problems by the analysis of the system impulse response. *Inverse Problems in Science and Engineering*, Vol. 22, 297–308, 2014.

20. S. Loehle and U. Fuchs: Heat flux calibration measurement using noninteger system identification method for cooled surfaces. *AIAA Journal of Thermophysics and Heat Transfer*, Vol. 25, No. 2, 213–217, 2011.

21. F. Hufgard, S. Loehle, T. Hermann, S. Schweikert, M. McGilvray, J. von Wolfersdorf, J. Steelant and S. Fasoulas: Analysis of porous materials for transpiration cooled heat flux sensor development. *In: International Conference on High-Speed Vehicle Science Technology, CEAS*, 2018 Moscow, Russia.

22. S. Loehle, S. Schweikert and J. von Wolfersdorf: Method for heat flux determination of a transpiration cooled wall from pressure data. *Journal of Thermophysics and Heat Transfer*, Vol. 30, No. 3, 567–572, 2016.

Part D

Measuring in
Two-Phase Flow

15

Challenges and Advances in Measuring Temperatures at Liquid–Solid Interfaces

Ana S. Moita
Universidade de Lisboa

Miguel R. O. Panão
University of Coimbra

Pedro Pontes
Universidade de Lisboa

Emanuele Teodori
ASML Holding N.V

António L. N. Moreira
Universidade de Lisboa

CONTENTS

15.1 Introduction

Transport phenomena occurring at liquid–heated solid interfaces govern many flows that are present in a variety of daily activities and industrial applications. Droplet/spray–wall interactions and pool/flow boiling are probably two of the most known examples of complex flows, present in a variety of applications, for which the accurate description of the fluid flow and heat transfer mechanisms occurring at the liquid–solid interface can provide a significant contribution to understanding and fully describing these flows. Droplet and spray impingement are relevant in numerous applications such as in fuel injection systems, for IC engines [1], and in several cooling systems [2–4]. Characterizing the heat transfer processes occurring due to a single droplet impacting on a heated solid surface is a complex task, which depends on the intricate relationship between the heat transfer processes and droplet dynamics [1,5]. At impact, a thin liquid film flows radially around the stagnation point (impact point), depicting a variable thickness which affects the temperature distribution in the liquid and on the solid, within the wetted area. In this case, the temperature at the liquid–solid interface strongly depends on the **effusivity** ($\beta = \sqrt{\rho c_p k}$, where ρ is the density, c_p the specific heat, and k the thermal conductivity. Subscripts l and w are related to the liquid and the surface, respectively). For the sake of simplicity, Seki et al. [6] proposed a definition for the uniform and constant contact temperature, T_c, as

$$T_c = \frac{\beta_w T_w + \beta_l T_l}{\beta_w + \beta_l} \tag{15.1}$$

This formulation is still a valid approximation, when one needs to estimate the temperature at droplet–surface interface, but the legitimacy of this approximation is, in fact, limited to the early stages of droplet impact, i.e., at $t^* = t/(D_0/U_0) < 2$, where D_0 and U_0 are the initial droplet diameter and velocity, respectively [7], and when the surface materials have high values of thermal effusivity (of the order of 7–$14 \times 10^3 \text{J/K s}^{0.5}\text{m}^2$, as typically observed in metals). Both dynamic and heat transfer mechanisms are further known to be affected by wettability.

Additional complexity occurs as the droplet evaporates and/or the liquid film spreading on the surface starts to boil. The characteristic spatial resolution required to **describe** the thermal processes occurring, in this case, is of the order of microns, while the typical spreading process of a droplet lasts tens of milliseconds. Such demanding temporal and spatial scales are also characteristic of spray impingement studies, which must further deal with polydispersity (different sizes and velocities of the droplets) and interaction mechanisms. (The various droplets composing the spray may interact between them during impact and spreading or may impact on a liquid film formed due to the deposition of the spray [1].)

Pool boiling studies, on the other hand, were initially focused on reconstructing pool boiling curves, which would determine the performance of the system to be studied, based on a plot of the power imposed to the heated surface and the superheat $\Delta T = T_{\text{surface}} - T_{\text{sat}}$, where T_{sat} is the saturation temperature of the liquid at a given pressure [8]. These studies required a precise evaluation of the heat power/heat flux imposed on the heating surface and accurate measurements of the surface temperature and the liquid temperature close to the surface.

However, the development of the nucleation theory and of the mechanistic models, such as those of Han and Griffith and of Kurul and Podowski [9–11], which consider a significant contribution of the quenching and induced liquid flow near the surface, at bubble formation and departure to the total boiling heat transfer, highlighted the need to further detail the entire process of nucleation and bubble dynamics. Such detailed characterization requires surface temperature measurements within characteristic length scales of the order of microns (or smaller) and typical timescales ranging from milliseconds to a few seconds.

Under these scenarios, the challenge of measuring any quantity at the liquid–solid interface (e.g., temperature) starts right at choosing a suitable technique. Furthermore, a single measurement technique is unlikely to be able to provide all the quantities required to characterize both thermal and fluid dynamic processes, which is the reason for the combination of different techniques in most recent studies [12–14].

Measurement techniques and particularly thermometry techniques can be distinguished mainly as intrusive, semi-intrusive, or non-intrusive [15]. Intrusive techniques involve physical contact with the sample to be investigated and include thermistors and thermocouples. Semi-intrusive and non-intrusive techniques are contactless, but while the former requires a small modification of the sample, such as the addition of a dye or a tracer, the latter relies on intrinsic temperature-dependent properties of the medium (e.g., refractive index, density, viscosity, adsorption and emission of light). Examples of non-intrusive techniques include interferometric techniques, Raman spectroscopy, and thermography.

While in several reviews on this topic [16], intrusive techniques are immediately discarded to provide useful information on liquid–solid interface temperatures, given their limitations regarding uncertainty and low temporal and spatial resolutions, they are one of the most common and inexpensive solutions that are available for the majority of researchers. On the other hand, despite their vast potential, contactless techniques are often more complex, are expensive, and require specific calibration and post-processing procedures to obtain reliable data.

In line with this, **the present review** introduces a span of possible techniques to characterize the thermal processes occurring at liquid–solid interfaces. The first step is to identify the limitation of contact techniques and review the most recent and reliable methods to obtain temperature and heat flux measurements based on thermocouples.

Afterward, semi-invasive and noninvasive techniques are shortly revised in two subsections introducing the cutting-edge contactless methods and how they can be used to obtain reliable results. A particular case study will also be provided, in contactless techniques, to detail the calibration and post-processing procedures followed, which allowed obtaining reliable results. **Infrared thermography** is probably one of the most commonly used non-intrusive techniques, and despite the limited range of applications, the example provided for time-resolved thermography with a high spatial resolution shows the kind of quantitative information obtained. If not adequately addressed during the measuring campaigns, most limitations are due to calibration and post-processing issues.

As various techniques are summarized, recognizing their main advantages and disadvantages for the applications discussed here, challenges and prospects are identified and proposed at the end of the chapter.

15.2 Intrusive Techniques: Thermistors and Thermocouples

15.2.1 Overview and State-of-the-Art Solutions

Thermocouples, thermistors, and similar sensors can measure temperatures within a wide range. However, they only provide local and discrete measurements, being unable to provide complete information on when temperature gradients and temperature fields are required. Geometry and flow characteristics make it particularly challenging to take advantage of the use of thermocouples to accurately characterize liquid flows such as pool and flow boiling. However, they have been widely used for several decades to obtain surface temperature measurements in liquid–surface interfacial flows. These surface temperature measurements are then used to extrapolate the liquid temperature at interfacial flows.

Thermocouples are sensors made from joining two wires of dissimilar conductive materials for which a temperature difference between their junctions produces an electromotive force. The voltage difference generated associated with the Seebeck effect is small in the order of millivolts.

The pair of electric conductors defines the type of thermocouples; for example, the pair chromel and alumel is used in thermocouples of type K. Besides determining the range of temperatures that the thermocouples can measure, the pair of materials chosen also affects the precision of measurements. Hence, although a wide variety of applications uses type K thermocouples due to the comprehensive range of temperatures covered (–200°C to +1,350°C) and their robustness in terms of accuracy, type T thermocouples (copper–constantan) are the most accurate.

Besides the materials, the diameter of the junction and the positioning significantly affect the accuracy and even the response time. Thermistors (and RTDs—resistance temperature detectors) are a type of resistor whose resistance is temperature dependent, more than in standard resistors. They are usually more accurate than thermocouples, but they are also more expensive. Another alternative is heat flux sensors. Recent studies have clearly shown the improved accuracy in the reconstruction of boiling curves with the use of micro-heat flux sensors [17–18]. These studies allowed observing details on the onset of boiling, mainly when boiling occurs on superhydrophobic surfaces, where boiling is triggered just at 1°C–3°C superheat. However, the use of these sensors still depends on the geometry and flow conditions.

A different scenario occurs in studies addressing spray and droplet impingement.

Indeed, while visualization has been the preferred technique to obtain dynamic information on liquid–solid interfaces, e.g., droplet dynamics, since the ground-breaking work of Lord Worthington [19–20], thermocouples are the most usual technique to measure the temperature variation on the impacting surface, during droplet contact. Wachters and Westerling [20] were pioneers in investigating the heat transfer on droplets impacting on heated solid walls. A heat flux meter was assembled to control the surface temperature. The heating assembly was a solid gold cylinder of 5 mm height and 15 mm diameter inserted into a small circular silver vessel. Wachters and Westerling [21] focused their attention on droplet dynamics using high-speed photography. Seki et al. [6] pioneered the measurement of temperature variation in millimetric droplets of water and ethanol, using a temperature-sensitive nickel film resistor ($0.7 \times 1.5\,mm^2$). This film was separated from the heated surface by an electrically insulating layer of SiO, which was also used to cover the outer surface of the resistor. With this apparatus, Seki et al. [6] were able to capture temperature oscillations on the surface during the impact of single droplets. Labeish [22] improved the accuracy obtained for the transient surface temperature measurements. Labeish's setup had a thermoelectrode with a diameter of 0.2–0.3 mm fixed in the aperture of the heated surface, which served as the other thermoelectrode. Labeish [22] tried different material combinations for the surface and thermoelectrodes, namely nickel–nichrome, stainless steel–nichrome, copper–constantan, and silver–nichrome. The junction obtained by polishing the surface has the shape of a ring with a diameter equal to that of the thermoelectrode.

After Labeish [22], several authors followed a similar strategy to capture the transient temperature variation on surfaces at droplet/spray impact. Most of them position the thermocouples at the impact region [23], as their size usually restricts the spatial resolution to hundreds of microns, and, as aforementioned, the thermocouples only provide discrete temperature measurements at a given point. The fact that thermocouples are intrusive sensors will also affect the flow and wetting properties of the surfaces.

With the advances in micro-fabrication, a few experiments tested smaller probes. The primary goal was obtaining more detailed information on the surface–liquid temperature within a more significant number of discrete locations, with an improved spatial resolution. Hence, for instance, Betz et al. [24] used an array of ~100-μm-wide parallel long thin-film gold lines to measure the evolution of the surface temperature distribution when a water drop slowly evaporated from the surface kept at 60 K before droplet impingement. However, since this system would solely provide an average temperature along each line, Betz et al. [24] used numerical simulations to analyze the radial distribution of the surface temperature under the impacting droplets. Kim et al. [25,26], Xue and Qiu [27], and Paik [28] used similar arrays of micro-sensors. Spatial and temporal resolutions could be improved up to 50 μm and 100 ns, respectively [29].

Shi et al. [30] further pushed the limit regarding spatial resolution, using nanoscale thermocouple junctions to obtain thermal microscopy measurements of surface temperature distribution. A resolution as high as 1×10^{-3} K was reported by Shi et al. [30], using a unique modulation approach [31]. Still, these sensors often cannot assure the required temporal resolution to capture the onset of phase change [32].

On the other hand, other authors report resolutions for temperature measurements of the order of 1×10^{-3} K, using micro-fabricated suspended thin-film Pt serpentine line resistance thermometers and sophisticated differential measurement schemes [33,34].

Despite these efforts, limitations on spatial and temporal resolutions are still an issue, apart from the topics mentioned above related to the way the contact techniques always affect the flows [16]. Also, the costs associated with these micro- and nano-sensors can be significant. Hence, non-intrusive techniques are preferred, especially when temporal and spatial scales are significantly small, and the flows have a high complexity degree, such as those dealing with liquid phase change.

Nevertheless, commercial thermocouples are probably the most usual and cost-effective solution that is available to experimentalists. Once their limitations are identified, one may take advantage of them to obtain useful information on temperature and heat fluxes at liquid–solid interfaces.

The following paragraphs revise various thermocouple solutions and the procedures that should be followed in the calibration and in the data reduction procedures to obtain reliable data.

15.2.2 Local Measurement of Surface Temperature with High Temporal Resolution

15.2.2.1 Eroding-Type Fast Response Thermocouples

Aziz and Chandra [35] used fast response thermocouples of eroding type, produced by Nanmac, which reported response times of the order of 10 μs. The purpose was to measure the temperature evolution of the impact of molten droplets on a solid stainless steel substrate. The eroding fast response K-type thermocouples consist of two fine ribbons of chromel and alumel, electrically separated by an insulating layer of mica and enclosed in a stainless steel sheath (see Figure 15.1).

The junction made through erosion lies between the two ribbons. Thus, its size is small enough to ensure the fast response obtained from calibration. The thermocouple is inserted into the stainless steel substrate and positioned at the top surface close to the point of drop impact.

In surface thermocouples, the calibration of the thermal effusivity or thermal product is an essential parameter for the calculation of the instantaneous heat flux. For eroding thermocouples, with an exposed junction, Buttsworth [36] proposed two methods for different timescales.

The simplest method is to impinge a single water droplet heated to T_w on the thermocouple junction initially at T_s, which will step into a contact temperature T_c lower than T_w. Assuming the droplet does not rebound, constant thermal properties of both thermocouple and droplet, and one-dimensional heat conduction for the step change in temperature, there is a relation between these temperature values and a term dependent on the thermal effusivity of water (β_w) and thermocouple (β_s):

$$\frac{T_c - T_s}{T_w - T_s} = \frac{\beta_w}{\beta_w + \beta_s} \tag{15.2}$$

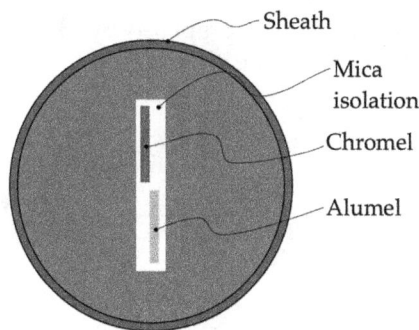

FIGURE 15.1
Schematic of the eroding-type thermocouples.

With the measurement of temperature values, and provided that β_w is known, one determines the effective value for the thermocouple effusivity (β_s). Comparing the calibrated values with those obtained for a weighted average from the data available for the thermocouple's constituting materials (chromel, Cr, and alumel, Al) reported by Caldwell [37], namely for the specific heat and thermal conductivity as a function of temperature,

$$c_{p,Al}\left[J/kg\,K\right]=0.07512T+500.8$$

$$k_{Al}\left[W/m\,K\right]=0.02981T+18.42$$

$$c_{p,Cr}\left[J/kg\,K\right]=0.1786T+394.3$$

$$k_{Cr}\left[W/m\,K\right]=0.01912T+12.11$$

Despite its fast response, there are a few limitations to these sensors. The main one is its size—$O(10^{-3})$ m, followed by the spatial resolution of local measurements. Moreover, due to its insertion into the substrate, it also limits the number of materials used for the sheath, which should be of the same material as the substrate. Another limitation of the eroding-type thermocouples is the erosion effect exerted by the impact of the droplet or spray, which may damage the junction, leading to more maintenance work than expected.

15.2.2.2 Thin-Film Thermocouple

Considering the limitation of the eroding-type thermocouples, Heichal et al. [38] proposed a different approach based on a thin film of conductive material. Their purpose was to obtain response times lower than 1 µs with the ability to withstand hot fluids and being built in smooth or rough surfaces of any conductive material. Finally, these thermocouples should be simple to fabricate and maintain.

Figure 15.2 illustrates the design proposed by Heichal et al. [38]. In this design, the substrate is one of the thermocouple materials, while a fine

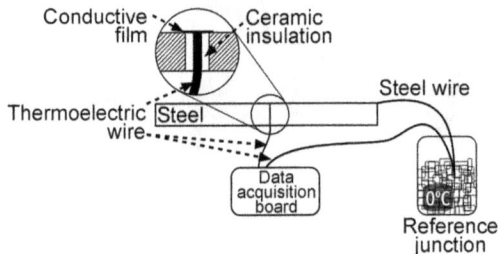

FIGURE 15.2
Diagram of a thin-film thermocouple design based on Heichal et al. [38].

thermoelectric wire, encapsulated by ceramic insulation, is the other. A thin film of highly conductive material deposited on the top surface ensures the electrical connection between the thermocouple materials and works as their junction. Heichal et al. [38] introduced the second junction, operating as the reference, in an ice bath at 0°C. While the eroding-type thermocouples require thermal isolation from the hole used to insert them in the substrate **(for example, steel in Figure 15.2)**, and the material must be the same as the substrate to ensure appropriate measurement of the surface temperature, in thin-film thermocouples, the conductive film provides an accurate measure of the surface temperature. Moreover, the fact that the film thickness can be as thin as 0.1–2 μm allows a short response time.

The conductive film is an essential element in this sensor. Once applied, the resistance should lie between 30 and 70 Ω. Heichal et al. [38] recommended using metallic films made of silver and platinum inks applied with a brush. Also, copper films are used by placing a drop of copper sulfate on the junction and applying a 12 V DC potential, using the substrate as the cathode and the copper wire in the droplet as the anode. In their calibration, the voltage produced did not depend on the material of the conductive film (see an example in Figure 15.3).

Concerning calibration, it is essential to choose the wire material that produces a higher output voltage with the substrate. In the case of SS 303, a constantan wire provided the maximum output, but in the case of other substrate materials, it is convenient to test which of the conventional wire materials (NICr60, copper, iron, constantan, chromel, alumel, and platinum) produces the highest output voltage.

Due to the low voltages obtained, it is essential to make sure connections do not increase the noise in the signal, since the wires may act as antennae capturing ambient interference of electromagnetic (EMI) or radio

FIGURE 15.3
Details of the conductive film made of silver ink (micro-tip conductive pen, SPI Supplies).

frequency (RFI) origin. The total resistance of film, wires, and substrate should be below 70 Ω; otherwise, it may amplify the noise.

The calibration of thin-film thermocouples involves two features: (1) the relation between the temperature and the voltage produced between junctions and (2) the response time.

The calibration of the measurement junction may use a small furnace with controlled temperature, as in Heichal et al. [38], or a second calibrated temperature sensor placed next to the thermocouple under calibration and use a thermal bath with controlled temperature, as in Díaz et al. [39]. The bath is adequate for lower temperature ranges (20°C–80°C). Afterward, a simple curve fitting allows obtaining the calibration polynomial curves, $T = f(V)$.

The response time of the thin-film thermocouples depends on the timescale associated with the conductive film. For example, in the case of silver, assuming one-dimensional transient heat conduction on a film with a thickness on the order of $1\,\mu m$, the timescale, $\tau_f \sim L_f^2/\alpha$, results in values on the order of nanoseconds. This short timescale means the highly conductive film will produce a significant influence on the thermocouple's response time. Therefore, to generate a unit impulse response function (UIRF), the best approximation is using a laser pulse of a few nanoseconds. Heichal et al. [38] followed this approach, and the results confirmed the timescales predicted, which are significantly lower than the timescales usually present in heat transfer phenomena at the liquid–solid interface, as in drop or spray impact events. After measuring the surface temperature, the goal is to retrieve information on the instantaneous heat flux, as explored in the next section.

15.2.3 Heat Flux Exchanged in Liquid–Solid Interfaces

Assuming one-dimensional (z) heat conduction, the primary challenge in calculating the heat flux in liquid–solid interfaces is under a transient heat transfer. Namely, it allows quantifying the evolution of the heat transfer coefficient, which is one of the most relevant parameters to understand the physics of the heat transfer phenomena involved and develop adequate correlations for modeling purposes. There are two recurrent experimental conditions. The first is to assume that heat transfer is similar to a semi-infinite solid, and the other is to consider a finite slab with an imposed heat flux.

15.2.3.1 Transient Heat Transfer Assuming a Semi-Infinite Solid

If we approach the transient heat transfer occurring at the liquid–solid interface similar to the surface of a semi-infinite solid, the assessment of this assumption depends on a scale analysis made to the heat conduction equation, further reducing to a relation between the penetration scale (L_c) of a perturbation at the surface, the substrate thermal diffusivity ($\alpha = k/(\rho c_p)$), and the timescale of heat transfer events (t):

$$L_c = \sqrt{\alpha t} \tag{15.3}$$

If this scale is lower than the surface thickness, the semi-infinite solid approach is valid. Assuming an initial condition of uniform temperature distribution on the substrate, if we consider the temperature difference between the measured surface temperature, $T_s(z, t)$, and its initial value, $\theta(t) = T_s(z, t) - T_s(z,0)$, the heat conduction energy equation becomes:

$$\frac{\partial \theta}{\partial t} = \alpha \frac{\partial^2 \theta}{\partial z^2} \tag{15.4}$$

In this case, there is no imposed heat flux. The initial and boundary conditions are as follows:

$$\theta(z,0) = 0 \tag{15.5a}$$

$$-k_s \frac{d\theta(0,t)}{dz} = q''(0,t) \tag{15.5b}$$

$$\theta(L_s,t) = 0 \tag{15.5c}$$

with k_s the substrate thermal conductivity, $q''(0,t) \left[W/m^2 \right]$ the heat flux removed by forced convection at the liquid–solid interface ($z = 0$), and L_s the characteristic length of the substrate. The approach to solving Eq. (15.4) follows that of Reichelt et al. [40], which applies Laplace transformations to the differential terms as follows:

$$\left[\begin{array}{l} L\left(\dfrac{\partial \theta(z,t)}{\partial t} \right) = s\Theta(z,s) - \Theta(z,0) \\[3mm] L\left(\dfrac{\partial^2 \theta(z,t)}{\partial z^2} \right) = \dfrac{\partial^2 \Theta(z,s)}{\partial z^2} \end{array} \right. \tag{15.6}$$

resulting in:

$$s\Theta(z,s) - \Theta(z,0) = \alpha \frac{d^2\Theta(z,s)}{dz^2} \tag{15.7}$$

which the general solution corresponds to:

$$\Theta(z,s) = Ae^{z\sqrt{s/\alpha}} + Be^{-z\sqrt{s/\alpha}} \tag{15.8}$$

For the boundary conditions in (15.4b) and (15.4c), the solution of (15.8) is:

$$\Theta(z,s) = \frac{q''(z,s)}{k} \sqrt{\frac{\alpha}{s}} e^{-z/\sqrt{\alpha}} \tag{15.9}$$

The top surface at $z = 0$ is the liquid–solid interface. Thus, in this location, after some mathematical manipulation using the convolution theorem with: $F_1(s) = \dfrac{d\theta(0,t)}{dt}$ and $F_2 = 1/\sqrt{s}$, Eq. (15.8) becomes:

$$q''(0,t) = \frac{k_s}{\sqrt{\pi\alpha_s}} \int_0^t \frac{d\theta(0,t^*)}{dt^*} \cdot \frac{1}{\sqrt{t-t^*}} dt^* \tag{15.10}$$

If we normalize the time, t, by the timescale of an energy perturbation into the entire characteristic length of the substrate, the result is the Fourier number, $\tau = \dfrac{\alpha t}{L_s^2}$, and the integral is numerically calculated based on discrete-time intervals between τ_i and $\tau_i + \Delta\tau$, with $\Delta\tau$ the inverse of the temperature data acquisition frequency, resulting in:

$$q''(0,\tau_i) = \frac{k_s}{\sqrt{\pi}L_s} \sum_{k=0}^{i-1} \int_{\tau_k}^{\tau_{k+1}} \frac{dT_s}{d\tau^*}(\tau^*) \cdot \frac{1}{\sqrt{\tau-\tau^*}} d\tau^* \tag{15.11}$$

The derivative inside the integral term expanded in a Taylor series within the interval $[\tau_k, \tau_{k+1}]$ is further resolved by Reichelt et al. [40] to obtain the instantaneous heat flux numerically as:

$$q''(0,\tau_i) = 2\frac{k_s}{L_s} \sqrt{\frac{\Delta\tau}{\pi}} \sum_{k=0}^{i-1} \left[\left(T'_{s,k} + T''_{s,k}\Delta\tau \left(i - \frac{2k+1}{2} \right) \right) R_{i,k} - T''_{s,k} \frac{\Delta\tau}{3} S_{i,k} \right] \tag{15.12}$$

$$\begin{cases} T'_{s,k} = \dfrac{T_{s,k+1} - T_{s,k}}{\Delta\tau} \\[2mm] T''_{s,k} = \dfrac{(T_{s,k+2} - T_{s,k+1}) - (T_{s,k} - T_{s,k-1})}{2\Delta\tau^2} \\[2mm] R_{i,k} = (i-k)^{1/2} - (i-k-1)^{1/2} \\[2mm] S_{i,k} = (i-k)^{3/2} - (i-k-1)^{3/2} \end{cases}$$

Reichelt et al. [40] showed a good agreement between the analytical solution of a sinusoidal temperature variation for a semi-infinite body with frequencies above 500 Hz, and the numerical solution proposed above.

Most experiments with drop or spray impact heat transfer obtain changes well above this frequency. Thus, we expect inaccuracies to be negligible. The limitation of this method is for frequencies of the order of 0.1 Hz, as pointed out by Chen and Nguang [41].

15.2.3.2 Transient Heat Transfer in a Finite Slab with an Imposed Heat Flux

When the experimental conditions consider an imposed heat flux, q''_{imp}, there is a different solution for calculating the transient heat flux from surface temperature measurements. The problem is still one-dimensional, and the energy equation expressing this problem is:

$$\frac{\partial \theta}{\partial t} = \left(\frac{k_s}{\beta}\right)^2 \frac{\partial^2 \theta}{\partial z^2} + \frac{k_s}{L_s \beta^2} q''_{imp} \tag{15.13}$$

where $\theta(z,t) = T_s(z,t) - T_s(L_s,0)$ and β is the thermocouple effusivity made of the same material as the substrate or similar thermophysical properties. The energy equation includes the thermal effusivity based on the evidence provided by Cossali et al. [42] of its importance in the temperature variation at the liquid–solid interface heat transfer. The approach followed in this case, introduced in Panão and Moreira [43], differs from the one proposed by Reichelt et al. [40], using $\left(\dfrac{k_s}{\beta}\right)^2$ instead of α, which allows later to simplify specific terms. The initial and boundary conditions considered in this case were:

$$\theta(z,0) = \theta(L_s,0)\big|_{=0} - \frac{q''_{imp}}{k_s}(L_s - z) \tag{15.14a}$$

$$\frac{d\theta(0,t)}{dz} = -\frac{q''(0,t)}{k_s} \tag{15.14b}$$

$$\frac{d\theta(L_s,t)}{dz} = -\frac{q''_{imp}}{k_s} \tag{15.14c}$$

In the initial condition, the term $\theta(L_s, 0)$ is null and the boundary condition in Eq. (15.13c) corresponds to the imposed heat flux at the bottom surface of the slab. Applying the Laplace transformations to solve the energy equation, the final general solution is:

$$\theta(z,s) = A e^{z\sqrt{s/(k_s/\beta)^2}} + B e^{-z\sqrt{s/(k_s/\beta)^2}} + \frac{q''_{imp}}{s^2}\left(\frac{L_s - z}{k_s} + \frac{k_s}{L_s \beta^2}\right) \tag{15.15}$$

Using the initial and boundary conditions in Eq. (15.13), the solution at the top surface ($z = 0$) results in:

$$\theta(0,s) = \frac{q''(0,s)}{\beta\sqrt{s}} \frac{e^{2\xi\sqrt{s}}+1}{e^{2\xi\sqrt{s}}-1} + \frac{q''_{imp}}{\beta s^{3/2}}\left(\frac{1}{\sqrt{s}}\left(\frac{1}{\xi}-\xi\right) + \frac{1}{s}\frac{\left(e^{\xi\sqrt{s}}-1\right)^2}{e^{2\xi\sqrt{s}}-1} - \frac{2e^{\xi\sqrt{s}}}{e^{2\xi\sqrt{s}}-1}\right) \quad (15.16)$$

with $\xi = \dfrac{L_s\beta}{k_s}$. Expressing the relation in Eq. (15.15) for the surface heat flux q'' $(0, s)$ requires an analysis on the exponential functions elevated to $\sqrt{s}$, which directly reflects the frequency of temperature variations, according to Chen and Nguang [41]. In fact, for values of this frequency between 0.5 and 1 kHz, as in most liquid–solid interface heat transfer events, one may approximate the solution for q'' $(0, s)$ to the boundary where $\sqrt{s} \to \infty$. When evaluating the limits of the fraction terms containing the exponential functions in this boundary, Eq. (15.15) expressed for the surface heat flux simplifies to:

$$q''(0,s) = \beta s\theta(0,s)\frac{1}{\sqrt{s}} - q''_{imp}\left(\frac{1}{s^{3/2}}\left(\frac{1}{\xi}-\xi\right) + \frac{1}{s^2}\right) \quad (15.17)$$

When the frequency associated with temperature variations is small, for example, in the absence of any heat transfer at the liquid–solid interface, it means $\sqrt{s} \to 0$; thus, the limits of the fractions with the exponential functions lead to $q''(0,s) = q''_{imp}$, which shows the consistency of the formulation above.

The application of the convolution theorem to find the analytical solution of (15.17), after some mathematical manipulation, results in:

$$q''(0,t) = \frac{\beta}{\sqrt{\pi}} \int_0^t \frac{d\theta(0,\tau)}{d\tau} \cdot \frac{1}{\sqrt{t-\tau}} d\tau - q''_{imp}\left(t + \frac{2}{\xi}\sqrt{\frac{t}{\pi}}\right) \quad (15.18)$$

It is noteworthy how the solution presented in (15.18) reduces to the one proposed by Reichelt et al. [40] in the absence of any heat sources, $q''_{imp} = 0$. Also, the solution in (15.17) shows the effect of q''_{imp} on the surface heat flux increases for thinner substrates ($L_s \to 0$) and in the unfolding of heat transfer in time. Finally, the numerical implementation of Eq. (15.18) to obtain the instantaneous surface heat flux follows the approach developed in Schultz and Jones [44] as:

$$q''(0,t_n) = \frac{2\beta_{t_n}}{\sqrt{\pi\delta t}} \sum_{i=1}^{n} \frac{\theta(0,t_i)-\theta(0,t_{i-1})}{\sqrt{n-i}+\sqrt{n-i-1}} - q''_{imp}\left(t_n + \frac{2}{\xi_{t_n}}\sqrt{\frac{t_n}{\pi}}\right) \quad (15.19)$$

with $\xi_{t_n} = \dfrac{L_s\beta_{t_n}}{k_s}$ and $t_i = i \cdot \delta t$ considering δt as the sampling time retrieved from the sampling rate in the data acquisition system.

15.3 Semi-Intrusive and Non-Intrusive Techniques

15.3.1 Overview

Semi-intrusive and non-intrusive methods have evolved significantly within the last decade, with relevant applications to interfacial flows at the meso- and microscales [16]. Despite their advantages, such as allowing simultaneous measurements of temperatures in two-dimensional fields, intact measures close to the actual value, and no disturbance during measurements [16], they can be quite complex to implement and require adequate calibration and data reduction procedures to obtain reliable measurements.

Conventional techniques such as fluorescence thermometry are semi-intrusive, as they require the use of a tracer. In fluorescence thermometry, electrons in a fluorophore undergo a transition from the valence band to the conduction band, as they suffer from external stimulation. As they return to the valence band, an amount of energy is released, known as fluorescence. Usually, the intensity of the lifetime of mixing fluorophore with the liquid phase is temperature dependent. Based on a preexistent relation (calibration), the intensity (or lifetime) values convert to temperature, and the choice of the fluorophore is of significant importance in the final result. Also, before performing any measurement, several issues related to quenching, bleaching, the intensity of the light source, among others, must be the subject of a careful sensitivity analysis [45].

In the interferometric techniques, two coherent beams are superimposed, forming an interference pattern. Since this pattern is usually affected by the refractive index of the materials or medium, measuring the phase differences of light waves caused by spatial variations in the refractive index allows inferring the temperature [16]. Another example of a non-intrusive technique is thermochromic liquid crystals (TLCs), which are often used to measure the temperature in liquids or solid surfaces. This technique addresses the use of materials that can reflect specific colors at certain temperatures. The process of color change is reversible and repeatable if no irreversible change or damage occurs to the TLCs. The composition of TLCs determined the relation between the color and the temperature, covering temperatures between 30°C and 120°C, with discrete bands of 0.5°C–20°C. The characteristic response time of TLCs is nearly 10 ms, which is perfect for many fluid flows, although it is usually not enough for more complex interfacial flows such as droplet–wall interactions or nucleate boiling (onset). X-ray and g-ray tomography, Raman (micro-)spectroscopy, Brillouin light scattering (BLS) [46], and rainbow thermometry are other non-intrusive techniques which are worth exploring. They reach temperature resolutions of the order of 0.5–1 K, with a sub-micrometer spatial resolution and temporal resolution of ms. In this context, interesting techniques have been explored on droplet–wall interactions, particularly to describe the temperature variations and heat fluxes at nucleation. For instance, Yu et al. [47] introduced synchrotron X-ray imaging with high spatial (~2 µm) and temporal (~20,000 Hz) resolutions as a method to visualize the behavior and shape of liquid–vapor interfaces underneath nucleate bubbles on surfaces with engineered micro-pillars.

Fluorescence has intensively been explored to describe dynamic and heat transfer processes during bubble growth and detachment in the attempt to characterize the temperature variations at single bubble formation and detachment [48–50]. Some of these approaches included submicron thermal imaging using fluorescence microscopy [51].

Despite their interest, there is a need to further investigate these techniques and their adaptation to the particular characteristics of interfacial flows. For instance, at the liquid–solid–vapor interface, during bubble detachment, there are strong temperature gradients which alter the interfacial tensions, leading to highly heterogeneous distributions of the fluorophore concentration. Also, temperature measurements can be strongly affected by diffraction effects at the surface or the liquid–vapor interface.

In this context, infrared thermography is a well-established non-intrusive technique. Despite its diffraction-limited spatial resolution of the order of ~10 µm, which still disables a more in-depth analysis into the liquid–surface–gas interfaces, this technique can produce essential information on the temperature fields at the surface, intricate flows, such as pool and flow boiling, although associated with complementary techniques. Furthermore, calibration and data reduction require careful procedures due to the strong dependence of the obtained measurements on the geometry and emissivity. The following paragraphs briefly describe these issues to illustrate the use of thermography to characterize temperature fields and heat flux at liquid–solid interfaces.

15.3.2 Infrared (IR) Thermography with High Temporal and Spatial Resolutions

15.3.2.1 Working Principles and Influencing Parameters

All bodies emit radiation at an absolute temperature above zero kelvin. The intensity of this radiation depends on the direction, wavelength, and temperature of the body. The irradiated energy reaching a surface has three possible outcomes: reflection, transmission, or absorption [52]. Thermographic cameras receive the irradiation from the target object and determine its temperature from such an energy balance, which also considers the attenuation from the atmosphere. In fact, in almost all situations requiring the use of a thermographic camera, it is separated from its target by the atmosphere, which has a higher or lower transmittance in different wavelengths, thus causing attenuation. The atmosphere attenuation depends on the complexity of its composition. Attenuation must be considered when choosing the wavelength band to use, as well as the nature and temperature of the target [53]. Usually, concerning atmospheric attenuation, the LWIR band behaves better than the MWIR band [53]. However, one may always overcome this issue by using a selective range of wavelength to avoid a lower atmospheric transmittance. The performance of an IR thermographic camera depends on many factors, such as the thermal sensitivity, spatial resolution, acquisition frequency, temperature, and dynamic ranges, and the working IR band [3]. The spatial resolution is directly related to the lens used. Modern cameras with special detectors can currently achieve values of sensitivity lower than tens of mK and very high acquisition frequencies, higher than 1,000 fps.

However, reliable measurements require custom-made and carefully prepared calibration and post-processing processes, as discussed in the following subsection.

15.3.2.2 Calibration and Data Reduction Procedures

Sielaff [54] combined high-speed visualization with IR thermography to characterize bubble dynamics and the heat transfer processes occurring at bubble formation and release. **IR analysis was used to take the temperature variation on the heating surface during the processes of bubble growth and detachment.** Image post-processing is a well-established procedure, which, **after an appropriate calibration**, allows obtaining detailed temporal evolution on bubble growth, the velocity of the bubble contact line, and contact angles, as also recently described in Teodori et al. [55]. However, taking reliable quantitative information from IR analysis requires delicate calibration procedures, as well as adequate preparation of the camera settings.

Teodori et al. [12] performed the following procedures to obtain a full characterization of the fluid dynamics and heat transfer processes occurring during droplet–wall interactions. A high-speed infrared (IR) camera (ONCA-MWIR-InSb from Xenics—ONCA 4696 series) was placed below the heated target, while a high-speed camera (Phantom v4.2) was mounted to take the side views of the droplets. The designation of the infrared camera refers to the spectral range of the camera, mid-wave infrared, and the typology of the detectors, InSb (indium antimonide).

During impact, simultaneous but not synchronized high-speed video and high-speed thermographic images were taken to record the dynamic behavior of the droplets. **Again, one should stress that IR images provide information about temperature fields on the heating surface, but these are naturally related to the heat transfer processes occurring with the liquid and vapor. The use of thin surfaces assures a negligible difference between the two sides of the heating surface, thus allowing the estimation of even heat fluxes transferred between the liquid/vapor and the surface.** The acquisition frequency and resolution were 2,200 fps and 512×512 px^2 and 1,000 fps and 150×150 px^2 for the high-speed video and high-speed thermographic cameras, respectively. For each experimental condition considered here, five tests were performed to assure the reproducibility of the experiments.

Before each experiment, the camera required a software-based additional calibration consisting of the following steps:

- *Object–camera* distance evaluation: The distance between the object and the IR camera must be carefully evaluated since the transmittance of the atmosphere can influence temperature measurements.

- *Integration time*: To measure the transient behavior of the investigated phenomena requires setting the integration time of the camera according to the appropriate image acquisition rate in frames per second. The lower the integration time, the higher the frames per second. Moreover, the integration time defines the range of temperature that is measurable. The lower the integration time, the wider the temperature range, resulting in a noisier thermographic image.

- *Offset calibration*: Once the integration time is defined, the camera must be calibrated again to consider the change in this parameter. The offset calibration also avoids the occurrence of the "Narciso" effect, which is the reflection in the infrared image of the optics of the camera itself.

- *Image windowing*: In the case of high-speed recording, one has to limit the dimension of the recorded image by "windowing" the global image.

- *Atmosphere temperature*: The quantification of the transmittance requires measuring the temperature of the surrounding atmosphere (T_{amb}) before each test.

- *Ambient temperature*: The incident radiation due to the reflection from the surrounding environment results in inaccurate temperature measurement and, consequently, it must be evaluated and set as a parameter. An infrared-reflecting object placed in front of the camera at the same distance of the object whose temperature will be measured allows evaluating the radiation due to the ambient temperature (e.g., aluminum foil; reflectivity $\rho \approx 0.99$). Its emissivity must be set equal to 1 in the software, and its temperature is measured with the IR camera. The resulting ambient temperature is entered as an input parameter for the final measurement of the desired temperature.

- *Emissivity*: The steps to evaluate this parameter are as follows:
    1. Paint the object with black ink or glue a black thermographic tape on its top ($\varepsilon = 0.95 - 0.96$);
    2. Obtain the emissivity map by heating the object to a known temperature, record an infrared image, and set that image to be at the measured temperature. This image serves as an emissivity map.

The radial temperature profiles **observed at the bottom side of the heating surface** were obtained after post-processing the IR images using a custom-made MATLAB code, which allows converting the raw IR images to temperature data. For the calibration method, a custom-made cavity-based black body radiator device was designed and assembled. For each imposed temperature at the cavity and after achieving a stable condition for which the temperature is homogeneous in the cavity, the measured temperatures were converted into radiated energy flux (W/m²) performing an energy balance and plotted against the received intensity signal in ADUs (analog-to-digital units). Then, a polynomial curve is fitted to these data. This calibration is performed considering a pixel-by-pixel approach. **As for the post-processing procedure, it consists of several steps and a sequence of advanced and applied filters [12].**

15.3.2.3 Surface Temperature and Heat Flux Measurements at Liquid–Solid Interfaces

Theofanous et al. [56] pioneered the use of infrared-based visualization to characterize the temperature of a heated surface in pool boiling. Others followed the similar visualization exercises. However, the actual mechanisms explaining the high heat fluxes removed during pool boiling heat transfer are not yet fully described. Following the "convection analogy models" derived in the 1950s (e.g., [57]), more recent theories have recognized the importance of vaporization occurring at the liquid–vapor–solid interface. Currently, the two main models considered are the "micro-layer model" and the "contact line model." The micro-layer model assumes that a thin layer can be trapped at the wall from a growing bubble and evaporates, contributing to the total heat transferred. The existence of the evaporating micro-layer was first suggested by Snyder and Edwards [58] and subsequently developed as a model by Cooper and Lloyd [59].

On the other hand, the contact line model considers that heat transfer occurs in three central regions, as schematically defined in Figure 15.4 [60–61]: the liquid–vapor–solid interface, namely the area of the adsorbed liquid film, the micro-region, and the macro-region.

In the literature, there are contrasting opinions concerning the dependence of the heat flux removed on the stage of the bubble growth and departure, often defining the micro-region as a three-phase contact line [62]. Stephan and Hammer [60] reported the complete description of the solution of the equations characterizing the micro-region.

Infrared thermography has been used to shed some light on these models, as several authors tried to confirm the micro-layer and the contact line models experimentally. Hence, experimental observation of the surface temperature field resulting from the heat flux removed during micro-layer evaporation has recently been achieved [62–64]. In these studies, the boiling surface consisted of an ITO (indium tin oxide) coating of 1 μm thickness, electrically

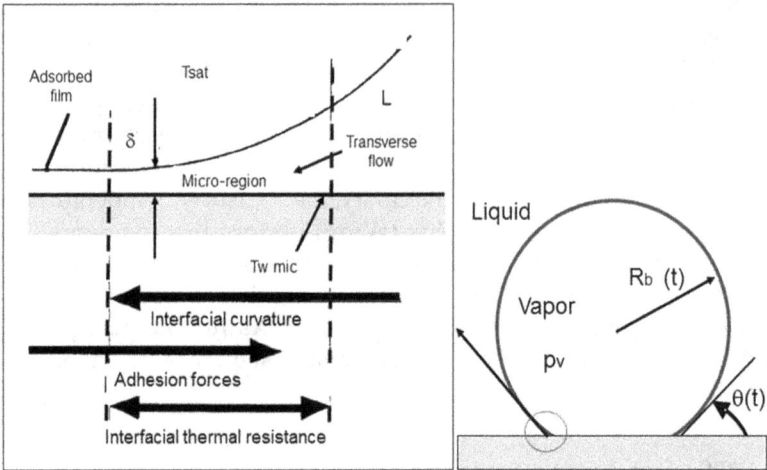

FIGURE 15.4
Identification of the main phenomena in the micro-region adopted according to the contact line model. (Image adapted from Ref. [60].)

heated and deposited on a sapphire substrate. Being the sapphire IR transparent and the ITO visible light transparent, high-speed visualizations and high-speed IR thermography were both performed at the backside of the ITO heater. Thermographic images, recorded with an acquisition frequency of 500 Hz, had, for the configuration used, a spatial resolution of 100 μm/pixel.

The thermographic images highlighted the existence of a region where temperature drops (high heat flux) in the bubble base outer ring and an area of high temperature (low heat flux) in the bubble base inner ring. In agreement with these observations, some studies [62–64] correlated these two regions with the existence of a dry patch at the inner ring of the bubble base and an evaporating micro-layer surrounding the dry patch, in agreement with the model of [59]. The same research group provided a further insight into the existence of the aforementioned regions in [65], where the authors developed a technique that allowed the definition of the position of the triple contact line (liquid–vapor–solid contact line) based on IR thermography, while the analysis of the interference patterns evaluated the thickness of the micro-layer. This kind of measurements further allowed authors such as Sato and Niceno [66] to benchmark CFD (computation fluid dynamics) solvers for the simulation of nucleate pool boiling.

On the other hand, some studies [54,67–70] integrated synchronized high-speed visualizations of the bubble motion from the front side with infrared thermography with a high temporal resolution, to measure the temperature field of the backside of the surface, where boiling was occurring. The surface used was a thin (20 μm) stainless steel foil electrically heated, structured to allow the existence of one single nucleation site, and was painted black for increased emissivity. The thin foil was chosen to reduce the thermal inertia of the surface itself and assume that the temperature measured at the bottom would be close to the one at the vapor–liquid–solid interface. The IR camera was acquiring images at 1,000 Hz with a spatial resolution of 30 μm/pixel. In general agreement, the authors noticed a ring of high heat flux at the triple contact line and negligible heat flux at the central region of the bubble, in agreement with the "contact line" model of [60–61].

Within this context, Wagner and Stephan reported typical temperature profiles obtained using time-resolved thermography [68] and were able to identify peaks of heat flux or dips in temperature in the micro-region (apparent contact line), confirming that the heat removal mechanism as involved in the bubble growth process occurs mainly in this region [62–72]. In [72], the boiling surface was a 25-μm-thick black-painted electrically heated titanium foil. With water as the liquid used, wall temperature measurements at the back of the foil using the IR camera synchronized with high-speed visualizations allow further insights into the bubble growth process from the front side. The acquisition frequency of the IR camera was 1,000 Hz with a spatial resolution of 40 μm/pixel. Contrary to the other authors, Golobic et al. [72] noticed temperature profiles with a valley at the central region of the bubble base and then a steady decrease along the bubble base radius. Further work of the group is reported in [73–74].

The "micro-layer" and "contact line" models describe the heat transfer mechanisms to occur in different ways. Each of them has followers, but there is yet no agreement, or irrefutable experimental proof on which is the correct one. This lack of understanding requires further work be performed toward the development of the current measurement techniques.

Teodori et al. [12] used a similar strategy, i.e., a 20-μm stainless steel thin foil heated by Joule effect, to relate fluid dynamics and heat transfer governing droplet impact and spreading on a heated wall. This combined analysis allowed detailing the temperature profiles on the surface and heat fluxes along droplet radius and relating them to the spreading lamella, including the local temperature variations associated with bubble entrapment at the impact region and the thickness of the lamella at the rim. These experimental results were useful to derive a semiempirical approach explaining the heat transfer between the surface and the droplet, involving the particular flow field inside the lamella, and benchmark a CFD code developed by the group of Dr. Georgoulas [75].

15.4 Final Remarks and Future Prospects

This chapter revised and discussed different solutions to measure and characterize temperatures and heat fluxes at liquid–solid interfaces. As summarized in Table 15.1, contact methods such as thermocouples have various

TABLE 15.1

Summary of Several Key Features for Intrusive, Semi-Intrusive, and Non-Intrusive Methods: Comparative Analysis

	Example	Temporal/Spatial Resolution	Advantages	Disadvantages
Intrusive	Thermocouples	μs/mm; temperature measurements within <°C	Easy to install, affordable, endure harsh environments	Intrusive Provides discrete measures in single positions Accuracy issues due to thermal resistance and positioning/assembly
	Heat flux sensors	ns/μm (considering the micro-sensors); temperature measurements of the order of 1×10^{-3} K	Accurate provision of (averaged) heat flux measurements	Intrusive Expensive Significant errors may be introduced due to the high thermal resistances that can be produced at sensor assembly
Semi-intrusive/non-intrusive	Laser (fluorescent-based) diagnostic techniques	ns/nm; measurements within accuracy better than mK	Non-intrusive Accurate Provide temperature/flow fields	Expensive Complex calibration/post-processing procedures Require adequate tracers
	Thermography	ms/μm; measurements within accuracy better than mK	Non-intrusive Accurate Provide temperature fields Allow evaluation of heat fluxes Do not require tracers	Expensive Complex calibration/post-processing procedures

limitations, but once one recognizes them and applies the appropriate calibration and data reduction procedures, they can provide useful insights into the heat transfer processes occurring at liquid–surface interfacial flows.

Non-intrusive techniques usually allow better spatial and temporal resolutions, but are much more complex to apply. In this case, these techniques are generally more expensive and require specific calibration and data reduction procedures. This issue is illustrated here for the application of infrared thermography with a high temporal and spatial resolution to the characterization of droplet–wall interactions and for the description of bubble dynamics in nucleate boiling flows. Future challenges arrive on the characterization of phase change fluid flows, as the high temperature gradients occurring at small spatial and temporal scales (of the order of microns and a few milliseconds or less, respectively) induce problems related to the distribution of the tracers and reflection at liquid–vapor interfaces, among others. Similar issues occur on thermometry techniques applied to microfluidics, which are probably the biggest challenge to overcome. Micro- and nano-sized sensors still have restrictions related to their spatial resolution being their assembly yet complex and expensive. In this context, non-intrusive techniques may be the answer, requiring additional research toward the development of the appropriate tracers, calibration, and processing procedures, in order to deliver reliable measurements.

Acknowledgments

The authors are grateful to Fundação para a Ciência e Tecnologia (FCT) for partially financing the research under the framework of the project JICAM/0003/2017, in the context of *Projecto 3599- Promover a Produção Científica, o Desenvolvimento Tecnológico e a Inovação*. A. S. Moita also acknowledges FCT for her contract under the framework of IF 2015 recruitment program (IF 00810–2015) and the exploratory project associated with this contract. Pedro Pontes also acknowledges the University of Lisbon for his PhD fellowship.

References

1. Moreira, A. L. N., Moita, A. S., Panão, M. R. 2010. Advances and challenges in explaining fuel spray impingement: How much of single droplet impact research is useful? *Prog Energy Combust Sci* 36:554–580.
2. Panão, M., Guerreiro, J., Moreira, A. K. N. 2012. Microprocessor cooling based on an intermittent multijet spray system. *Int J Heat Mass Trans* 55:2854–2863.

3. Moura, M., Teodori, E., Moita, A. S., Moreira, A. L. N. 2016. Cooling system with controlled pool boiling of dielectrics over micro-and-nano structured integrated heat spreaders. *Proceedings of IEEE ITherm Conference, The Intersociety Conference on Thermal and Thermomechanical Phenomena in Electronic Systems*, Las Vegas, NV, pp. 378–387.

4. Kim J. 2007. Spray cooling heat transfer: The state of the art. *Int J Heat Fluid Flow* 28(4):753–767.

5. Liang, G., Mudawar, I. 2017. Review of droplet impact on heated walls. *Int J Heat Mass Transfer* 106:103–126.

6. Seki, M., Kawamura, H., Sanokawa, K. 1978. Transient temperature profile of a hot wall due to an impinging liquid droplet. *J Heat Transfer* 100:167–169.

7. Moita, A. S., Moreira, A. L. N., Roisman, I. 2010. Heat transfer during drop impact onto a heated surface. *Proc ASME Int Heat Transfer Conf IHTC* 14(6):803–810.

8. Nukiyama, S. 1966. The maximum and minimum values of the heat Q transmitted from metal to boiling water under atmospheric pressure. *Int J Heat Mass Transfer* 9:1419–1433.

9. Han, C. Y., Griffith, P. 1962. The mechanism of heat transfer in nucleate boiling. Technical Report No. 7673–19, Depaertement of Mechanical Engineering M.I.T.

10. Kurul, N., Podowski, M. 1990. Multidimensional effects in forced convection subcooled boiling. *Proceedings of 9th International Heat Transfer Conference*, Jerusalem, pp. 21–25.

11. Malavasi, I., Teodori, E., Moita, A. S., Moreira, A. L. N., Marengo, M. 2018. Wettability effect on pool boiling: A review. In: *Encyclopaedia of Two-Phase Heat Transfer and Flow III: Macro and Micro Flow Boiling and Numerical Modeling Fundamentals (a 4-Volume Set)*, Vol. 4, Edited by J. Thome. World Scientific Publishing Co Pte Ltd, Singapore, pp. 1–62.

12. Teodori, E., Pontes, P., Moita, A. S., Moreira, A. L. N. 2018. Thermographic analysis of interfacial heat transfer mechanisms on droplet/wall interactions with high temporal and spatial resolution Exp. *Thermal Fluid Sci* 96:284–294.

13. Teodori, E., Moita, A. S., Pontes, P., Moura, M., Moreira, A. L. N., Bai, Y., Li, X., Liu, Y. 2017. Application of bioinspired superhydrophobic surfaces in two-phase heat transfer experiments. *J Bionic Eng* 14(3):506–519.

14. Gerardi, C., Buongiorno, J., Hu, L. W., McKrell, T. 2010. Study of bubble growth in water pool boiling through synchronized, infrared, thermometry and high-speed video. *Int J Heat Mass Transfer* 53:4185–4192.

15. Childs, P. R. N. 2016. Nanoscale thermometry and temperature measurement. In: *Thermometry at the Nanoscale: Techniques and Selected Applications*, Edited by L. D. Carlos, F. Palacio, F. P. Parada. RSC – The Royal Society of Chemistry, London.

16. Kim, M. M., Giry, A., Mastiani, M., Rodrigues, G. O., Reis, A., Mandin, P. 2015. Microscale thermometry: A review. *Microelectron Eng* 148:129–142.

17. Teodori, E., Valente, T., Malavasi, I., Moita, A. S., Marengo, M, Moreira, A. L. N. 2017. Effect of extreme wetting scenarios on pool boiling conditions. *Appl Thermal Eng* 115:1424–1437.

18. Malavasi, I., Bourdon, B., DiMarco, P., DeConinck, J., Marengo, M. 2015. Appearance of a low superheat "quasi-Leidenfrost" regime for boiling on superhydrophobic surfaces. *Int Commun Heat Mass Transfer* 63:1–7.

19. Worthington, A. M. 1876. On the form assumed by drops of liquids falling vertically on a horizontal plate. *Proc Royal Soc London* 25:261–271.

20. Worthington, A. M. 1877. A second paper on the form assumed by drops of liquids falling vertically on a horizontal plate. *Proc Royal Soc London* 25:498–503.

21. Wachters, L. H. J., Westerling, N. A. J. 1966. The heat transfer from horizontal plate to sessile water drops in the spheroidal state. *Chem Eng Sci* 21:923–936.

22. Labeish, V. G. 1994. Thermodynamic study of a drop impact against a heated surface. *Exp Thermal Fluid Sci* 8:181–184.

23. Pasandideh-Fard, M., Aziz, S., Chandra, S., Mostaghimi, J. 2001. Cooling effectiveness of a water drop impinging on a hot surface. *Int J Heat Fluid Flow* 22:201–210.

24. Betz, R., Jenkins, J., Kim, C.-J. C., Attinger, D. 2013. Boiling heat transfer on superhydrophilic, superhydrophobic, and superbiphilic surfaces. *Int J Heat Mass Transf* 57:733–774.

25. Lee, J., Kim, J., Kiger, K. T. 2001. Time- and space-resolved heat transfer characteristics of single droplet cooling using microscale heat arrays. *Int J Heat Fluid Flow* 22:188–200.

26. Kim, J.-H., Ahn, S. I., Kim, J. H., Zin, W.-C. 2007. Evaporation of water droplets on polymer surfaces. *Langmuir* 23:6163–6169.

27. Xue, Z., Qiu, H. 2005. Integrating micromachined fast response temperature sensor array in a glass microchannel. *Sens Actuators, A* 122:189–195.

28. Paik, S. W., Kihm, K. D., Lee, S. P., Pratt, D. M. 2007. Spatially and temporally resolved temperature measurements for slow evaporating sessile drops heated by a microfabricated heater array. *J Heat Transfer* 129:966.

29. Bhardwaj, R., Longtin, J. P., Attinger, D. 2010. Interfacial temperature measurements, high-speed visualization and finite-element simulations of droplet impact and evaporation on a solid surface. *Int J Heat Mass Transfer* 53(19–20):3683–3691.

30. Shi, L., Kwon, O., Miner, A. C., Majumdar, A. 2001. Design and batch fabrication of probes for sub-100 nm scanning thermal microscopy. *J Microelectromech Syst* 10:370–378.

31. Lee, W., Kim, K., Jeong, W., Zotti, L. A., Pauly, F., Cuevas, J. C., Reddy, P. 2013. Heat dissipation in atomic-scale junctions. *Nature* 498:209–212.

32. Buongiorno, J., Cahill, D. G., Hidrovo, C. H., Moghaddam, S., Schmidt, A. J., Shi, L. 2014. Micro- and nanoscale measurement methods for phase change heat transfer on planar and structured surfaces. *Nanoscale Microscale Thermophys Eng* 18:270–287.

33. Wingert, Z. C., Chen, Y., Kwon, S., Xiang, J., Chen, R. 2012. Ultra-sensitive thermal conductance measurement of one-dimensional nanostructures enhanced by differential bridge. *Rev Sci Instrum* 83:024901.

34. Weathers, A., Bi, K., Pettes, M. T., Shi, L. 2013. Reexamination of thermal transport measurements of a low-thermal conductance nanowire with a suspended micro-device. *Rev Sci Instrum* 84:084903.

35. Aziz, S. D., Chandra, S. 2000. Impact, recoil and splashing of molten metal droplets. *Int J Heat Mass Transfer* 43(16):2841–2857.

36. Buttsworth, D. R. 2001. Assessment of effective thermal product of surface junction thermocouples on millisecond and microsecond time scales. *Exp Thermal Fluid Sci* 25(6):409–420.

37. Caldwell, F. R. 1962. Thermocouple materials. In Temperature; Its Measurement and Control in Science and Industry. Reinhold, New York, vol. 2, pp. 81–134.

38. Heichal, Y., Chandra, S., Bordatchev, E. 2005. A fast-response thin film thermocouple to measure rapid surface temperature changes. *Exp Thermal Fluid Sci* 30(2):153–159.

39. Díaz, A. J., Ortega, A. 2013. Investigation of a gas-propelled liquid droplet imping- ing onto a heated surface. *Int J Heat Mass Transfer* 67(C):1181–1190.

40. Reichelt, L., Meingast, U., Renz, U. 2002. Calculating transient wall heat flux from measure- ments of surface temperature. *Int J Heat Mass Transfer* 45(3):579–584.

41. Chen, X. D., Nguang, S. K. 2003. The theoretical basis of heat flux sensor pen. *Adv Decis Sci* 7(1):1–10.

42. Cossali, G. E., Marengo, M., Santini, M. 2005. Single-drop empirical models for spray impact on solid walls: A review. *Atomization Sprays* 15(6):699–736.

43. Panão, M. R. O., Moreira, A. L. N. 2009. Intermittent spray cooling: A new technology for controlling surface temperature. *Int J Heat Fluid Flow* 30(1):117–130.

44. Schultz, D. L., Jones, T. V. 1973. Heat-transfer measurements in short-duration hypersonic facilities, Agardograph No.165.

45. Abram, C., Fond, B., Beyrau, F. 2018. Temperature measurement techniques for gas and liquid flows using thermographic phosphor tracer particles. *Prog Energy Combust Sci* 64:93–156.

46. Berne, B. J., Pecora, R. 2000. *Dynamics Light Scattering*. Dover Publications, New York.

47. Yu, D. I., Kwak, H. J., Noh, H., Park, H. S., Fezzaa, K., Kim, M. H. 2018. Synchrotron x-ray imaging visualization study of capillary-induced flow and critical heat flux on surfaces with engineered micropillars. *Sci Adv* 4:e1701571.

48. Jaque, D., Vetrone, F. 2012. Luminescence nanothermometry. *Nanoscale* 4(15):4301e26.

49. Kosseifi, N., Biwole, P. H., Mathis, C., Rousseaux, G., Boyer, S. A. E., Yoshikawa, H. N., et al. 2013. Application of two-color LIF thermometry to nucleate boiling. *J Mater Sci Eng B* 3:281e90.

50. Quinto-Su, P. A., Suzuki, M., Ohl, C.-D. 2014. Fast temperature measurement following single laser-induced cavitation inside a microfluidic gap. *Sci Rep* 4:5445.

51. Sedmak, I., Urbancic, I., Podlipec, R., Strancar, J., Mortier, M., Golobic I. 2016. Submicron thermal imaging of a nucleate boiling process using fluorescence microscopy. *Energy* 109:436–445.

52. Incropera, F. P., DeWitt, D., Bergman, T. L., Lavine, A. S. 2007. *Fundamentals of Heat and Mass Transfer*, 6th Edn. John Wiley & Sons, New York.

53. Astarita, T., Carlomagno, G. M. 2013. *Infrared Thermography for Thermo-Fuid-Dynamics*. Springer-Verlag, Berlin Heidelberg, Germany.

54. Sielaff, A. 2014. Experimental investigation of single bubbles and bubble interactions in nucleate boiling. Damstrast: Technische Univeristat Darmstradt (PhD Thesis Library). Retrieved from http://tuprints.ulb.tu-damstradt.de/3703.

55. Teodori, E., Valente, T., Malavasi, I., Moita, A. S., Marengo, M, Moreira, A. L. N. 2017. Effect of extreme wetting scenarios on pool boiling conditions. *Appl Thermal Eng* 115:1424–1437.

56. Theofanous, T. G., Tu, J. P., Dinh, A. T., Dinh, T. N. 2002. The boiling crisis phenomenon. *Exp Thermal Fluid Sci* 26(6–7), P.I:775–792, P.II:793–810.

57. Rohsenow, W. M. 1952. A method of correlating heat transfer data for surface boiling of liquids. *Trans ASME* 74:969–976.
58. Snyder, N., Edwards, D. 1956. Summary of conference on bubble dynamics and boiling heat transfer. *Memo Jet Propul Lab*Memo (20–137):14–15.
59. Cooper, M., Lloyd, A. 1969. The microlayer in nucleate pool boiling. *Int J Heat Mass Transfer* 12:895–913.
60. Stephan, P., Hammer, J. 1994. A new model for nucleate boiling heat transfer. *Heat Mass Transfer* 30(2):119–125.
61. Kim, J. 2009. Review of nucleate pool boiling heat transfer mechanisms. *Int J Multiphase Flow* 35:1067–1076.
62. Gerardi, C., Buongiorno, J., McKrell, T. 2010. Study of bubble growth in water pool boiling through synchronized, infrared thermometry and high-speed video. *Int J Heat Mass Transfer* 53(19):4185–4192.
63. Duan, X., Buongiorno, J., McKrell, T. 2011. Integrated particle imaging velocimetry and infrared thermometry for high resolution measurement of subcooled nucleate pool boiling. *The 14th International Meeting on Nuclear Reactor Thermal hydraulics, NURETH 14*, Toronto, Canada.
64. Duan, X., Philips, B., McKrell, T., Buongiorno, J. 2013. Synchronized high-speed video, infrared thermometry, and particle image velocimetry data for validation of interface-tracking simulations of nucleate boiling phenomena. *Exp Heat Transfer* 26(2–3):169–197.
65. Kim, H., Buongiorno, J. 2011. Detection of liquid–vapor–solid triple contact line in two-phase heat transfer phenomena using high-speed infrared thermometry. *Int J Multiphase Flow* 37:166–172.
66. Sato, Y., Niceno, B. 2015. A depletable micro-layer model for nucleate pool boiling. *J Comput Phys* 300:20–52.
67. Wagner, E., Stephan, P., Koeppen, O., Auracher, H. 2007. High resolution temperature measurements as moving vapor/liquid and vapor/liquid/solid interfaces during bubble growth in nucleate boiling. *Proceedings of 4th International Berlin Workshop on Transport Phenomena with Moving Boundaries*, Berlin, Germany.
68. Wagner, E., Stephan, P. 2009. High resolution measurements at nucleate boiling of pure FC-84 and FC-3284 and its binary mixtures. *ASME J Heat Transfer* 131:1–12.
69. Schweizer, N., Stephan, P. 2009. Experimental study of bubble behaviour and local heat flux in pool boiling under variable gravitational conditions. *Multiphase Sci Technol* 21(4):329–350.
70. Stephan, P., Sielaff, A., Fischer, S., Dietl, J., Herbert, S. 2013. A contribution of the basic understanding of nucleate boiling phenomena: Generic experiments and numerical simulations. *Thermal Sci Eng* 21(2):39–57.
71. Golobic, I., Petkovsek, J., Baselj, M., Papez, A. 2009. Experimental determination of transient wall temperature distributions close to growing vapor bubbles. *Heat Mass Transfer* 45:857–866.
72. Golobic, I., Petkovsek, J., Kenning, D. 2012. Bubble growth and horizontal coalescence in saturated pool boiling on a titanium foil, investigated by high-speed IR thermography. *Int J Heat Mass Transf* 55:1385–1402.
73. Zupančič, M., Steinbucher, M., Sedmak, I., Golobič, I. 2015. Enhanced pool-boiling heat transfer on laser-made hydrophobic/superhydrophilic polydimethylsiloxane-silica patterned surfaces. *Appl Thermal Eng* 91:288–297.

74. Petrovsek, J., Heng, Y., Zupancic, M., Gjerkes, H., Cimerman, F., Golobic, I. 2016. IR thermographic investigation of nucleate pool boiling at high heat flux. *Int J Refrig* 61:27–139.
75. Teodori, E., Pontes, P., Moita, A., Georgoulas, A., Marengo, M., Moreira, A. 2017. Sensible heat transfer during droplet cooling: Experimental and numerical analysis. *Energies* 10(6):790(27p).

16

Measuring Heat Transfer Coefficient during Condensation Inside Channels

Davide Del Col, Stefano Bortolin, and Marco Azzolin
University of Padova

CONTENTS

16.1 Introduction

Convective condensation inside channels is a heat transfer process often encountered in engineering applications (e.g., refrigeration, air-conditioning, loop heat pipes), but due to the complex phenomena involved (vapor shear stress, thermal conduction in the liquid film, turbulence, surface tension, gravity, mass transport, waves at the interface, wall wettability), it is not yet fully understood. For these reasons, measurements of the condensation heat transfer coefficient may be needed in industrial applications (e.g., to characterize the heat transfer performance of a given fluid in a specific channel) or for a more fundamental investigation (e.g., to study the surface tension or the gravity effect) and the development of accurate prediction methods. Depending on the desired accuracy and on the experimental conditions, several techniques can be used for the measurement of the heat transfer coefficient.

The present chapter would help the reader in the selection and implementation of different experimental techniques starting from the experience matured in this field at the University of Padova.

The condensation heat transfer coefficient α_c is defined as the ratio of the heat flux q to the difference between the saturation temperature T_{sat} and the wall temperature T_{wall}, as reported in Eq. (16.1):

$$\alpha_c = q/(T_{sat} - T_{wall}) \tag{16.1}$$

Considering the specific case of in-tube condensation, the heat transfer coefficient is expected to change along the length of the channel (α_c is a function of the axial coordinate z) and this is due, first of all, to the vapor quality variation. A mean value of the heat transfer coefficient (e.g., measured with the fluid evolving from saturated vapor to saturated liquid) is not generally helpful. Therefore, two different types of measurement that are able to account for the heat transfer coefficient variation along the channel must be considered: a quasi-local approach that is based on a relatively small variation of the vapor quality along the channel and a local approach that allows measuring the heat transfer coefficient at different axial positions along the channel. It must be pointed out that, at a given axial position, due to gravity or surface tension effects (in case of non-circular channels), the thickness of the liquid film may not be uniform and thus the heat transfer coefficient can vary along the perimeter of the channel cross section (Toninelli et al. 2017; Bortolin et al. 2014). This type of measurement would request an evaluation of the wall temperature field in correspondence of the channel cross section which is usually not available. For this reason, hereinafter, the heat transfer coefficient is to be considered cross-section-averaged.

When tackling the issue of performing condensation measurements, one of the main problems is related to the determination of the heat flux. In fact, unlike in the case of flow boiling where the heat transfer rate can be accurately measured based on the direct determination of electrical power, in condensation, the heat flow rate must be extracted from the fluid and, to achieve this, a secondary fluid (e.g., water) is often employed. As a consequence, the heat flux must be indirectly measured from the temperature variation and from the mass flow rate of this secondary fluid. A less common alternative is the use of Peltier elements and heat flux sensors, but this technique will not be considered here. The presence of a secondary fluid on the external side complicates the installation of the temperature sensors inside the channel wall, and this must be considered in the design of the test section. Experimental techniques involving a secondary fluid can be distinguished between those that provide a direct measurement of the wall temperature and those that are based on the preliminary knowledge of the secondary fluid heat transfer coefficient (based on Wilson plot method).

16.2 Approach Based on Wilson Plot Method

When measuring condensation heat transfer coefficients, one challenge is the measurement of the channel wall temperature. This is usually done by placing a temperature sensor (e.g., thermocouple) inside a small hole machined in the wall of the tube. The critical aspects are the reduced wall thickness of the tube usually available and the presence of a secondary fluid on the external side (see Sections 16.3 and 16.4). The Wilson plot method can be applied to avoid measuring the wall temperature.

Let us now consider a test section consisting of a tube-in-tube heat exchanger: The primary fluid flows inside the internal tube (diameter D_i), and it partially condenses from vapor quality x_{in} to vapor quality x_{out}; a secondary fluid (e.g., water) flows in counterflow in the external annulus to remove the condensation heat. The goal is to determine the internal condensation heat transfer coefficient α_c from the measurement of the overall heat transfer coefficient U_o and from the knowledge of the external heat transfer coefficient α_o (without measuring the wall temperature). An application of the present technique to the measurement of the condensation heat transfer coefficient with R22 flowing inside a micro-fin tube can be found in Del Col et al. (2002).

The overall heat transfer coefficient U_o (based on the external area A_o) is given by:

$$U_o = \frac{q}{\Delta T_{lm}} = \frac{Q}{A_o \Delta T_{lm}} \tag{16.2}$$

where ΔT_{ml} is the log-mean temperature difference determined from the annulus-side inlet and outlet temperatures and from the inlet and outlet saturation temperatures. It must be pointed out that the log-mean temperature difference can be used under the hypothesis of uniform overall heat transfer coefficient along the channel and, thus, a limited variation of vapor quality (typically, $\Delta x \sim 0.1$–0.2) must be maintained during the tests. A small vapor quality change leads to a quasi-local measurement satisfying the hypothesis of "almost uniform" heat transfer coefficient in the test section, but from the other side, reducing Δx means also reducing the heat flow rate Q. The heat flow rate, together with the vapor quality variation of the condensing fluid, can be obtained from a heat balance between the two fluids as reported below:

$$Q = \dot{m}_2 c \left(T_{2,out} - T_{2,in} \right) = \dot{m}_1 \left(h_{1,in} - h_{1,out} \right) = \dot{m}_1 h_{LV} \Delta x \tag{16.3}$$

where $\dot{m}_1$ and $\dot{m}_2$ are, respectively, the mass flow rate of the primary condensing fluid and the secondary fluid, $T_{2,out}$ and $T_{2,in}$ are the temperature of the secondary fluid at the outlet and inlet of the test section, $h_{2,in}$ and $h_{2,out}$ are the specific enthalpy of the condensing fluid at the inlet and outlet of the tube, and h_{LV} is the latent heat. In Eq. (16.3), it is assumed that the secondary fluid is an uncompressible liquid having a constant specific heat c and negligible pressure drop. It is evident that, since $\dot{m}_2$ cannot be reduced below a minimum value (due to the uncertainty of the mass flow meter and also due to the need to maintain the flow turbulent, as it will be clear in the following), a small vapor quality variation leads to a reduced temperature difference and thus to a higher uncertainty of the heat flow rate. Therefore, the vapor quality variation is a compromise between a small variation of the condensation heat transfer coefficient and a feasible measurement of the heat flow rate. It emerges as this technique is not suited for small-diameter channels and low mass flow rates of the condensing fluid (a small heat flow rate exchanged causes a large vapor quality variation).

In conclusion, the determination of the overall heat transfer coefficient using Eqs. (16.2) and (16.3) requires the measurement of the following quantities: mass flow rate of the two fluids, temperature difference of the secondary fluid, and saturation temperature of the primary fluid. For the measurement of the temperature of the secondary fluid, static mixers must be installed prior to the temperature wells at the inlet and outlet of the test section to guarantee an effective measurement of the adiabatic mixing temperature. The saturation temperature can be measured in adiabatic sectors prior to and after the test section with thermocouples placed on the external side of the wall. Pressure ports located prior to and after the channel allow comparing the measured saturation temperature with the saturation temperature obtained from the pressure. Considering Eq. (16.3), the inlet specific enthalpy must be also known. Usually, a pre-section is installed before the actual test section that brings the fluid from the superheated state to the saturated state (at the desired inlet vapor quality) exchanging heat with a cold fluid.

Assuming no fouling resistance, the condensation heat transfer coefficient can be obtained from

$$\alpha_c = \frac{1}{\left[\dfrac{1}{U_o} - \dfrac{1}{\alpha_o} - \dfrac{D_o \ln(D_o/D_i)}{2\lambda_{\text{wall}}}\right]\dfrac{A_i}{A_o}} \qquad (16.4)$$

where α_o is the heat transfer coefficient on the external side, D_o is the outside diameter of the inner tube, λ_{wall} is the wall thermal conductivity, and A_i/A_o is the ratio between the internal and external areas of the tube. To obtain the condensation heat transfer coefficient α_c, the external heat transfer coefficient α_o must be known and it can be determined using the Wilson plot technique.

16.2.1 Modified Wilson Plot Method

The Wilson plot method is a tool for the determination of the heat transfer coefficient in different convective heat transfer processes. It allows finding appropriate coefficients for the heat transfer coefficient correlations starting from sets of experimental data taken without the direct measurement of the surface temperature. A general review of the Wilson plot method with the modifications proposed by researchers in the past years can be found in Fernández-Seara et al. (2007), and a critical discussion on Wilson plot technique is reported in Rose (2004). The method can be used for a variety of situations, from the specific case of in-tube condensation to the case of heat transfer in plate heat exchangers (Mancin et al. 2013; Longo 2009) and shell-and-tube heat exchangers. The Wilson plot method was developed by Wilson in 1915 (Wilson 1915), and in this section, it is proposed with a modification that adopts a correlation for the convective heat transfer coefficient on the external side (annulus) as a power function of the Reynolds (Re) and Prandtl (Pr) numbers instead of the fluid velocity as in the original work by Wilson.

Consider now a set of experiments performed with the same tube-in-tube heat exchanger test section mentioned in Section 16.2 and using water (or a different liquid) on both the internal and external sides. The flow rate of the water on the external side is varied, while the flow conditions (inlet temperature and mass flow rate) on the internal side are maintained constant. Suppose that for the annulus (external side), the equation of the heat transfer coefficient α_o can be written in the following form:

$$\mathrm{Nu}_o = \frac{\alpha_o D_h}{\lambda_{\text{water}}} = C\mathrm{Re}_o^n \mathrm{Pr}_o^{1/3} \left(\frac{\mu_o}{\mu_{o,\text{wall}}} \right)^{0.14} \tag{16.5}$$

where Nu_o is the Nusselt number on the external side, D_h is the hydraulic diameter of the annulus, λ_{water} is the water thermal conductivity, μ_o is the dynamic viscosity of the water at the bulk temperature of the fluid, and $\mu_{o,\text{wall}}$ is the water dynamic viscosity calculated at the wall temperature. All the properties are calculated at the mean temperature between inlet and outlet of the annulus.

The total thermal resistance $1/(U_o A_o)$ can be measured (Eq. 16.2), and it is equal to the sum of the inside $(1/(\alpha_i A_i))$, wall (R_{wall}), and annulus-side $(1/(\alpha_o A_o))$ thermal resistances

$$\frac{1}{U_o A_o} = \frac{1}{\alpha_i A_i} + R_{\text{wall}} + \frac{1}{\alpha_o A_o} \tag{16.6}$$

Since the internal heat transfer coefficient α_i is held constant during the water–water tests and the wall resistance does not vary, substituting Eq. (16.5) in Eq. (16.6), we obtain:

$$\frac{1}{U_o A_o} = C_1 \mathrm{Re}_o^{-n} \mathrm{Pr}_o^{-1/3} \left(\frac{\mu_o}{\mu_{o,\text{wall}}} \right)^{-0.14} + C_2; \quad C_1 = \frac{D_h}{C A_o \lambda_{\text{water}}}; \quad C_2 = R_{\text{wall}} + \frac{1}{\alpha_i A_i}$$

$$\tag{16.7}$$

where C_1 and C_2 are constants for the test series. Equation (16.7) is linear in the form $Y = C_1 X + C_2$ and the experimental data points can be plotted in a Wilson plot (Figure 16.1) with abscissa $X = \mathrm{Re}_o^{-n} \mathrm{Pr}_o^{-1/3}(\mu_o/\mu_{o,wall})^{-0.14}$ and ordinate $1/(U_o A_o)$. The plot of the data should show the experimental points lying on a straight line having intercept C_2 and slope C_1. The interpolating equation (and thus the coefficients C_1 and C_2) can be determined by the weighted least squares (WLS) regression method (see Section 16.4.2.1) that accounts also for the uncertainty of each experimental point. Once C_1 is determined, both the constant C (Eq. 16.7) and the external heat transfer coefficient α_o (Eq. 16.5) can be obtained.

Regarding the value of the exponent n in Eq. (16.5), it can be assumed from previous data or it can be regarded as an unknown constant and found by minimization of the sum of square residuals. In this case, since the overall resistance is not linear with n, an iterative scheme must be followed.

In conclusion, for an effective application of the modified Wilson plot method described here, the following aspects must be considered:

- A valid form of the correlation for the heat transfer coefficient Eq. (16.5) must be available.
- All the experimental tests must fall in the same regime (e.g., turbulent) where the heat transfer correlation is valid.
- The overall heat transfer coefficient must be measured with good accuracy.
- The fouling resistance must be negligible or constant during the experimental campaign.
- The two thermal resistances (internal side and external side) should have the same order of magnitude.
- A large database encompassing the secondary fluid conditions that are expected during condensation tests is needed.

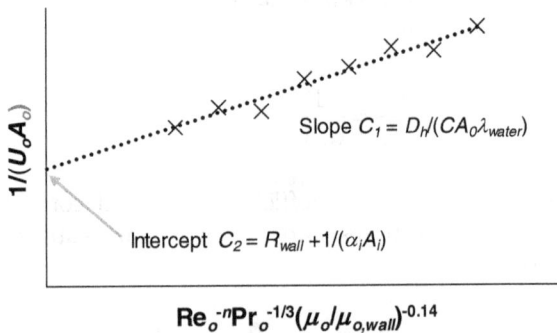

FIGURE 16.1
Wilson plot considering Eq. (16.5) as a functional form for the heat transfer coefficient on the external side and a constant thermal resistance on the internal side.

16.2.2 Evaluation of the Experimental Uncertainty

The uncertainty analysis of the heat transfer coefficient α_c can be performed according to JCGM Guide (JCGM 2008). In this section, only the main aspects will be discussed and the reader can refer to JCGM (2008) for more details. The experimental uncertainty is made of two parts: The first component is the Type A uncertainty (u_A) that derives from repeated observations (usually, for each test condition, several measurements are performed with the test rig operating in steady-state conditions), whereas the second one is the Type B uncertainty (u_B) that derives from instruments calibration and manufacturer's specifications. The combined standard uncertainty u_C of a generic measured parameter θ (it could be a temperature, a mass flow rate, a heat transfer coefficient, etc.) results from the Type A and Type B components according to Eq. (16.8).

$$u_C(\theta) = \sqrt{u_A^2(\theta) + u_B^2(\theta)} \tag{16.8}$$

When a quantity ξ (e.g., heat transfer coefficient) is not directly measured, but it can be expressed as a function f of N input quantities $\theta_1, \theta_2, ..., \theta_N$, as in the case of heat transfer coefficient (Eq. 16.4) or vapor quality variation (Eq. 16.3), its combined standard uncertainty is determined applying the law of propagation of uncertainty:

$$u_C(\xi) = \sqrt{\sum_{i=1}^{N}\left(\frac{\delta f}{\delta \theta_i}\right)^2 u_C^2(\theta_i) + 2\sum_{i=1}^{N-1}\sum_{j=i+1}^{N}\frac{\delta f}{\delta \theta_i}\frac{\delta f}{\delta \theta_j}\text{cov}(\theta_i,\theta_j)} \tag{16.9}$$

where the second part of the expression under square root can be neglected for uncorrelated input quantities.

Considering the condensation heat transfer coefficient (Eq. 16.4), the major contributions are due to the uncertainty of the overall heat transfer coefficient U_o and the external heat transfer coefficient α_o, whereas the contributions due to geometry (areas and diameters) could be negligible for conventional-diameter channels. The combined uncertainty on the measured overall heat transfer coefficient can be determined by applying the law of propagation of uncertainty to Eqs. (16.2) and (16.3) once the Type A and B contributions related to each measured quantity (temperatures and mass flow rates) are known. For common engineering fluids, the contributions due to the uncertainty of the thermodynamic properties (e.g., specific heat and specific enthalpies) can be neglected.

More attention is required during the estimation of the uncertainty related to the annulus-side heat transfer coefficient. In fact, in this case, the heat transfer coefficient α_o is calculated with Eq. (16.5) and the constant C is, in turn, obtained using the Wilson plot method and calculating the slope C_1 of the straight line by the weighted least squares (WLS) regression method. Therefore, the uncertainty in the constant C_1 must be obtained as a first step and then the uncertainty on the annulus-side heat transfer coefficient can be calculated applying again the law of propagation of uncertainty. Assuming that the form of Eq. (16.5) well describes the heat transfer coefficient in annuli (this is a prerequisite for the application of the Wilson plot method), each experimental point shown in Figure 16.1 will be characterized by its own experimental uncertainty and the WLS regression method will allow determining the value of the coefficient C_1 (different weights are given to the data points depending on their respective uncertainty) together with its associated variance and thus uncertainty. The weighted least squares (WLS) regression method is described in detail in Section 16.4.2.1.

Once all contributions have been evaluated, the expanded uncertainty of the heat transfer coefficient is obtained by multiplying the combined standard uncertainty by a coverage factor k (e.g., $k = 2$ for an interval having a level of confidence of approximately 95%).

A last comment must be made regarding the use of the Wilson plot method that is evident from the uncertainty analysis on the condensation heat transfer coefficient. Neglecting the wall thermal resistance and supposing that the overall heat transfer coefficient and the annulus heat transfer coefficient have the same uncertainty, and that the internal area is equal to the external area $A_o = A_i$, it can be found that the uncertainty of the condensation heat transfer coefficient $u(\alpha_c)$ is proportional to the term $(\alpha_o - U_0)^{-2}$. So, if $\alpha_c \gg \alpha_o$, U_0 approaches α_o and $u(\alpha_c)$ can become very high. Therefore, as previously mentioned, it is important that the thermal resistances on the internal and external sides present comparable values.

16.3 Direct Wall Temperature Measurement Approach

A different experimental technique for the evaluation of the heat transfer coefficient (HTC) during convective condensation inside tubes is based on the direct measurement of the channel wall temperature by embedding some thermocouples in the wall. In this case, it will be possible to determine the condensation "quasi-local" heat transfer coefficient overcoming the limitations of the Wilson plot method.

Let us consider a test section made of a tube-in-tube heat exchanger in which a primary fluid (e.g., refrigerant) condenses in the inner tube and a secondary fluid (e.g., water) flows in the external annulus. The condensation heat transfer coefficient α_c can be determined from the measurement of the heat flow rate Q, the wall temperatures T_{wall}, and saturation temperatures T_{sat}:

$$\alpha_c = \frac{Q/A}{\left[\left(T_{sat,in} - T_{wall,in}\right) - \left(T_{sat,out} - T_{wall,out}\right)\right]/\ln\dfrac{\left(T_{sat,in} - T_{wall,in}\right)}{\left(T_{sat,out} - T_{wall,out}\right)}} \qquad (16.10)$$

where the subscripts in and out refer, respectively, to the inlet and outlet of the test section.

The objective here is to accurately measure the wall temperature to directly obtain the condensation heat transfer coefficient. The wall temperature has to be measured at the inlet and outlet of the test section possibly at different angular locations at the same axial position. This is important to get a mean value of the wall temperature, since the two-phase flow can be not axisymmetric (e.g., when stratified flow is present) and thus the heat transfer coefficient (and the wall temperature) can vary along the perimeter of the channel cross section. This technique was applied in Cavallini et al. (2001) to measure the condensation heat transfer coefficient of numerous refrigerants inside an 8-mm-diameter tube. In their work, the inner tube was instrumented with 8 copper–constantan thermocouples embedded in the wall to measure the surface temperature (Figure 16.2). They placed four thermocouples 100 mm after the inlet of the measuring section and four thermocouples 100 mm before the outlet. The average wall temperature (that is needed in Eq. (16.10)) was determined as:

$$T_{\text{wall,in}} = \frac{\sum_{i=1}^{n} T_{\text{wall,in},i}}{n} \qquad T_{\text{wall,out}} = \frac{\sum_{i=1}^{n} T_{\text{wall,out},i}}{n} \qquad (16.11)$$

With four thermocouples embedded in the tube wall, they obtained a representative mean wall temperature to determine the cross-section-averaged heat transfer coefficient.

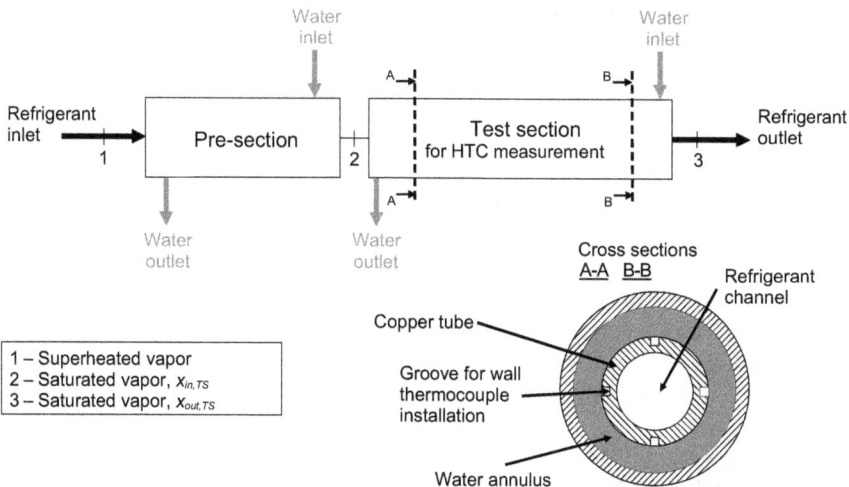

FIGURE 16.2
Drawing of the test section for condensation heat transfer coefficient measurements (Cavallini et al. 2001).

One concern about the fabrication of thermocouples is that the diameter of the wires should be as small as possible to avoid the effects of heat conduction along the wires and to realize small junctions. Moreover, the accommodation of the wall thermocouples is made so that the thermocouple wires do not cross the coolant path and at the same time the junctions are in contact only with the wall of the channel, avoiding any influence of the water flow on the wall temperature measurement. The thermocouples wires, following the grooves machined on the outer side of the internal tube, exit from the two ends of the test section without crossing the water channel and thus minimizing the possible errors due to heat conduction along the thermocouple wires.

In Eq. (16.10), the heat flow rate Q can be evaluated on the water side measuring the water mass flow rate $\dot{m}_{water}$ and the water temperature T_{water} at the inlet/outlet of the test section. It is clear that the accurate measurements of the water temperatures are fundamental for the precise evaluation of the heat flow rate. Thus, it is important that, before the temperature sensors, the water flow is mixed to avoid possible temperature inhomogeneity in the flow and make a correct measurement of the adiabatic mixing temperature. For this reason, mixers should be placed on the water path before the water inlets and after the water outlets. Moreover, if the temperature difference across the heat exchanger is small (e.g., 2 K or less), it is preferable to measure the temperature difference using a multi-junction thermopile or using high-accuracy thermistors.

In Eq. (16.10), the temperature of the condensing fluid at the inlet/outlet of the test section can be determined with temperature sensors (e.g., thermocouples can be inserted into the condensing fluid) or, if the fluid is in saturated conditions, with pressure sensors.

Once the mean wall temperatures at the inlet and outlet of the test section are measured and the saturation temperature is determined, the heat transfer coefficient can be calculated using Eq. (16.10). The heat transfer coefficient measured with this technique is a quasi-local value at the mean vapor quality x_{mean}.

$$x_{mean} = \frac{x_{in,TS} + x_{out,TS}}{2} \qquad (16.12)$$

where $x_{in,TS}$ is the vapor quality at the inlet of the test section and $x_{out,TS}$ is the vapor quality at the exit (after partial condensation). The vapor quality at the inlet of the test section $x_{in,TS}$ is controlled using a pre-section (tube-in-tube heat exchanger) where the refrigerant is desuperheated and partially condensed (Figure 16.2). It is important to notice that, with this technique, the vapor quality change in the test section $\Delta x = (x_{in,TS} - x_{out,TS})$ should be limited to maximum 0.15–0.2 to keep the hypothesis of "quasi-uniform" heat transfer coefficient and heat flux in the test section.

A similar technique was applied to measure the heat transfer coefficient in a tube with smaller internal diameter (below 4 mm) in Azzolin et al. (2018).

Differently from the previous application, the test section was divided into several subsections to obtain more than one heat transfer coefficient at different vapor qualities in the same test run. Each subsection looks like a miniaturization of the test section presented in Figure 16.2. The condensing primary fluid (e.g., refrigerant) flows in the internal tube, and the secondary fluid (water in this case) flows in the external tube.

Some recommendations for the design of the heat transfer test section can be summarized as follows:

- High heat transfer coefficient and enhanced surface area on the external path: This allows obtaining a low heat transfer resistance on the water side, reducing the temperature difference between the channel wall and the coolant flow. The higher temperature difference will be on the internal side, between the condensing fluid and the wall. This will be favorable to a precise measurement of the condensation heat transfer coefficient because it reduces the uncertainty contribution of the saturation-minus-wall temperature difference.

- Low transversal thermal resistance of the channel wall and low axial heat conduction: These seem to be two opposite features because a high thermal conductivity of the test tube will increase the axial heat flux. A high thermal conductivity of the wall will lead to a lower error due to deviation in temperature sensor positioning, but to avoid high axial heat flux, the wall thickness should be kept as small as possible.

- Good coolant mixing: For a precise measurement of the water temperature, a good mixing of the water is necessary where the sensor is located.

The realization of the test section proposed by Azzolin et al. (2018) started from a thick copper tube that was machined to obtain fins and grooves on the external side. The fins, as mentioned above, have different objectives:

- To increase the heat transfer area.
- To realize a higher external heat transfer coefficient.
- To allow inserting the wall thermocouples inside holes machined on the copper without wires crossing the coolant path.

In the design phase, it is important to select the geometry of the external path (i.e., the number of fins, distance between the fins, and position of the thermocouples). Thus, some numerical simulations can be run, with different geometries, imposing the condensation heat transfer coefficient as boundary condition (an estimation can be obtained from correlations available in the literature). As a result of the simulations, the temperature field in the tube wall and in the water path can be calculated. From this numerical analysis, it is possible to select the best geometry for the coolant path and the positions for thermocouples. For the case presented in Azzolin et al. (2018), it is obtained that nine fins per subsection can be a good compromise between technical feasibility and accuracy (the error between the calculated and the imposed heat transfer coefficient was below 1.5% in any case). Thus, in the final design, each subsection has nine fins with two grooves to create the water passage and between two consequent fins, the grooves are rotated by 90°. In each subsection, six thermocouples are embedded in the copper wall in the 2nd, 5th, and 8th fins, on both the sides where the copper tube has not been externally machined (Figure 16.3). This distribution of wall thermocouples allows evaluating the quasi-local condensation heat transfer coefficient α_c in each subsection with two different methods:

- With the mean logarithmic temperature difference as reported in Eq. (16.10): $T_{wall,in}$ and $T_{wall,out}$ are obtained, respectively, from the wall temperatures measured with the thermocouples (TCs) in the 2nd fin (TCs 1-2) and in the 8th fin (TCs 5-6).
- Using Eq. (16.13): A mean value of the saturation temperature $T_{sat,m}$ in the subsection is considered, and the mean wall temperature $T_{wall,m}$ is obtained by the two thermocouples located in the 5th fin (TCs 3-4).

$$\alpha_c = \frac{Q}{A\left(T_{sat,m} - T_{wall,m}\right)} \tag{16.13}$$

Since the saturation pressure (and thus the corresponding saturation temperature) is measured at the inlet and outlet of the test section, the saturation temperatures $T_{sat,in}$, $T_{sat,out}$, $T_{sat,m}$ required, for each subsection, in Eqs. (16.10) and (16.13) can be obtained using the following methods:

- Linear interpolation from the two measured saturation pressure values at the inlet and outlet of the test section (if the pressure drop is limited along the channel).

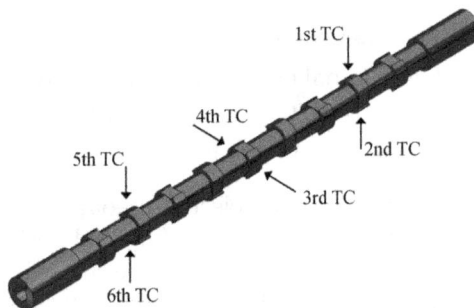

FIGURE 16.3
View of the copper tube machined on the external side; the arrows show the locations of the wall thermocouples in the subsection.

- Estimation of the pressure gradient (due to both frictional and momentum change contributions) in each subsection by selected correlations; the calculated pressure gradient is then multiplied by a corrective factor in order to match the total calculated pressure drop in the test section to the measured value of pressure drop (see also Section 16.4.1).

For the evaluation of the mean wall temperature $T_{wall,m}$ in Eq. 16.13, Meyer and Ewim (2018) proposed a method that can be adopted when wall temperatures are measured at different axial positions along the test section, even if they are not equally spaced. With this procedure, the mean wall temperature $T_{wall,m}$ is obtained from the trapezium integration technique of the thermocouple measurements at the different axial locations.

The heat flow rate, specific enthalpies of the condensing fluid, and vapor qualities are computed following the same procedure previously described, but having the foresight to apply it in each subsection.

To complete the data reduction with this technique, the following contributions should be considered:

- Dissipated heat flow rate: Even if the test section is well insulated, there can be a heat flow rate dissipated toward the ambient. In principle, this heat flux can be reduced keeping the water temperature close to the ambient, but obviously, this implies restrictions on the working conditions during tests. To evaluate the dissipated heat flow rate, some tests can be performed varying the water-to-ambient temperature difference while the primary side of the test section is vacuumed.

- Position of the thermocouples: The effective wall temperature sensor is located inside the tube wall, but at a certain distance from the internal channel surface. The measured temperature can be corrected to account for the conduction resistance in the metal wall.

- Effect of the axial conduction: Due to the heat conduction along the metal tube, some heat can be extracted from the condensing fluid in the short adiabatic zones at the edges of the test section. The axial heat flux depends on the operating conditions, and therefore, numerical simulations can be performed to account for its contribution.

16.4 Local Measurement

The previous techniques are applicable to measuring heat transfer coefficient in conventional and small-diameter channels down to 3 mm. When further reducing the channel dimensions (diameters below 3 mm), the cross-sectional area becomes very small and so the mass flow rate. For instance, in the case of 1-mm-diameter channel, the small flow rates (due to the reduced cross-flow area) result in small heat transfer rates; it becomes very difficult to limit the vapor quality change in the test section and to get "quasi-local" values of the heat transfer coefficient. Since the heat duty must be measured on the secondary fluid (water) side, a small heat flow rate means also a reduced temperature difference on the coolant side, leading to a high value of the experimental uncertainty. To overcome this issue, Garimella (2004) developed a "thermal amplification technique": The test section is cooled using water flowing in a closed (primary) loop at a high flow rate, thus ensuring that the condensation side presents the governing thermal resistance. Heat exchange between this primary loop and a secondary cooling water stream at a much lower flow rate is used to obtain a large temperature difference, which in turn is used to measure the condensation duty.

A different approach was proposed by Matkovic et al. (2009), which allows the measurement of local heat transfer coefficients in a single mini-channel. In Matkovic et al. (2009), the test section was designed to measure local heat transfer coefficients and to reduce the nominal experimental uncertainty associated with the measuring technique. This technique, in the recent years, has been applied to measure the condensation heat transfer coefficient of pure fluids (Del Col, Bortolato, et al. 2015; Del Col et al. 2017) and zeotropic mixtures (Del Col, Azzolin, et al. 2015) and also to study the effect of the shape of the mini-channel during convective condensation (Del Col et al. 2011).

The test section is made of two parts: a pre-section where desuperheating and partial condensation occur, and the actual measuring section where the heat transfer coefficient is measured. The test section was obtained by proper machining an 8-mm-external-diameter copper rod having a 1-mm internal bore. The geometry of the external path (Figure 16.4a) was designed following the tips reported in Section 16.3. This complex geometry allows increasing the external heat transfer area, thus drastically decreasing the related heat transfer resistance. Furthermore, the complex flow passage causes mixing of the coolant throughout the channel by continuously changing the flow direction and disturbing the boundary layer. High heat transfer coefficients are thus obtained even at moderate mass flow rates (beneficial for precise heat flux measurements). Besides, perfect local mixing of the coolant is a necessary condition to link the temperature variation along the coolant flow to the heat flow rate transferred.

The goal of the measuring sector is to perform local heat transfer coefficient measurements; thus, it is fundamental to place numerous temperature sensors along the sector. The number of temperature sensors installed comes out as a compromise between an appropriate temperature

(a)

(b)

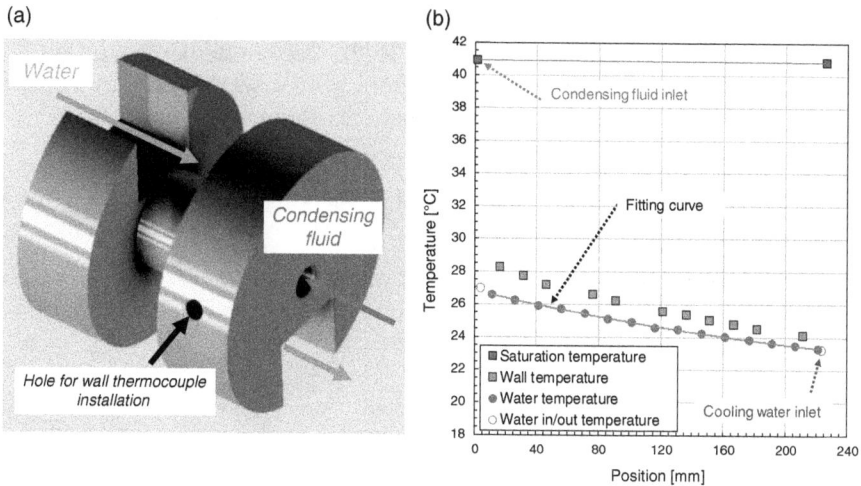

FIGURE 16.4
(a) Details of the test section showing the water path and the location of the wall thermocouple;
(b) Temperature profiles measured during a condensation test run.

profile description and a feasible design. As a result of CFD modeling, in the measuring section,

- a certain number (15) of thermocouples has been installed in the coolant channel, to measure the water temperature in correspondence of the fin carving and
- a certain number (13) of thermocouples has been embedded in the wall.

Furthermore, two thermocouples and a thermopile are installed in the water circuit at the inlet and outlet of the measuring sector and at the inlet/outlet of the pre-section to evaluate the total heat flux exchanged.

The test section is connected to the test rig by means of two stainless steel capillaries. The stainless steel capillary tubes have three major roles:

- Achieving thermal separation between the desuperheater (pre-section) and the measuring sector and between the measuring sector and the outlet to the test section.
- Providing adiabatic sectors where the measurement of the saturation temperature of the fluid can be done with a good accuracy on the outer tube surface.
- Accommodating the pressure ports for the evaluation of the saturation temperature.

A typical profile of wall and water temperature measurements for a condensation test is reported in Figure 16.4b: The condensing fluid enters the measuring sector in saturated conditions with a saturation temperature equal to 41°C, while the cooling water enters in countercurrent at 23°C.

16.4.1 Determination of the Local Heat Transfer Coefficient and Vapor Quality

For the determination of the local heat transfer coefficient (Eq. 16.14), three parameters are needed: the local saturation temperature $T_{sat}(z)$, the local wall temperature $T_{wall}(z)$, and the local heat flux $q(z)$.

$$\alpha_c(z) = \frac{q(z)}{T_{sat}(z) - T_{wall}(z)} \qquad (16.14)$$

In Eq. (16.14), z is the axial coordinate along the tube and is oriented with the flow of the condensing fluid.

The saturation temperature can be obtained in the adiabatic segments at the inlet/outlet of the mini-channel from the values of the measured pressure, and it can be checked with the value measured by the temperature sensors. The local saturation temperature can be obtained with two methods:

- Using a linear interpolation from the two measured saturation pressure values.
- Through the estimation of the local pressure gradient accounting for both frictional pressure drop and pressure recovery due to momentum variations during condensation. These different contributions can be evaluated with proper correlations, as shown in Del Col, Bortolato et al. (2015), and then multiplying the local calculated pressure gradient by a corrective factor in order to match the total calculated pressure drop to the value of measured pressure drop in the measuring sector.

In this case, the main point of the data reduction procedure is the calculation of the local heat flux. In the present measuring sector, there are several temperature sensors (in this case, thermocouples) placed in contact with the water along the external path. The local heat flux $q(z)$ is proportional to the derivative of the water temperature along the measuring sector:

$$q(z) = -\dot{m}_{water}\ c_{water}\ \frac{1}{\pi D_i}\ \frac{dT_{water}(z)}{dz} \qquad (16.15)$$

where $dT_{water}(z)/dz$ is the derivative of the polynomial equation interpolating the water temperatures measured by the thermocouples. The first step is to decide how many thermocouples should be considered for the interpolation.

Bearing in mind the energy balance in the measuring sector, the coolant temperature variation is directly associated with the corresponding enthalpy variation of the condensing fluid, so the thermodynamic vapor quality x_i corresponding to the *ith* thermocouple placed in the water flow is:

$$x_i = x_{in,MS} - \frac{\dot{m}_{water}}{\dot{m}_{ref}\, h_{LV}} c_{water} \left(t_{water,out} - t_{water,i} \right) \tag{16.16}$$

The vapor quality at the inlet of the measuring sector $x_{in,MS}$ is calculated from the energy balance on the water side in the preconditioning sector starting from the enthalpy of the superheated vapor. By using Eq. (16.16), it is possible to determine all the water thermocouples and the z locations corresponding to positive values of vapor quality.

Different interpolating equations can be considered. To minimize the uncertainty of heat transfer coefficients, the following order of preference is suggested: a second-order polynomial, an exponential equation with three parameters in the form $t(z) = a_0 + a_1 e^{-z/a_2}$, and a third-order polynomial (or higher order of the polynomial if needed). For each function, the equation parameters are calculated by means of WLS regression method. Different criteria can be used to select the form of the fitting equation: The first criterion is physical, and the second one is statistical. The physical criterion is based on the experimental uncertainty of the temperature measurements. Specifically, all the values of the calculated water temperatures with the fitting equation have to be within the expanded experimental uncertainty of the corresponding thermocouple readings. If this check test fails, the next fitting equation in the order of preference will be considered, until the criterion is satisfied. In the second criterion, the coefficients of determination R-square (R^2) and adjusted R-square (R^2_{adj}) are utilized to assess the fitting procedure (Rawlings et al. 1998). Unlike R^2, R^2_{adj} increases with the number of parameters of the fitting equation only if the new term significantly improves the model. Finally, in order to assure the accuracy and repeatability of the results, the heat transfer coefficients have to be insensitive to the method of interpolation, that is to say the variation in heat transfer coefficients using the fitting equation that meets the conditions of the statistical and uncertainty criteria and the next admissible equation in the order of preference should be within the experimental uncertainty.

Once the interpolating equation for the water temperature is established, the local heat flux (Eq. 16.15), the local heat transfer coefficient (Eq. 16.14), and the local vapor quality $x(z)$ (Eq. 16.17) are calculated:

$$x(z) = x_{in,MS} - \frac{\dot{m}_{water}}{\dot{m}_{ref}\, h_{LG}} c_{water} \left[t_{water,out} - t_{water}(z) \right] \tag{16.17}$$

16.4.2 Evaluation of the Experimental Uncertainty

In the present section, the problem of the estimation of the uncertainty on the local heat transfer coefficient will be addressed. As reported in Eq. (16.14), the local heat transfer coefficient is not directly measured, but it comes from three different quantities: the local heat flux $q(z)$, the local saturation temperature $T_{sat}(z)$, and the wall temperature at a given z-axial location $T_{wall}(z)$. The first step is the estimation of the Type A and Type B uncertainties for each of these three quantities, and then, as a second step, the combined standard uncertainty on the heat transfer coefficient can be obtained using the law of propagation of uncertainty (Eq. 16.9). The expanded uncertainty on the heat transfer coefficient can be obtained by multiplying the combined standard uncertainty by the desired coverage factor.

In the experimental technique presented here, the local wall temperature is directly measured (by thermocouples embedded in the wall) and, therefore, its uncertainty comes from the calibration procedure (Type B) and from repeated temperature measurements during the condensation test (e.g., 60 reading in 60 seconds when operating in steady-state conditions). It must be mentioned that, to obtain relatively low values of uncertainty (as the ones reported in Figure 16.5), the use of an ice point reference to maintain the cold junctions of thermocouples at a constant temperature is suggested and on-site calibration of thermocouples is needed. The calibration procedure consists in circulating the secondary fluid (e.g., water) under adiabatic conditions in the measuring section (the inner tube is in vacuum) and using two high-precision thermistors installed at the inlet and at the outlet of the water channel. A correction function for each thermocouple can be implemented by comparing the temperature measured by the thermocouple against the reference temperature gaged by the thermistors.

As previously reported, the local saturation temperature is estimated starting from the measured pressure values at the inlet and outlet. Consequently, the experimental uncertainty for the local saturation temperature can be calculated from a linear combination of the experimental uncertainty of the inlet and outlet saturation temperatures, derived from the pressure measurements. The uncertainty of the saturation temperature derived by pressure measurement can be determined by the difference between the value of saturation temperature at the measured pressure p and the value of saturation temperature at a pressure equal to $p + u_C(p)$, where $u_C(p)$ is the uncertainty on the pressure measurement.

The third term to be computed is the experimental uncertainty of the local heat flux $q(z)$. The local heat flux, as reported in Eq. (16.15), depends on the mass flow rate of the secondary fluid (water), the channel diameter, and the derivative of the secondary fluid temperature. For small-diameter channels, the contribution of the diameter uncertainty cannot be neglected and a proper characterization of the cross section of the channel (e.g., by analysis of images taken with optical microscope) is recommended. A particular aspect is the determination of the uncertainty related to the temperature gradient $dT_{water}(z)/dz$, and for its estimation, the WLS regression method can be used. A detailed description on this procedure is also provided in Del Col et al. (2013).

16.4.2.1 The Weighted Least Squares (WLS) Method

Considering a set of experimental data (as in the case of Figure 16.4b where a number of water temperature measurements is obtained at different axial positions z) and given a fitting equation that depends linearly on M+1 parameters (e.g., a second-order polynomial), the weighted least squares method can be used to determine all the parameters of this fitting equation with their respective uncertainties. Therefore, with reference to our specific problem, the WLS method can be used to estimate the uncertainty on the secondary fluid temperature gradient. In fact, once the coefficients of the selected fitting function (and their uncertainties) are calculated by the WLS method, the uncertainty on the temperature gradient can be obtained by calculating the derivative of the fitting function and then applying the law of propagation of uncertainty (Eq. 16.9).

Let us now focus on the WLS method and on how to calculate the fitting coefficients and their uncertainties. In Figure 16.4b, the fifteen measured water temperatures are interpolated using a second-order polynomial function (three coefficients to be determined), but for a more general application, the procedure is presented here considering a number N of experimental data to be fitted (each identified by the coordinates (z_i, y_i)) with a polynomial function in the form

$$y(z) = \sum_{k=0}^{M} a_k z^k \qquad (16.18)$$

In the specific case of Figure 16.4b, z_i are the axial locations of the water thermocouples and y_i are the corresponding water temperatures measured. The objective is to determine the M+1 coefficients a_k that minimize the following merit function:

$$\chi^2 = \sum_{i=1}^{N} \left[\frac{y_i - y(z_i)}{u_C(y_i)} \right]^2 \qquad (16.19)$$

where $u_c(y_i)$ is the combined uncertainty of each measured quantity y_i (the water temperature in our case). When applying the WLS method, the uncertainty related to the independent variable z (e.g., water thermocouples' position in the channel) is neglected. This assumption is valid if the water-side heat transfer area between two consecutive thermocouples is larger (18 times in this case) compared to the heat transfer area of a single groove where a water thermocouple is positioned (Figure 16.4a) and if the temperature variation inside the groove is lower than the uncertainty of the thermocouple. Therefore, a possible axial positioning error of the thermocouple placed inside the fin cut does not affect the result of the measurement.

The minimum of the merit figure Eq. (16.19) occurs where its derivative with respect to all parameters a_0, ..., a_M is equal to zero, and this condition yields the following matrix equation:

$$\left(A^T \times A \right) \times a = A^T x \, b \tag{16.20}$$

where A is an $N \times (M+1)$ matrix whose elements are obtained as reported in Eq. 16.21:

$$A_{ij} = \left(\frac{\delta y / \delta a_{j-1}}{u_C(y_i)} \right) \tag{16.21}$$

b is a vector of N constant terms defined as the ratio between the ith measured quantity (e.g., temperature measured at the ith location) and the corresponding uncertainty

$$b_i = \left(\frac{y_i}{u_C(y_i)} \right) \tag{16.22}$$

and finally, a is the vector of the $M+1$ coefficients of the fitting equation (three coefficients in the case of a second-order polynomial) to be determined.

The covariance matrix $C = (A^T \times A)^{-1}$ is related to the standard uncertainty of the estimated parameters a: The diagonal elements C_{jj} are the square uncertainties of the fitted parameters a, while the off-diagonal elements C_{jl} are the covariances between the estimated coefficients a_j and a_l, dubbed cov(a_j, a_l).

Once square uncertainties and covariances have been obtained, by applying the law of propagation of uncertainty (Eq. 16.9) to Eq. (16.15), the uncertainty on the heat flux can be obtained. It must be mentioned that the parameters a_0, ..., a_M are usually correlated and, therefore, all the terms in Eq. (16.9) must be considered.

16.4.2.2 Results from the Uncertainty Analysis

An example of the results that can be obtained by applying the aforementioned procedure is reported in Figure 16.5, where the percentage uncertainty on the local heat flux and on the local heat transfer coefficient is plotted versus the axial position. The results in Figure 16.5 were obtained starting from the temperature profiles reported in Figure 16.4b (propylene condensation inside a 1-mm channel at 40°C saturation temperature and 200 kg m^{-2}s^{-1} mass flux) and considering a coverage factor $k = 2$. It can be observed that, in such experimental conditions, the main contribution to the

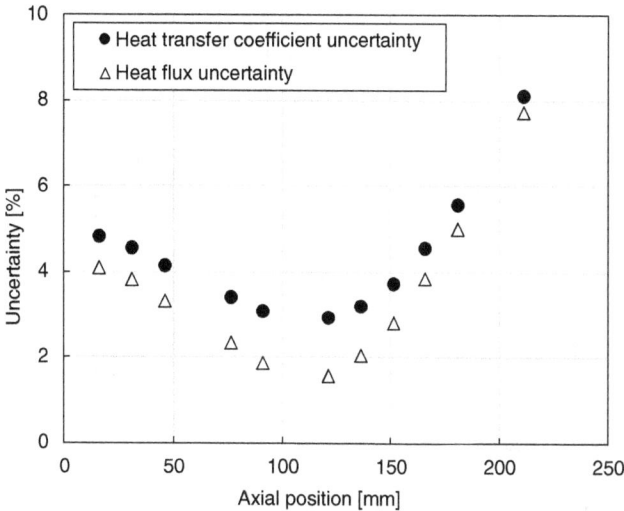

FIGURE 16.5
Evaluation of the expanded ($k = 2$) experimental uncertainty on the local heat transfer coefficient and on the local heat flux during a condensation test with propylene inside a 1-mm-diameter channel. The saturation temperature is equal to 40°C, and the mass flux is 200 kg m^{-2}s^{-1}.

heat transfer coefficient uncertainty is due to the uncertainty on the local heat flux. Furthermore, the experimental uncertainty on the heat flux displays a minimum located near to the center of the channel and it assumes larger values at the edges of the test section due to the interpolation procedure.

A last consideration must be done regarding the uncertainty of the local vapor quality. As shown in Section 16.4.1, at each axial position corresponds a value of the heat transfer coefficient and also a value of the vapor quality. The local vapor quality is obtained from Eq. (16.17), and this equation involves the following quantities: mass flow measurements (primary and secondary fluids), inlet vapor quality, and outlet and local temperatures of the secondary fluid. Once the fitting equation is obtained for the secondary fluid (as illustrated above), the combined uncertainty of the local vapor quality $u_c(x)$ can be easily calculated applying the law of propagation of uncertainty to Eq. (16.17).

16.5 Conclusions

The present chapter presented some experimental techniques for the measurement of the condensation heat transfer coefficient inside channels using a secondary fluid for heat removal. In all the techniques, the heat flux is obtained from the energy balance on the side of the secondary fluid:

1. An average or "quasi-local" approach based on the Wilson plot method that, does not require a direct wall temperature measurement.

2. A "quasi-local" approach based on the direct measurement of the wall temperature. This method is applicable in conventional tubes (i.e. internal diameter equal to 6 mm and higher), and it can be extended to channels with an internal diameter of about 3 mm with some precautions. This approach accepts some variation of the vapor quality along the tube. Once the heat flux is known and the saturation-to-wall temperature difference is measured, the quasi-local heat transfer coefficient can be determined. This technique, compared to the Wilson plot method, adds some difficulties because it requires embedding temperature sensors in the wall, but allows a higher experimental accuracy.

3. A local approach to the measurement of the heat transfer coefficient in small-diameter channels. This technique requires huge efforts during the design and realization of the test section for the heat transfer measurements since, along the channel, several thermocouples should be installed on the water side and embedded in the tube wall. The local heat flux is obtained through the derivative of the water temperature profile measured along the test section, bearing in mind that a particular attention should be paid to the evaluation of the interpolating equation. This technique allows the measurement of the local heat transfer coefficient with high experimental accuracy, and it has been applied to single channels of different shapes with an internal diameter around 1 mm.

References

Azzolin, M., S. Bortolin, L.P. Le Nguyen, P. Lavieille, A. Glushchuk, P. Queeckers, M. Miscevic, C.S. Iorio, and D. Del Col. 2018. "Experimental investigation of in-tube condensation in microgravity." *International Communications in Heat and Mass Transfer* 96: 69–79.

Bortolin, S., E. Da Riva, and D. Del Col. 2014. "Condensation in a square minichannel: Application of the VOF method." *Heat Transfer Engineering* 35: 193–203.

Cavallini, A., G. Censi, D. Del Col, L. Doretti, G.A. Longo, and L. Rossetto. 2001. "Experimental investigation on condensation heat transfer and pressure drop of new HFC refrigerants (R134a, R125, R32, R410A, R236ea) in a horizontal smooth tube." *International Journal of Refrigeration* 24: 73–87.

Del Col, D., M. Azzolin, S. Bortolin, and A. Berto. 2017. "Experimental results and design procedures for minichannel condensers and evaporators using propylene." *International Journal of Refrigeration* 83: 23–38.

Del Col, D., M. Azzolin, S. Bortolin, and C. Zilio. 2015. "Two-phase pressure drop and condensation heat transfer of R32/R1234ze(E) non-azeotropic mixtures inside a single microchannel." *Science and Technology for the Built Environment* 21: 595–606.

Del Col, D., M. Bortolato, M. Azzolin, and S. Bortolin. 2015. "Condensation heat transfer and two-phase frictional pressure drop in a single minichannel with R1234ze(E) and other refrigerants." *International Journal of Refrigeration* 50: 87–103.

Del Col, D., S. Bortolin, A. Cavallini, and M. Matkovic. 2011. "Effect of cross sectional shape during condensation in a single square minichannel." *International Journal of Heat and Mass Transfer* 54: 3909–3920.

Del Col, D., S. Bortolin, D. Torresin, and A. Cavallini. 2013. "Flow boiling of R1234yf in a 1 mm diameter channel." *International Journal of Refrigeration* 36: 353–362.

Del Col, D., R.L. Webb, and R. Narayanamurthy. 2002. "Heat transfer mechanisms for condensation and vaporization inside a microfin tube." *Journal of Enhanced Heat Transfer* 9: 25–37.

Fernández-Seara, J., F.J. Uhía, J. Sieres, and A. Campo. 2007. "A general review of the Wilson plot method and its modifications to determine convection coefficients in heat exchange devices." *Applied Thermal Engineering* 27: 2745–2757.

Garimella, S. 2004. "Condensation flow mechanisms in microchannels: Basis for pressure drop and heat transfer models." *Heat Transfer Engineering* 25: 104–116.

JCGM, Joint Committee for Guides in Metrology. 2008. *Evaluation of Measurement Data - Guide to the Expression of Uncertainty in Measurement.* Sèvres: Bureau International des Poids et Mesures (BIPM). http://www.bipm.org/en/publications/guides/gum.html.

Longo, G.A. 2009. "R410A condensation inside a commercial brazed plate heat exchanger." *Experimental Thermal and Fluid Science* 33: 284–291.

Mancin, S., D. Del Col, and L. Rossetto. 2013. "R32 partial condensation inside a brazed plate heat exchanger." *International Journal of Refrigeration* 36: 601–611.

Matkovic, M., A. Cavallini, D. Del Col, and L. Rossetto. 2009. "Experimental study on condensation heat transfer inside a single circular minichannel." *International Journal of Heat and Mass Transfer* 52: 2311–2323.

Meyer, J.P., and D.R.E. Ewim. 2018. "Heat transfer coefficients during the condensation of low mass fluxes in smooth horizontal tubes." *International Journal of Multiphase Flow* 99: 485–499.

Rawlings, J.O, S.G. Pantula, and D.A. Dickey. 1998. *Applied Regression Analysis.* New York: Springer.

Rose, J.W. 2004. "Heat-transfer coefficients, Wilson plots and accuracy of thermal measurements." *Experimental Thermal and Fluid Science* 28: 77–86.

Toninelli, P., S. Bortolin, M. Azzolin, and D. Del Col. 2017. "Effects of geometry and fluid properties during condensation in minichannels: Experiments and Simulations." *Heat and Mass Transfer/Waerme- Und Stoffuebertragung* 55: 14–57.

Wilson, E.E. 1915. "A basis for rational design of heat transfer apparatus." *Transactions of the ASME* 37: 47–82.

17

Optical Measurement Techniques for Liquid–Vapor Phase Change Heat Transfer

Jocelyn Bonjour, Serge Cioulachtjian, Frédéric Lefèvre, Stéphane Lips, and Rémi Revellin

Univ Lyon, CNRS, INSA-Lyon, Université Claude Bernard Lyon 1

CONTENTS

17.1 Introduction

Phase change is an efficient heat transfer mode involved in various practical applications. To cite a few examples, this phenomenon is used in steam turbine power plants, refrigeration units, heat pumps, or microelectronics cooling systems. Whatever the scale, all the concerned applications require high heat transfer rates, homogeneous temperature fields, or a high compactness of thermal components. In all these technologies, a proper design should ensure not only the effectiveness, the reliability, and the efficiency, but also the safety of use of the systems. Pool boiling, flow boiling, and/or thermo-capillary phase change heat transfer must thus be sufficiently understood to become predictable. The complexity of the various multi-scaled fundamental phenomena taking part in liquid–vapor phase change heat transfer makes its understanding still challenging. Generally speaking, mass, momentum, and energy transfer between the two phases (liquid and vapor) either stagnant or flowing in a confined channel, with energy transfer from the walls, gives rise to a multiplicity of possible phase distributions. The latter are responsible for significant variations in the transfer rates of momentum, energy, and/or mass. For instance, heat is often transferred between the walls and the fluid by either nucleation followed by release of distinct bubbles, or diffusion into and through a liquid film. The thermo-hydraulic behavior of the two-phase fluid can only be analyzed if some important parameters are accurately measured, e.g., liquid film thickness, bubble size, bubble frequency and velocity, and wall temperature and flow patterns. For such measurements, different optical measurement techniques have successfully been developed and used at CETHIL during the last decades, which are as follows:

- Visualization and characterization of flow structures with high-speed video cameras, whose use enables the determination of flow patterns, bubble dynamics, or liquid film thickness during pool boiling, flow boiling, or Taylor bubble flow.
- Confocal microscopy used to measure the shape of the meniscus within the capillary structure of a heat pipe and the thickness of the condensation film on the wall.
- The determination of the temperature field on the outer wall of heated tubes, ensured by means of an infrared camera observing a sapphire tube coated with ITO. The latter enables a total transparency of the heated area in the visible spectrum while heating.

This chapter aims at showing how these optical methods—used successfully in many other fields of physics and engineering—can bring new insights into phase change heat transfer if they are properly implemented. This will also exemplify how new important results can be obtained or new phenomena can be discovered in the field of liquid–vapor phase change heat transfer.

17.2 High-Speed Videography for the Analysis of Boiling

It is the essence of *mechanical engineering* to determine the *mechanisms* governing the phenomena that the *engineer* intends to employ at his profit. This obviously holds true for the branch of mechanical engineering that focuses on liquid–vapor heat transfer. When boiling and condensation became subjects of interest for the engineers (i.e., during the industrial revolution driven by steam engines), the researchers were first observing the phenomena (bubble dynamics, two-phase flow, etc.), describing them as precisely as possible, prior to interpreting them through force analysis, momentum exchange, heat transfer, etc. One of the most ancient reported works on phase change heat transfer that rely on this methodological approach is probably the manuscript published by J.G. Leidenfrost in 1756 ("*De aquae communis nonnullis qualitatibus tractatus,*" which can be translated as "A tract about some qualities of common water"). The development of photography and then videography, up to modern high-speed videography for which the frame rate can reach 100,000 fps (frames per second), was obviously of greatest help to the researchers. This section aims at introducing good practices when using these techniques for the analysis of boiling. It will also exemplify how these techniques were successfully used to make some progresses in the understanding of the phenomena involved in pool boiling and flow boiling.

17.2.1 Good Practices, Experimental Biases, and Artifacts

Although observing liquid–vapor flows to describe them seems intuitive, many experimental precautions must be taken to make sure that the images and video sequences are meaningful. A number of biases may affect the quality of the results.

Of course, a good optical access must be guaranteed, even though the vessel in which boiling takes place is generally gas- and liquid-tight to avoid any interaction with the external air (e.g., to control the saturation temperature and pressure, or because the absence of dissolved gas in the liquid is generally required to reach an acceptable repeatability of the experiments). This means that the optical access must be carefully assessed during the design of the experimental setup; that is, transparent windows (viewports) or tubes must be employed with an adequate use of seals such as O-rings or flat seals. Then, an important choice concerns the illumination of the region of interest. The light source must be strong enough to obtain images with a high contrast, without affecting the thermal conditions in the vessel. The development of LED technology over the last decade was particularly helpful to limit the undesirable thermal effects, while previously, light spots or laser lightning was often shown to affect the results of the experiments by noticeably heating up the fluid. For instance, such LED spots can reach more than 53,000 cd/m² with a uniformity better than 98.7% over a square area of 100 × 100 mm² without any thermal effect on the fluid.

In the case of tubular geometry (typical of flow boiling), a background light illumination is generally recommended (Figure 17.1a): The observed tube is placed between the light source and the video camera, which increases the contrast of the images. The observations in such conditions are the basis of the identification of the numerous types of flow patterns described in the literature. An example of some typical flow configurations are given in Figure 17.1b. It is sometimes advisable to use a mirror to record simultaneously the top and the side views on the same video sequence (Figure 17.1c).

In the case of flat surface (typical of pool boiling on a heated wall), if the light is not diffuse, the axes of the light source and of the camera are generally oriented quasi-normally to the wall (both the camera and the light source are facing the wall), or parallel to the plane of the wall but perpendicular to one another (lateral illumination), or parallel to the plane of the wall but aligned parallel to one another (background illumination). In the former case, the quality of the image depends on the reflection of the light on the wall. The diffuse light or the two first arrangements allow the observation of the overall behavior of the two-phase flow associated with the liquid–vapor phase change phenomena. Such a configuration was successfully used

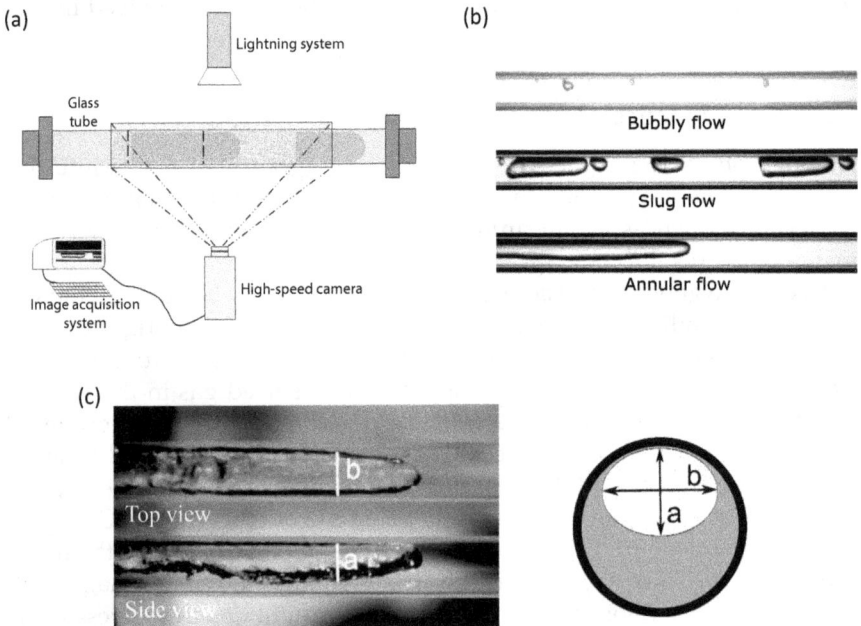

FIGURE 17.1
Flow boiling: (a) typical experimental configuration of background lightning for the study of flow boiling (Charnay et al. 2014); (b) examples of flow patterns (Charnay et al. 2014); (c) simultaneous video recording of top and side views (Revellin et al. 2012).

to determine the various flow patterns during pool boiling, for instance in the early work of Gaertner (1965) who distinguished the regimes of discrete bubbles, vapor columns, vapor mushrooms, and vapor patches. Its limitation lies in the fact that the phenomena occurring in the foreground hide those occurring in the background and that the image is blurred out of the zone corresponding to the focal distance (e.g., Figure 17.2a where the forefront and background bubbles appear blurry). Background illumination was thus efficiently used to study boiling on thin heaters (e.g., wires (cf. Figure 17.2b) and ribbons) or to focus on single-bubble dynamics (cf. Figure 17.2c).

Because of the differences in the refraction indexes of the glasses and the fluid involved (particularly because of the gap in the refraction indexes between the liquid and the vapor), because of the curvatures of the bubbles, and of the tubes if any, the images are usually strongly deformed and usually include complex reflections. In the classical case of annular flow (Figure 17.1b, bottom), the liquid film developing along the wall and the tube wall itself cannot sometimes be distinguished one from another as they may form a single-light-colored zone. For simple enough geometries, optical models can be developed to determine the shape and size of an object from its characteristics on the image. Otherwise, a calibrated grid or an object of calibrated size must be visible on the images to be used as a reference.

Another source of image deformation is the so-called "mirage effect" that occurs when light rays are refracted by passing through a medium of nonconstant optical index of refraction. As the optical index usually varies with temperature, mirage effects are particularly pronounced in the vicinity of heated walls. This phenomenon has probably led to a number of misinterpretations of the results on the mechanisms of triggering of critical heat flux or on the micro-layer theory (according to which the bubbles grow because of the evaporation of the thin liquid layer formed between the bubble and the wall), because even the existence of the micro-layer was disputed on unclear images. To highlight the importance of the mirage effect, experiments were performed by placing a 1-mm-diameter spherical plastic ball on a surface immersed in a fluid (Siedel, Cioulachtjian, and Bonjour 2008). A picture was first taken without heating the surface (thus without any mirage effect, cf. Figure 17.3a). The surface

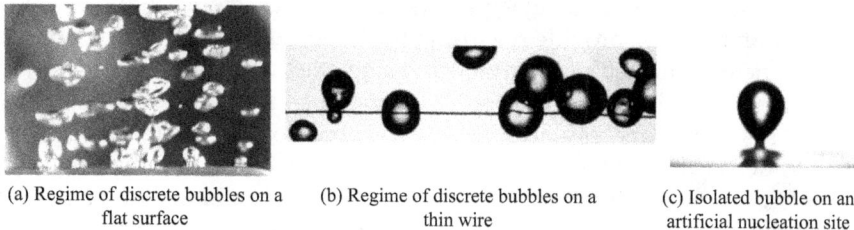

| (a) Regime of discrete bubbles on a flat surface | (b) Regime of discrete bubbles on a thin wire | (c) Isolated bubble on an artificial nucleation site |

FIGURE 17.2
Typical examples of images for pool boiling regime identification (a, b) or for bubble dynamics analysis (c). Lateral diffuse illumination (a) or background illumination (b, c).

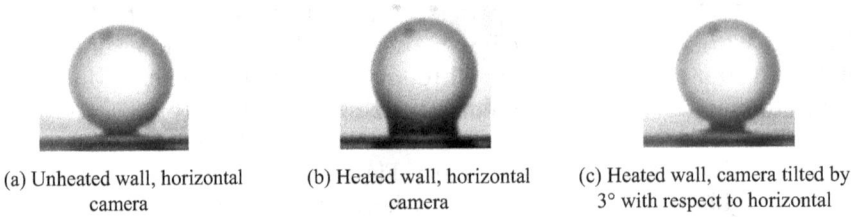

| (a) Unheated wall, horizontal camera | (b) Heated wall, horizontal camera | (c) Heated wall, camera tilted by 3° with respect to horizontal |

FIGURE 17.3
Illustration of the importance of the mirage effect in the vicinity of a heated wall (images (b) and (c): wall temperature 10 K above the liquid temperature).

was then heated as if in the experimental conditions of boiling. If no particular care was taken, the ball would look the same as in Figure 17.3b, where a strong mirage effect alters the reality and leaves the impression of a very large bubble foot. To get rid of this effect, the camera was slightly tilted to allow a 3° angle with the horizontal plane. The optical path was thus modified to cross a much thinner superheated layer of liquid, without distorting much the image. (Compare Figure 17.3a and 17.3c.)

17.2.2 Recent Progresses in the Understanding of Bubble Dynamics Owing to High-Speed Videography

Since the pioneering work of Nukiyama (1934), many progresses have been made to understand the fundamental processes of boiling (bubble nucleation, growth, and detachment). To isolate these fundamental phenomena, a research track has developed over time to study boiling from isolated controlled nucleation site. This allows observing the life cycle of individual bubbles, sometimes with interactions with the preceding or successive bubbles. This gave birth to an abundant amount of publications based on various optical techniques, taking the benefit of ever-improving high-speed video cameras with increased space and/or time resolution. A few major findings concerning the bubble growth laws, bubble frequency, or bubble waiting time are given here, as a synthesis of two PhD theses defended in the research group of the authors (Siedel 2012; Michaïe 2018). As a whole, the work of Siedel became an experimental reference owing to the detailed description he could provide regarding the bubble dynamics during boiling of pentane on a heated wall on which a single artificial nucleation site had been created. With millions of bubble images such as the video sequence of Figure 17.4, it became essential to develop an automatic image processing code to detect the bubble contours (Figure 17.5), to determine the evolution of their volume as a function of time (Figure 17.6). The image processing is as follows: The bubble contour is first determined by locating the maximum gray gradient using the method known as "Sobel method" (Sobel, 1990). Then, the light zone at the center of the bubble's image (that is the image, through the bubble, of the background viewport providing lightning) is

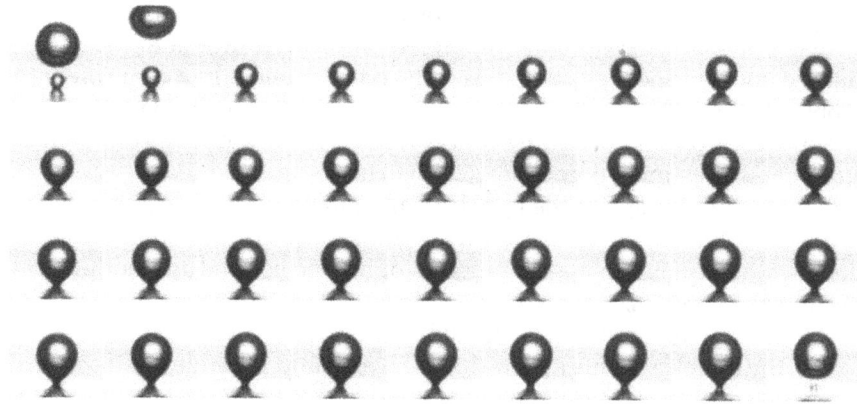

FIGURE 17.4
Typical video sequence of a single bubble growth.

detected and removed, as well as any other reflections of light. Finally, the bubble's volume is measured as if the bubble was a stack of 1-pixel-thick vapor cylinders (Siedel, 2012). This method has the advantage of taking into account the whole contour rather than simplifying the bubble as a sphere or a truncated spheroid as it is sometimes the case in the literature, which was

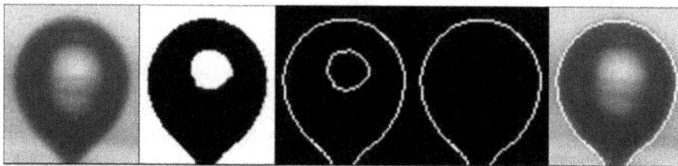

FIGURE 17.5
Various steps of the detection of the bubble contour and volume measurement.

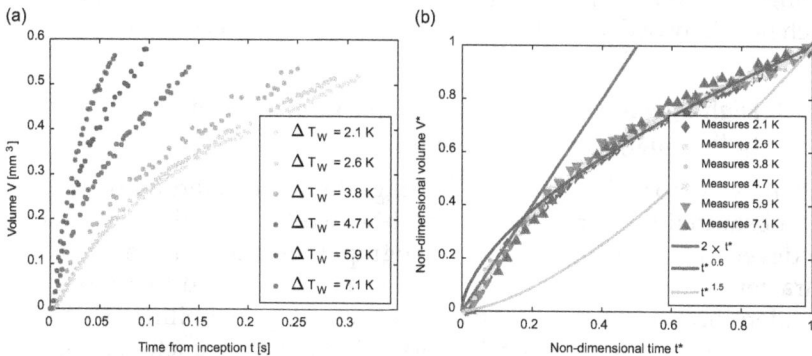

FIGURE 17.6
Growth law of an isolated bubble during boiling of pentane at 36°C. (a) Volume vs. time as a function of wall superheat. (b) Dimensionless variables.

shown to lead to as much as 20% error in the volume determination when there is a neck at the base of the bubble. Another advantage of this method is that the bubble is not considered as fully axisymmetric: In a case where the bubble is slightly tilting, the error in the volume determination remains low. Under these considerations, the uncertainty on the bubble's volume measurement was estimated by Siedel (2012) to be about 3%.

Because of the shape of the bubbles (far from being spherical) and owing to the image processing method that results in accurate volume measurement, it was decided to reason in terms of bubble volume and not in terms of equivalent radius, which is often done in the literature. As shown in Figure 17.6a, which represents the bubble volume (V) vs. time from inception (t), the bubble volume at detachment was found to be almost insensitive to the wall superheat (difference between the wall and liquid temperatures) even though it affects the instantaneous growth rate. All the growth laws could be reconciled into a single dimensionless growth curve (Figure 17.6b) that fits the trend of a dimensionless volume $V*$ (ratio of volume to volume at detachment) vs. dimensionless time $t*$ (ratio of time to time at detachment) as: $V* = 2.t*$ for $t* < 0.2$ and $V* = t*^{0.6}$ for $t* > 0.2$. These highly time-resolved measurements could demonstrate that the classical models of bubble growth cannot be applied to boiling: These models, essentially based on hydrodynamic considerations, yield to growth laws fitting $R_{eq} \sim t^{0.5}$, i.e., $V* \sim t*^{1.5}$ for long times (whose trend is plotted in Figure 17.6b for the sake of comparison with the actual dimensionless growth law).

When the observed objects are too small, optical magnifiers (e.g., zoom lenses, telephotos, and bellows) are of common use. Beyond these devices, images of particularly high resolution were obtained by using a short-field telescope involving parabolic mirrors, but with a focal distance of about 1 m, to reach resolutions up to a few μm/pixel. Yet, an alternative approach was developed by Michaïe (2018). It was indeed demonstrated that instead of magnifying the image, observing large bubbles (up to 10 cm) obtained during boiling close to the triple point (water: $T = 0.01°C$ and $P = 6$ mbar) can bring much new knowledge on the fundamental mechanisms of pool boiling.

17.2.3 High-Speed Videography: A Major Contribution to the Analysis of Flow Boiling

Maybe even more than for pool boiling, the understanding of flow boiling substantially improved over the last two to three decades owing to the development of high-speed videography. Because of the (overall) 1D character of the flow, the visualization can indeed lead to a more objective identification of the flow patterns than for pool boiling. This identification, along with an accurate thermo-hydraulic characterization, is an evident basis for modeling (often on a time-averaged approach) two-phase flows during flow boiling. Since the 1980s or 1990s, a huge amount of data has thus been published under various forms, such as flow pattern maps and prediction tools for pressure drops or for heat transfer coefficients. The corresponding database was progressively extended to high or low pressure, to high or low velocities, to more and more fluids, etc. One subject, among others, became a subject of scientific debate: the effect of the size of the tube in which flow boiling takes place, and its classification into the families of macro-, mini-, or microchannels. Various criteria were proposed, based on manufacturing technology, on force balances (with dimensionless numbers, for instance the Bond number to account for the capillary and gravity forces), and on heat transfer mechanisms. A recent contribution to this debate has been made by Layssac through his PhD thesis in the research group of the authors (Layssac 2018; Layssac et al. 2017). The importance of accurately quantifying the symmetry (or eccentricity) of the flow to determine the flow patterns, involved forces, and thermo-hydraulic characteristics was demonstrated. A method for quantifying the eccentricity of the flow had previously been developed (Donniacuo et al. 2015). For the case of annular flow, it was based on flow visualization and image processing, taking into account the refraction of light in the tube and in the fluid to determine the thickness of the top and bottom liquid films (Figure 17.7). In the present example, in which a round tube is used, a correction must be applied to take into account the deformation of the image. All details are available in Donniacuo et al. (2015) and are not given here for the sake of clarity.

All examples briefly presented in this section demonstrate that the continuous improvement in the performance of the high-speed cameras, in terms of both spatial and temporal resolutions, leads to a continuous improvement in the understanding of the two-phase systems, just by enabling us to visualize the phenomena occurring in these systems. This trend will probably carry on in the future, and new exciting results can still be expected, thanks to high-speed videography.

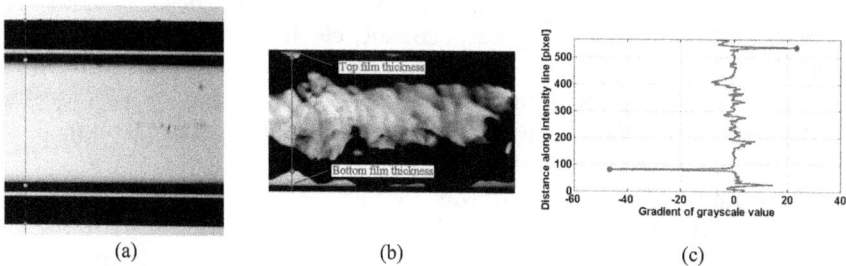

FIGURE 17.7
Main steps for the determination of the film thickness (a) single phase flow: determination of the outer and inner wall boundaries; (b) visualization of the liquid films and vapor core (walls are cropped); (c) determination of the film thicknesses from grayscale value gradients.

17.3 Liquid–Vapor Interface Characterization Using Confocal Microscopy

Besides high-speed videography, other optical techniques were also implemented by the authors. Among them, confocal microscopy was intensively used to measure the thickness and the shape of liquid films within several passive two-phase capillary devices, better known as heat pipes (Lefèvre et al. 2010; Lips, Lefèvre, and Bonjour 2011, 2010b, 2010a, 2010c; Rullière, Lefèvre, and Lallemand 2007). A STIL© confocal measuring probe enabled measuring film thicknesses in the range of 2–300 μm throughout partially transparent heat pipes. The shape of the meniscus inside grooved capillary structures of various geometries (longitudinal or radial grooves, crossed-grooves, etc.) was determined as well as the shape and the thickness of condensation films on the fins of the grooved surfaces. These parameters have a huge influence on the thermal and hydrodynamic performance: Both the evaporation and condensation heat transfer are strongly linked to the shape and the thickness of these films, and both the capillary pressure and the pressure drops within the grooves of a heat pipe depend on the curvature of the liquid–vapor interface. Therefore, if one wants to characterize properly the thermal characteristics of these devices, it is critically important to get an accurate measurement of these films.

After a brief introduction on the different techniques available to measure the thickness of liquid films, this section presents the principle of confocal microscopy and the experimental methodology that was settled to produce proper and accurate experimental data with this measurement technique. Finally, some examples are described to show the capability and the limits of confocal microscopy to measure the thickness of liquid films.

17.3.1 Introduction

Several experimental techniques were developed to measure liquid film thickness (Tibiriçá, do Nascimento, and Ribatski 2010). These methods can be classified into three categories: acoustic, electrical, and optical. Even if acoustic and electrical methods are non-intrusive, they are not convenient for the measurement of liquid films of varying thickness, which are the kind of liquid films formed within a capillary structure. An optical method seems, therefore, more convenient for that purpose.

Shadowgraphy is a classical way of measuring liquid film thickness through a fully transparent device. Nonetheless, building a fully transparent device is not always possible, even for research purposes. In the case of a heat pipe, the capillary structure is usually metallic, which prevents observations through it. Therefore, it is only possible to acquire measurements from above. Furthermore, as the system requires working in two-phase conditions, it is mandatory to close it tightly. A transparent plate can be used for that purpose. Several optical methods require the insemination of the liquid with particles, which is not convenient in a two-phase system where a fluid evaporates on a heated section and condenses on a cooled section. Indeed, in working conditions, these particles remained trapped in the evaporator section, because they are not able to evaporate. Therefore, measurements are possible only in the heated section of the device. Other reflection-based techniques could be used for that purpose, but the presence of the transparent window makes the optical path complicated to analyze.

Chromatic confocal microscopy is a non-intrusive reflection method, which enables obtaining measurements orthogonally through a transparent window and is therefore appropriate for measurements of liquid films within a capillary structure. It has to be noted that nowadays, this technique is widely used for the acquisition of liquid film thickness in two-phase or in ambient conditions.

17.3.2 Functional Principle

The functional principle of confocal microscopy is presented in Figure 17.8. A point white-light source passes through an axial chromatic aberration lens. The image of this source is a continuum of monochromatic points distributed along the optical axis. If an interface is present in the measurement space, it reflects only the monochromatic image located at this very place. Therefore, by analyzing the wavelength of this image using a spectrometer, it is possible to determine the location of this interface very precisely.

The confocal microscope used by the authors was developed by STIL SA. It is the MICROMESURE2 measuring station associated with the CHR 150 controller. The vertical position of the optical device does not change along

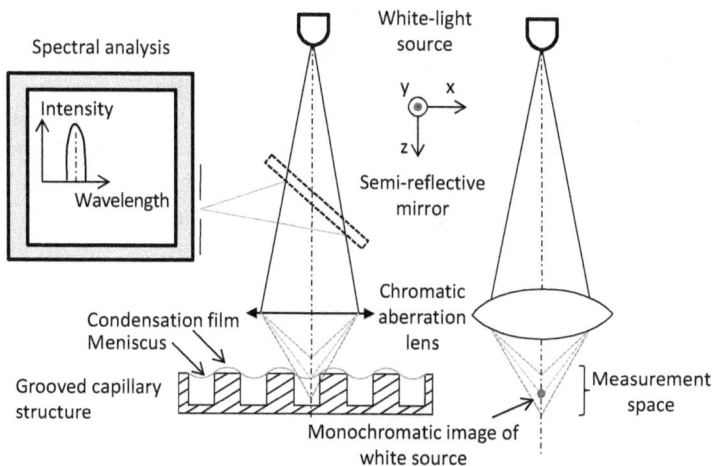

FIGURE 17.8
Working principle of chromatic confocal microscopy.

the experiment, while a motorized table allows the translation on the x–y plane. The measuring field is 350 µm, and the working distance of the pen is 12.7 mm, which is a strong constraint that needs to be taken into account when designing an experimental bench. The pen is calibrated by the manufacturer so that each part of the surface located within the 350 µm height of the measuring field is recorded with accuracy in the z-direction equal to 60 nm. The diameter of the light beam is equal to 3 µm, which corresponds also to the accuracy in the x- and y-directions. The surface to be inspected has to be placed on a plate motorized along the x- and y-axes in order to completely record the surface. The discretization step varies from 1 to 100 µm and outlines the acquisition time.

The surface to be scanned is placed directly under the optical pen. The calibrated chromatic aberration lens is protected within the optical pen by a glass porthole, called a reference window. When a transparent plate is located in between, which is the case in two-phase applications, the use of a new reference window is necessary to maintain the system calibration characteristics. The machining of such a window was made by the company STIL SA to take into account the optical characteristics of a 5-mm-thick transparent plate to be crossed in our applications. It is also possible to calibrate the optical system with a perfectly smooth and rectilinear inclined surface having an accurately known tilted angle.

Observations through a transparent window in two-phase conditions require that the temperature of the window be slightly higher than that of the saturation temperature in order to avoid condensation, which would prevent the measurement. This can be achieved with a flat heater placed against the window or by heating the ambient air surrounding the window.

17.3.3 Experimental Implementation of the Optical Method and Examples

Confocal microscopy was frequently used in the studies of the research group of the authors to measure the shape of the liquid–vapor interface within grooved capillary structures of flat plate heat pipes (FPHPs). The height and width of the grooves were in the range 50–500 µm. The capillary forces are dominant in this geometrical range, and the shape of the liquid–vapor is described by the Young–Laplace law, which links the pressure of the liquid and the pressure of the vapor to the curvature of the meniscus inside the capillary structure. Therefore, if the pressure of the vapor is known, measuring the curvature of the meniscus enables the local estimation of the pressure of the liquid within the capillary structure. In working conditions, the liquid flows from the condenser to the evaporator of the heat pipe, due to the variation of the meniscus curvature along the capillary structure. Therefore, by measuring the meniscus curvature at different locations, it is possible to estimate the liquid pressure variation along the heat pipe.

Figure 17.9a shows an example of measurements obtained at the condenser and at the evaporator sections of a copper FPHP filled with pentane: Four

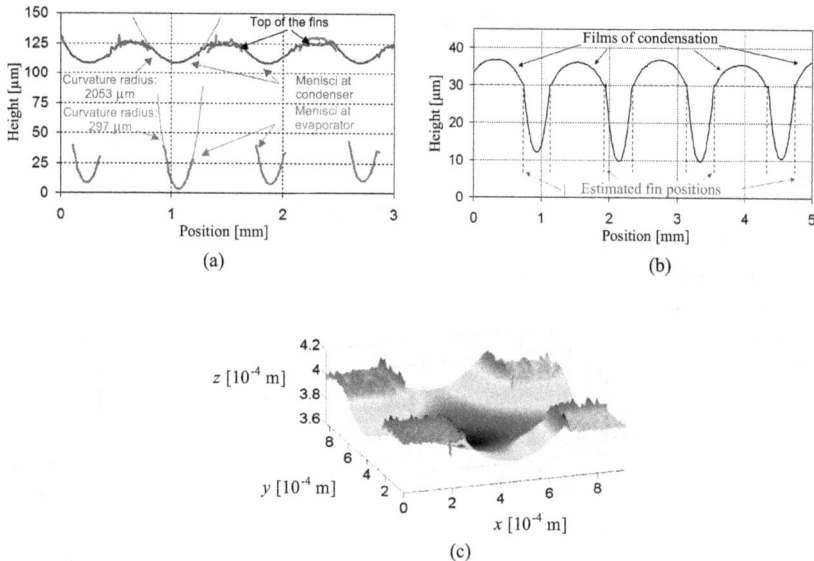

FIGURE 17.9
Some examples of liquid–vapor interface visualization in heat pipes (a) example of measurements at the condenser and the evaporator of a grooved copper plate; (b) example of measurements at the condenser of a grooved silicon plate; (c) example of measurements above the evaporator of a cross-grooved copper plate.

square grooves of side 400 μm, separated by 400 μm fins, are observed. Smooth-curved menisci and rough copper fins can easily be distinguished in both sections. At the evaporator, where the radius of curvature is small, the junction between the meniscus and the top of the fins is not visible, unlike at the condenser. When a monochromatic image of the white light intercepts a specular surface with a steep slope with respect to the horizontal, it is reflected outside the measurement space and prevents the measurement. The limiting angle given by the manufacturer is equal to 27°. The limitation of the observable slope of the surface is a strong constraint for the observation of capillary structures of very small dimensions (<100 μm approximately). Indeed, the part of the meniscus, which can be measured by the microscope, is then particularly small and does not enable determining the shape of the meniscus with a good accuracy. Nonetheless, for larger grooves, even if the entire meniscus cannot be measured, the radius of curvature can be estimated using the data collected in the center of the groove, where the slope of the meniscus is nearly flat. Besides, in all our measurements, the circle fitted using the experimental data is always attached to the top of the fins.

In order to estimate the accuracy of the optical method used for measuring the liquid pressure inside a groove, a grooved plate was tilted in nonworking conditions with a known controlled angle. In these conditions, the variation of the liquid pressure within a groove is directly linked to the hydrostatic pressure. By comparing the local hydrostatic pressure to the liquid pressure estimated by measuring the radius of curvature inside a groove, the accuracy of the method was found to be in the order of 1 Pa.

Another important parameter in two-phase systems is the thickness of liquid films, especially at the condenser, which has a strong influence on the overall thermal resistance of the system. In a grooved heat pipe, these films are formed on the top of the fins, where the cold wall is in closest contact with the vapor. Contrary to the liquid within the groove, which is generally thick (of the order of 100 μm), the liquid film on the fins is very thin (of the order of a few μm), which makes it difficult to measure. Indeed, two reflecting interfaces are thus present in the measuring field of the microscope and the spectral analysis of the reflected light becomes more delicate. The ability to measure this film depends on its thickness and the nature of the surface of the fin. On a copper fin, the measured signal is extremely noisy (Figure 17.9a). Since copper is particularly reflective, the system detects either the liquid–vapor interface or the solid–liquid interface, seen through the liquid thickness. Therefore, it is not possible to measure these films on a raw copper surface. On the contrary, on a silicon fin, the observed interface appears smooth because the silicon is much less reflective (Figure 17.9b). Films with thicknesses down to 2 μm can be measured in these conditions. The accuracy of the measurement is therefore not sufficient to study the junction zone between the meniscus in the groove and that of the condensate on the top of the fins.

In Figure 17.9a and b, FPHPs with longitudinal grooves were studied. In this type of capillary structure, the menisci are mostly cylindrical and, therefore, the surface is characterized by one single radius of curvature. Figure 17.9c presents an example of measurement obtained with a more complex capillary structure made of cross-groove. Peaks appear at the edges of the cross-grooves, which are due to the machining of the grooves within the copper plate. In this structure, the menisci are not cylindrical and it is necessary to determine the two main radii of curvature at each point of the surface to estimate the mean curvature. This requires realizing surface measurements with a huge resolution (of the order of five minutes for a surface of 1 mm × 1 mm with a 10-μm acquisition step).

The acquisition frequency of the confocal microscope ranges from 300 to 1000 Hz. It can thus also be used to characterize temporally the position of the liquid–vapor interface. Figure 17.10 presents an example of such a measurement for a square groove of 400 μm width exposed to a transversal AC electric field: The altitude of the center of the meniscus is plotted versus time. The temporal variation of the electric field (Figure 17.10a) induces a temporal variation of the electric forces acting on the meniscus and thus a temporal variation of the position of the meniscus within the groove (Figure 17.10b). A frequency sweeping procedure was implemented, which enables determining the resonance frequency of this type of system. The temporal resolution of the signal also enables performing a Fourier analysis to determine the spectral content of the signal (Figure 17.10c) and to get important information

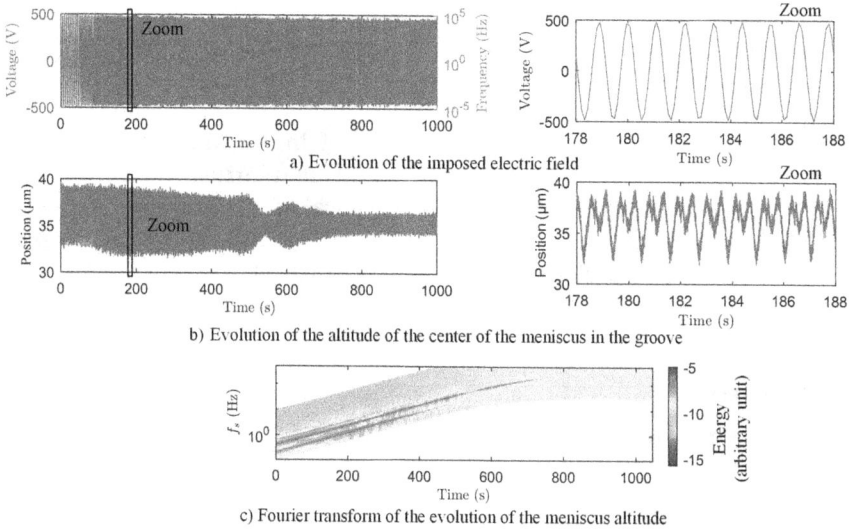

a) Evolution of the imposed electric field

b) Evolution of the altitude of the center of the meniscus in the groove

c) Fourier transform of the evolution of the meniscus altitude

FIGURE 17.10
Frequency sweeping and data post-processing of a meniscus exposed to an AC electric field: solicitation voltage on the left axis and time-dependent frequency on the right axis (a), mechanical response of the liquid interface (b), Fourier transform of the signal delivered by the confocal microscope (c).

on the physical phenomena involved during the experiments. In this example, the developed method enables showing that an AC electric field generates different resonance effects on the meniscus at the scales of the groove width or of the groove length, depending on the frequency of the solicitation (Cardin et al. 2018). This can strongly affect the thermal and hydrodynamic performance of a two-phase flow system.

In conclusion, confocal microscopy is a powerful technique for measuring the shape of menisci and liquid films in two-phase devices. This measurement enables the indirect estimation of some hydraulic and thermal characteristics of these devices that are not easily accessible with other classical sensors. However, several limitations of this technique have been highlighted in this section: Inclined liquid–vapor interfaces or thin liquid film interfaces near a reflecting wall are particularly difficult to obtain. Using a pen with a measuring field of 350 μm enabled measuring the film of thickness down to 2 μm. It is possible to reduce this limitation by using a pen with a smaller measuring field (20 μm for example), but as confocal microscopy is an optical method, the minimum resolution cannot be lower than the wavelength of the white light. To overcome this physical limitation, we have recently implemented another promising optical method based on surface plasmon resonance. The surface plasmon resonance corresponds to an evanescent wave at the interface of a corrugated metal (for example, a triangular grating of depth 100nm and of period 1 μm) and a dielectric fluid. The resonance is due to a collective excitation of the electrons at the surface, when it is illuminated by a white light. The wavelength of resonance changes when any physical or bio-chemical event takes place in a very thin region close to the metallic surface. Therefore, it is, for example, possible to apply this technique to observe the first nuclei of condensation on a cooled plate. Film thicknesses of the order of a few nanometers were measurable, and this non-intrusive technique was also successfully used to measure the very local temperature at the liquid–solid interface (Ibrahim et al. 2018, 2017). This measurement technique opens up important opportunities in the field of two-phase flow researches.

17.4 Wall Temperature Measurement of a Transparent Tube during Flow Boiling Using IR Camera

Both high-speed videography and confocal microscopy are non-intrusive methods, which enable determining the shape of the liquid–vapor interfaces in two-phase systems. Non-intrusive temperature measurements are also of major importance when studying liquid–vapor phase change heat trans-fer. Furthermore, the huge heat transfer coefficients between the wall and the fluid require accurate measurement techniques for their determination because the temperature difference between the wall and the fluid is small. Common temperature sensors, such as thermocouples or resistive thermal detectors, can be used, but their proper implantation is not so easy. They must be as small as possible for not disturbing the system, and their implemen-tation must be performed carefully, for instance by etching grooves in the wall, to ensure a temperature of the sensor very close to that of the wall with limiting influence of the fluid in contact with the wall. Therefore, this type of sensor is often limited to metallic walls.

For the study of two-phase flows during evaporation or condensation in tubes, transparent systems are often required to visualize the flow structures and are thus incompatible with the use of thermocouples. To address this issue and to be able to both observe the flow and measure the wall tempera-ture, a couple of research teams developed a non-intrusive wall temperature method for transparent systems based on IR imaging. This section details this method and the various steps of calibration and post-processing required to achieving an accurate wall temperature measurement.

17.4.1 Functional Principle of Measurement with IR Camera

Temperature measurement by means of an IR camera is widely used in many applications and is easy to implement when dealing with materials for which the assumption of a black or gray body can be used. The raw measure-ment of the camera has to be adjusted knowing the emissivity of the surface

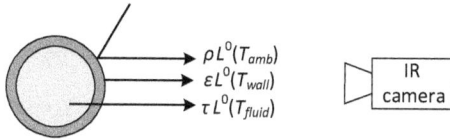

FIGURE 17.11
Summary of the various contributions affecting the camera response in the case of two-phase flow in a tube. T_{amb}, T_{wall}, and T_{fluid} are, respectively, the ambient, wall, and fluid temperatures. ε, ρ, and τ are, respectively, the wall emissivity, reflectivity, and transmissivity, which depend on the tube material and thickness. $L^0(T)$ is the luminance of a black body at temperature T.

to get itstemperature. However, when the emissivity of the surface is low, other phenomena need to be taken into account. The IR camera actually measures the global heat flux that it receives within its given spectral range. Figure 17.11 shows the various contributions affecting the response of the IR camera in the case of the study of two-phase flow in a tube. The bulk emission is the signal containing the information of the wall temperature, but also the reflection component from the ambient environment and optionally the transmission component through the wall that can induce important measurement bias. One must also note that the emissivity, transmissivity, and reflectivity of the materials are often directional, spectral, and temperature dependent. Therefore, the calibration procedure needs to be robust enough to take into account these effects.

17.4.2 Experimental Configuration and Data Acquisition

The example considered in the present section is the wall temperature measurement of a transparent evaporator used to study R245fa flow boiling for various mass fluxes, vapor qualities, saturation temperatures, and orientations. The evaporator is composed of a 200 ± 1.0 mm long sapphire tube which is externally coated with a layer of indium tin oxide (ITO). Note that the sapphire tube and the ITO coating are transparent in the visible wavelengths, whereas the ITO coating is optically thick in the IR wavelengths. These properties enable the visualization of the flow along with the wall temperature measurement of the ITO. The inner tube diameter is equal to 1.6 mm ± 0.15 mm (Figure 17.12). In the present study, the resistivity of the ITO coating is equal to $450\ \Omega \pm 10\ \Omega$ corresponding to a thickness equal to about 1 μm. The choice of the ITO coating thickness was motivated by a trade-off between its optical thickness and its resistivity. The two-phase flow can thus be studied both in the visible spectral domain by a high-speed camera and in the IR spectral domain by a thermal camera, the two being synchronized by the acquisition system. More details on the experimental setup can be found in Layssac, Lips, and Revellin (2019, 2018).

The wall temperature of the outer tube is measured with an IR camera FLIR SC 7000 using MATLAB to post-process the data. The IR camera is

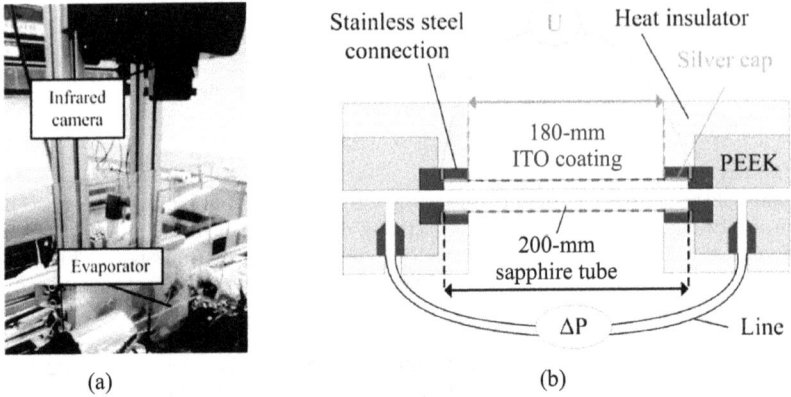

FIGURE 17.12
Photograph (a) and schematic (b) of the test section.

located at a distance of 300 mm from the tube. With this distance and considering a resolution of 512×640 pixels, the maximal zoom enables a precision characterized by a conversion factor of 10 pixels/mm. The electric signal is then coded in 14 bits, which enables obtaining a frame with grayscale values between 0 and 16383. These values are converted into black body temperatures with the software Altair as shown in Figure 17.13a.

The emissivity of the ITO coating being directional, the black body temperature of the tube is not constant in a cross section (Figure 17.13b). This behavior is typical of metallic coating for which the emissivity is higher for angles of emission close to 90°C. To take this effect into account, an averaging of the black body temperature field for orthoradial angles of the sapphire tube between 60° and 120° is performed and enables obtaining longitudinal profiles of black body temperature.

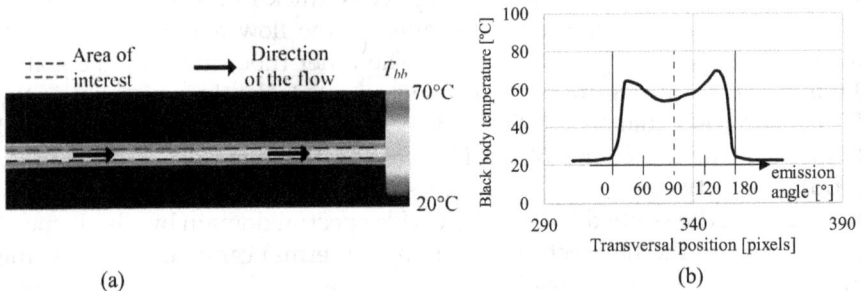

FIGURE 17.13
Example of raw temperature measurements (a) black body temperature field; (b) black body temperature profile in a cross section of the test section.

17.4.3 Post-Processing Analysis

Figure 17.14 presents an example of black body temperature profile for three real temperatures. The fluctuations of the black body temperature are due to the heterogeneity of the ITO coating, and a specific calibration is thus required to determine the apparent mean emissivity of each section of the tube. To determine the real wall temperature from the black body temperature, the effect of both the ambient temperature and the sensitivity of the camera must be taken into account. Considering the direct emission of the ITO coating and the reflection of the environment of the tube, the digital level (DL) of each pixel of the camera is a function of the sensitivity of the IR camera sensors F, the outer wall and ambient temperatures T_{ow} and T_{amb}, respectively, and the emissivity of the ITO coating:

$$DL = K\sigma_{sb}\left[\in F(T_{ow})T_{ow}^4 + (1-\varepsilon)F(T_{amb})T_{amb}^4\right] \qquad (17.1)$$

with K a multiplying factor, ε the emissivity of the ITO, and σ_{sb} the Stefan–Boltzmann constant. The first term in the brackets corresponds to the bulk emission of the coating, and the second term corresponds to the reflection of the surroundings. The sensibility of the camera sensor F corresponds to the part of the heat flux effectively measured by the camera in its spectral range. It depends on the relative spectral response of the camera ξ given by the manufacturer (Figure 17.15):

$$F(T) = \frac{1}{\sigma T^4}\int_{\lambda=0}^{+\infty}\xi(\lambda)M_\lambda^0(T)d\lambda \qquad (17.2)$$

The digital level may also be expressed considering the black body temperature T_{bb} given by the infrared camera, corresponding to a case where there is no reflection of the ITO coating:

$$DL = K\sigma_{sb}F(T_{bb})T_{bb}^4 \qquad (17.3)$$

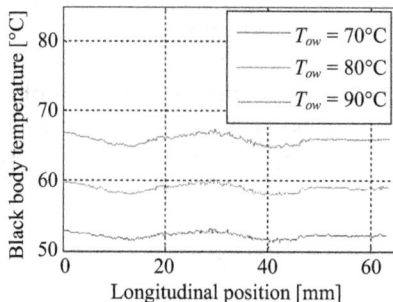

FIGURE 17.14
Example of mean black body temperature profile along the test tube.

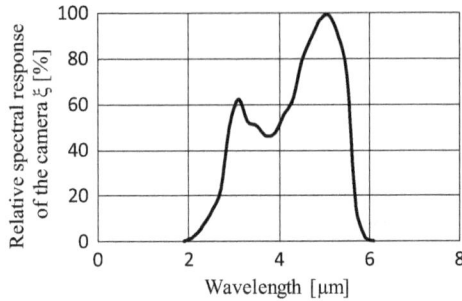

FIGURE 17.15
Spectral response of the camera.

The introduction of Eq. (17.1) into Eq. (17.3) enables deriving the following expression of the outer wall temperature of the tube:

$$T_{ow} = \left[\frac{1}{F(T_{ow})} \left[F(T_{amb})T_{amb}^4 + \frac{1}{\varepsilon} \left[F(T_{bb})T_{bb}^4 - F(T_{amb})T_{amb}^4 \right] \right] \right]^{\frac{1}{4}} \quad (17.4)$$

In the setup configuration, the ambient temperature T_{amb} is given by three thermocouples located on three polycarbonate plates placed on each side of the test section, which are assumed to be representative of the radiative environment of the tube. The black body temperature T_{bb} is directly given by the IR camera, and $F(T)$ can be calculated by means of Eq. (17.2). If the emissivity of the ITO coating is known, Eq. (17.4) can be solved by an iterative procedure as the term $F(T_{ow})$ depends on the real outer wall temperature.

17.4.4 Determination of the Wall Emissivity by Means of a Calibration Procedure

In practice, the emissivity of the ITO coating is unknown a priori. A calibration procedure is then required to determine the apparent wall emissivity as a function of temperature for each section of the test tube. To do so, a high-velocity single-phase flow was set in the tube (no heat flux imposed) in order to impose a known internal wall temperature. The external wall temperature is calculated with a specific local thermal model taking into account the heat losses toward the surroundings. It was not possible to insert a thermocouple inside the tube for the calibration because it may have damaged the ITO coating.

For each condition of known outer wall temperature T_{ow}, the emissivity ε is calculated by measuring the corresponding ambient temperature T_{amb} and black body temperature T_{bb} (given by the IR camera):

$$\varepsilon(T_{pe}) = \frac{F(T_{bb})T_{bb}^4 - F(T_{amb})T_{amb}^4}{F(T_{ow})T_{ow}^4 - F(T_{amb})T_{amb}^4} \quad (17.5)$$

An example of the apparent mean emissivity along the test section is presented in Figure 17.16 for three different temperatures. The spatial variations are due to the heterogeneity of the ITO coating. The low value of emissivity justifies the implementation of the complete method, and especially the necessity to take into account the sensibility of the camera (F), whereas a simple assumption of a gray body would not be enough to take into account the phenomena linked to the ambient temperature contribution. The uncertainty on the emissivity is presented in Figure 17.17 as a function of wall temperature. It is calculated considering the uncertainty on the wall temperature during the calibration (0.4 K), on the black body temperature measurement (0.1 K), and on the ambient temperature (0.25 K). Lowering the wall temperature increases the emissivity uncertainty. In our case, we choose to use the infrared thermography only for wall temperature higher than 80°C in order to get an uncertainty on the emissivity lower than 0.004.

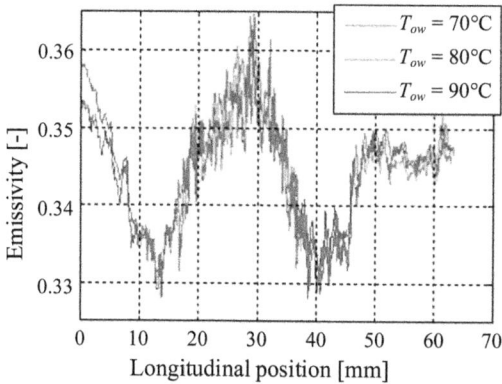

FIGURE 17.16
Variations of the ITO deposit emissivity with abscissa for three sapphire outer wall temperatures.

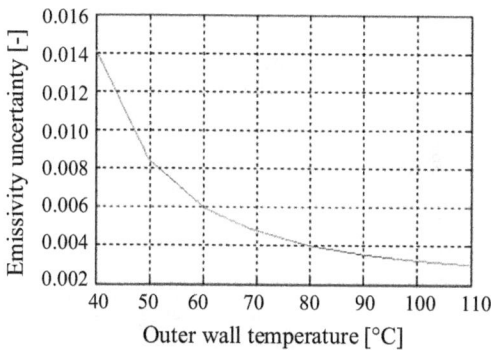

FIGURE 17.17
Evolution of the uncertainty of the mean emissivity of the wall as a function of its temperature.

Using this calibration procedure, the wall temperature can then be determined during the experiments, while heating the tube by Joule effect. The heat flux transferred to the fluid and the inner wall temperature are not measured directly. As a result, a local thermal model is required to calculate them, considering the heat losses of the evaporator (Layssac et al. 2019).

An example of temperature profile measured along the test section is presented in Figure 17.18 for a mass velocity of 152 kg/m²s and a vapor quality of 0.28. The high-frequency fluctuations are due to the noise of the black body temperature measurement and can be smoothed if one wants an average temperature. The global uncertainty of the measurement associated with the calibration procedure is presented in Figure 17.19. The uncertainty of the temperature measurement depends on the temperature as the part of the ambient contribution seen by the IR camera decreases when the wall temperature increases.

In conclusion, the infrared thermography can be used as a very interesting and non-intrusive tool for the wall temperature determination in convective two-phase flow studies.

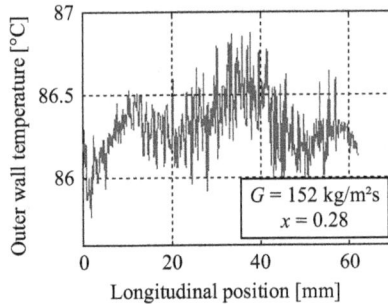

FIGURE 17.18
Example of spatial temperature profile of the wall of the test section.

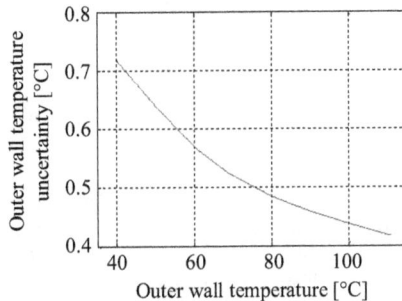

FIGURE 17.19
Uncertainty on the wall temperature determination.

However, important care must be taken to achieve an accurate temperature measurement. In the present example, the uncertainty is lower than 0.5 K for wall temperature higher than 75°C. Different possibilities can be used to increase the accuracy of the method:

- A thicker ITO coating would exhibit a higher emissivity and thus a lower final uncertainty on the temperature measurement.
- A better control of the ambient conditions would lead to a decrease in the uncertainty on the ambient temperature and thus on the temperature measurement.
- The calibration procedure could be improved to reduce the uncertainty of the real wall temperature during the calibration step, for example by trying to insert properly a thermocouple inside the tube. It is also possible to add a second transparent sapphire tube surrounding the test section in order to reduce the heat losses toward the surroundings.

Note that the calibration procedure can also be performed at the scale of each pixel. In the present study, the choice was made to perform the calibration at the scale of a section of the tube, as a trade-off between taking into account the longitudinal heterogeneity of the tube and smoothing the radial variation of emissivity. Anyway, whatever the studied configuration is, each research team needs to choose its own trade-off between the quality (cost) of the acquisition system, the control of the ambient conditions, the complexity of the calibration procedure and post-processing analysis, and the uncertainty of the final wall temperature determination.

17.5 Conclusion

In this chapter, three different measurement techniques implemented at CETHIL during the last decade for the study of phase change heat transfer have been presented:

High-speed videography for the analysis of boiling: This section introduced good practices when using high-speed and high-definition videography for the analysis of boiling. It also exemplified how this technique and the previous ones (photography and then videography) were successfully used to make some progresses in the understanding of the phenomena involved in pool boiling and flow boiling.

Film thickness measurements using confocal microscopy: This section described how this optical system can be used for measuring the shape of menisci and liquid films in two-phase devices. This measurement enables estimating indirectly some hydraulic and thermal characteristics of these devices that are not easily accessible with other classical sensors.

Temperature measurements using IR camera: This section dealt with the implementation of this non-intrusive tool for the wall temperature determination during flow boiling. Using infrared thermography enabled measuring the temperature of the wall while visualizing the flow, thanks to an ITO layer coated on a sapphire tube. However, since important care must be taken to achieve quantitative temperature measurement, the method and the various steps of the calibration and post-processing required for reaching an accurate wall temperature measurement were explained.

If the flow visualization is an important step toward the understanding of the phenomena governing phase change heat transfer, it is recommended to carry out temperature measurements at the same time and vice versa. In this way, hydrodynamic phenomena and thermal responses can be analyzed in parallel and accurate and robust prediction methods can be developed.

References

Cardin, N., L. Davoust, S. Lips, S. Siedel, and J. Bonjour. 2018. "Confocal microscopy as a means to investigate the EHD-driven formation of liquid-vapor interfaces along an array of microgrooves." In *19th International Symposium on the Application of Laser and Imaging Techniques to Fluid Mechanics*. Lisbon, Portugal.

Charnay R., Bonjour J. and Revellin R. 2014. "Experimental investigation of R-245fa flow boiling in minichannels at high saturation temperatures: Flow patterns and flow pattern maps", *International Journal of Heat and Fluid Flow*, Vol. 46, pp. 1–16.

Donniacuo, A., R. Charnay, R. Mastrullo, A.W. Mauro, and R. Revellin. 2015. "Film thickness measurements for annular flow in minichannels: Description of the optical technique and experimental results." *Experimental Thermal and Fluid Science*, Vol. 69, pp. 73–85 https://doi.org/10.1016/j.expthermflusci.2015.07.005.

Gaertner, R.F. 1965. "Photographic study of nucleate pool boiling on a horizontal surface." *Journal of Heat Transfer*, Vol. 87, pp. 17–29.

Ibrahim, J., M. Al Masri, C. Veillas, F. Celle, S. Cioulachtjian, I. Verrier, F. Lefèvre, O. Parriaux, and Y. Jourlin. 2017. "Condensation phenomenon detection through surface plasmon resonance." *Optics Express*, Vol. 25 (20), pp. 24189–98. https://doi.org/10.1364/OE.25.024189.

Ibrahim, J., M. Al Masri, I. Verrier, C. Veillas, F. Celle, S. Cioulachtjian, O. Parriaux, F. Lefèvre, and Y. Jourlin. 2018. "Temperature sensor through surface plasmon resonance." In *Optical Sensing and Detection V*, 10680:106800R. International Society for Optics and Photonics. https://doi.org/10.1117/12.2306243.

Layssac, T. 2018. "*Contribution à l'étude Phénoménologique de l'ébullition Convective En Mini-Canal.*" Thèse de doctorat, INSA de Lyon, Université de Lyon.

Layssac, T., C. Capo, S. Lips, A. W. Mauro, and R. Revellin. 2017. "Prediction of symmetry during intermittent and annular horizontal two-phase flows." *International Journal of Multiphase Flow*, Vol. 95, pp. 91–100.

Layssac, T., S. Lips, and R. Revellin. 2018. "Experimental study of flow boiling in an inclined mini-channel: Effect of inclination on flow pattern transitions and pressure drops." *Experimental Thermal and Fluid Science*, Vol. 98, pp. 621–33. https://doi.org/10.1016/j.expthermflusci.2018.07.004.

Layssac, T., S. Lips, and R. Revellin. 2019. "Effect of inclination on heat transfer coefficient during flow boiling in a mini-channel." *International Journal of Heat and Mass Transfer*, Vol. 132, pp. 508–18. https://doi.org/10.1016/j.ijheatmasstransfer.2018.12.001.

Lefèvre, F., R. Rullière, S. Lips, and J. Bonjour. 2010. "Confocal microscopy for capillary film measurements in a flat plate heat pipe." *Journal of Heat Transfer*, Vol. 132, p. 031502. https://doi.org/10.1115/1.4000057.

Lips, S., F. Lefèvre, and J. Bonjour. 2010a. "Thermohydraulic study of a flat plate heat pipe by means of confocal microscopy: Application to a 2D capillary structure." *Journal of Heat Transfer*, Vol. 132, p. 019008. https://doi.org/10.1115/1.4001930.

Lips, S., F. Lefèvre, and J. Bonjour. 2010b. "Combined effects of the filling ratio and the vapour space thickness on the performance of a flat plate heat pipe." *International Journal of Heat and Mass Transfer*, Vol. 53 (4), pp. 694–702. https://doi.org/10.1016/j.ijheatmasstransfer.2009.10.022.

Lips, S., F. Lefèvre, and J. Bonjour. 2010c. "Investigation of evaporation and condensation processes specific to grooved flat heat pipes." *Frontiers in Heat Pipes*, Vol. 1 (2), pp. 023001-1–023001-8. https://doi.org/10.5098/fhp.v1.2.3001.

Lips, S., F. Lefèvre, and J. Bonjour. 2011. "Physical mechanisms involved in grooved flat heat pipes: Experimental and numerical analyses." *International Journal of Thermal Sciences*, Vol. 50 (7), pp. 1243–52. https://doi.org/10.1016/j.ijthermalsci.2011.02.008.

Michaïe, S. 2018. *"Experimental Study of the Fundamental Phenomena Involved in Pool Boiling at Low Pressure."* Ph.D. thesis, Université de Lyon. https://tel.archives-ouvertes.fr/tel-01940955/document.

Nukiyama, S. 1934. "The maximum and minimum values of the heat Q transmitted from metal to boiling water under atmospheric pressure." *Journal of the Japanese Society of Mechanical Engineers*, Vol. 37, pp. 367–74.

Revellin R., Padilla M., Bonjour J., "Two-phase flow characteristics in singularities", *9th HEFAT International Conference*, Malta, 16–19 July 2012 (Invited keynote lecture).

Rullière, R., F. Lefèvre, and M. Lallemand. 2007. "Prediction of the maximum heat transfer capability of two-phase heat spreaders - experimental validation." *International Journal of Heat and Mass Transfer*, Vol. 50 (7–8), pp. 1255–62.

Siedel, S., S. Cioulachtjian, and J. Bonjour. 2008. "Experimental analysis of bubble growth, departure and interactions during pool boiling on artificial nucleation sites." *Experimental Thermal and Fluid Science*, Vol. 32 (8), pp. 1504–11. https://doi.org/10.1016/j.expthermflusci.2008.04.004.

Siedel, S. 2012. *"Bubble Dynamics and Boiling Heat Transfer: A Study in the Absence and in the Presence of Electric Fields."* Ph.D. Thesis, INSA Lyon.

Sobel, I. 1990. "An isotropic 3×3 image gradient operator," In H. Freeman, editor, *Machine Vision for Three-Dimensional Scenes*, pp. 376–379. New York, Academic Press.

Tibiriçá, C. B., F. Júlio do Nascimento, and G. Ribatski. 2010. "Film thickness measurement techniques applied to micro-scale two-phase flow systems." *Experimental Thermal and Fluid Science*, Vol. 34 (4), pp. 463–73. https://doi.org/10.1016/j.expthermflusci.2009.03.009.

18

Selected Problems of Experimental Investigations during Refrigerants Condensation in Minichannels

Tadeusz Bohdal, Henryk Charun, and Małgorzata Sikora
Koszalin University of Technology

CONTENTS

18.1 Introduction

The dynamic development of technologies in space technology and electronics has an impact on the global trend of machines and devices miniaturization in many areas. In some systems, there is a need to discharge a high heat flux density, larger than 1,000 W/cm² (Baummer et al. 2008). The use of conventional methods of heat collection has become ineffective or even impossible. In this case, the phase changes of refrigerants (condensation and boiling) are used, especially in small-diameter channels (Obhan and Garimella 2001). High values of heat transfer coefficient are obtained; they allow reducing the heat exchange area. Practical implementation of this method is done using compact minicondensers and minievaporators. To call the heat exchanger compact, it should be made from channels with a hydraulic diameter (with dimensions depending on the criteria) (Ohadi 2007, Mehendale et al. 2000, Kandlikar 2003) within 1–6 mm, usually in practice $d_h < 3$ mm.

In recent years, the number of articles about research methods and theoretical analysis in terms of boiling and condensation in small-diameter channels has significantly increased. The number of publications in the boiling range in minichannels is also higher. However, it is not possible to transfer in absolute terms the test results obtained during boiling in minichannel for the condensation process in this type of channels (Mikielewicz 1995, Thome et al. 2003a,b, Cavallini et al. 2005). A significant difficulty in designing and testing of compact minicondensers is the fact that there are no uniform models for the condensation of refrigerants (Sun and Mishima 2009, Garcia-Cascales et al. 2010). This chapter focuses on research problems related to experimental investigations of heat exchange and flow resistance during condensation of refrigerants in minichannels and also two-phase flow structures.

18.2 Some Research Problems of Compact Minicondensers

While designing compact refrigeration heat exchangers, two basic engineering problems should be solved. Firstly, determine, for a selected heat exchanger, the heat exchange surface and the driving power of the motion generators of the refrigerant that transmits the heat. This is usually done to calculate the values of heat transfer coefficient on both sides of the wall, during boiling or condensation of refrigerants in the flow and for factors mediating in heat transfer (water, air, brine, etc.). The driving power is determined on the basis of the knowledge of the mass flow value and the flow resistance of the refrigerant and intermediary factor. For relatively simple cases, the best known calculation correlations from the literature are used.

An effective way to determine the thermal and flow characteristics of boiling and condensation in minichannels is experimental research. In boiling tests in minichannels, the heat flow from the environment is directly supplied to the measuring section of the minichannel (often using electric heating). In the refrigerant condensation research in the minichannel, the heat flow must be transferred to the cooling medium, which is unfortunately a much more difficult problem. Therefore, indirect methods of collecting heat flux from the test section are usually used.

18.2.1 The Mechanism of Refrigeration Condensation in Minichannels and Basic Research Problems

The refrigerant condensation process in a compact minicondenser is similar to that in a conventional one. The superheated refrigerant vapor, after leaving the compressor discharge port, is directed to the condenser. In the first zone of condenser, the heat of the superheated steam is collected, and when the wall temperature of the minichannel reaches a value lower than the saturation temperature, the process of proper condensation begins. This process continues to achieve the saturated liquid state for the vapor quality $x = 0$. Condensation may be complete (in the range of vapor quality from $x = 1$ to $x = 0$) or incomplete. After the proper condensation zone, a single-phase process of the refrigerant liquid subcooling is carried out. It has experimentally been found for a compact condenser (Leducq et al. 2003) that the heat transfer zone for superheated steam can account for up to 15% of the total surface area of the exchanger, the area of the second zone (proper condensation) is 73%–80%, and that of the subcooling zone is 5%–12% of the heat exchanger area.

In both the conventional and compact condensers, the most important effect on heat exchange efficiency is due to the proper condensation zone, in which the heat transfer coefficient reaches high values, compared to single-phase zones of superheat and subcooling heat recovery. In the case of homogeneous refrigerant condensation, the isothermal and isobaric nature of the process is preserved, but during the condensation of zeotropic mixtures, the process is non-isothermic and proceeds with a temperature glide. For the condensation process in the conventional channel, the dominant influence is from gravity, inertial forces, and shear forces at the interface, causing changes in the two-phase flow structure.

The mechanism of energy and momentum exchange during refrigerants condensation in minichannels is more complicated than in convention channels. The interactions of viscous forces and surface tension play a fundamental role in it. This leads to the formation other two-phase flow structures than those in the conventional channels.

The basic experimental research problems of refrigerants condensation in minichannels are as follows:

1. Assessment of the mass, energy, and momentum exchange mechanisms during the condensation of refrigerants in minichannels.

2. Development of the maps of two-phase flow structures in a generalized version, especially for new pro-ecological refrigerants.

3. Determining the influence of the refrigerant type on the parameters of the condensation process and the results of thermal and flow investigations.

4. Determination of the influence of minichannel geometric parameters (including internal diameter, shape and length) on the condensation efficiency.

5. Indication of generalized and experimentally tested procedures for calculating the average and local values of heat transfer coefficient and flow resistance.

6. Development of theoretical basis for exchange and momentum energy for condensation in minichannels, with simultaneous explanation of differences in process mechanisms in conventional, mini-, and microchannels.

Some of the mentioned research problems are presented in the further part of the chapter. The importance of the research problem of the refrigerants condensation process in small-diameter channels is the reason for widely developed research cooperation of many science centers in this topic. An example is the cooperation of the University of Padova (Cavallini et al. 2005) and the University of Pretoria (Ewim and Meyer 2019, Ewim et al. 2018) in the field of condensation of new refrigerants, including zeotropic mixtures (Thome 2005).

18.2.2 Methodology of Experimental Investigations of Refrigerants Condensation in Minichannels

While developing the methodology of experimental research on condensation in mini-channels, modern research trends should be taken into account. Trends relate to the use of high-accuracy measuring devices. For example, in their study, Thome et al. (2003b) showed that in condensation investigations in the 1990s, Coriolis flow meters with a great accuracy were used. The condensation tests in channels were very difficult when the vapor quality is $x < 0.05$ or $x > 0.95$. Also in the literature, only few publications regarding condensation in minichannels, for mass flux density $G < 50$ kg/m^{2}s exist. Below this value, there are big problems with measuring accuracy.

Some parameters of condensation characteristics in minichannels can be measured directly (e.g., refrigerant, intermediate factor, or channel wall temperature, pressure, and differential pressure). Other parameters are determined by indirect methods (including heat flux from the measurement section, vapor quality x, void fraction ε, and average and local heat transfer coefficients α). One of the intermediate research methods is the *Wilson* method (Wilson 1915), with its later modifications, but its range is limited.

In most cases, the experimental test stand for the condensation of homogeneous refrigerants (pure refrigerant) can also be used in studies of mixtures of these refrigerants, including zeotropic mixtures (Macdonald and Garimella 2016).

18.2.3 Experimental Investigations of Thermal and Flow Characteristics of Condensation

Based on many years of research by the authors, a laboratory test stand was proposed, whose schematic diagram is shown in Figure 18.1.

The test stand shown in Figure 18.1 allows performing experimental measurements of condensation of various homogeneous refrigerants (pure) and their mixtures (including zeotropic mixtures) in single-pipe minichannels and in multipots in the range of the inner diameter $d = 0.21$–3.30 mm. A detailed description of the methodology and the results of the authors' research in the scope of condensation characteristics in minichannels is provided in the works of Bohdal et al. (2012, 2015) and Shin and Kim (2004a,b).

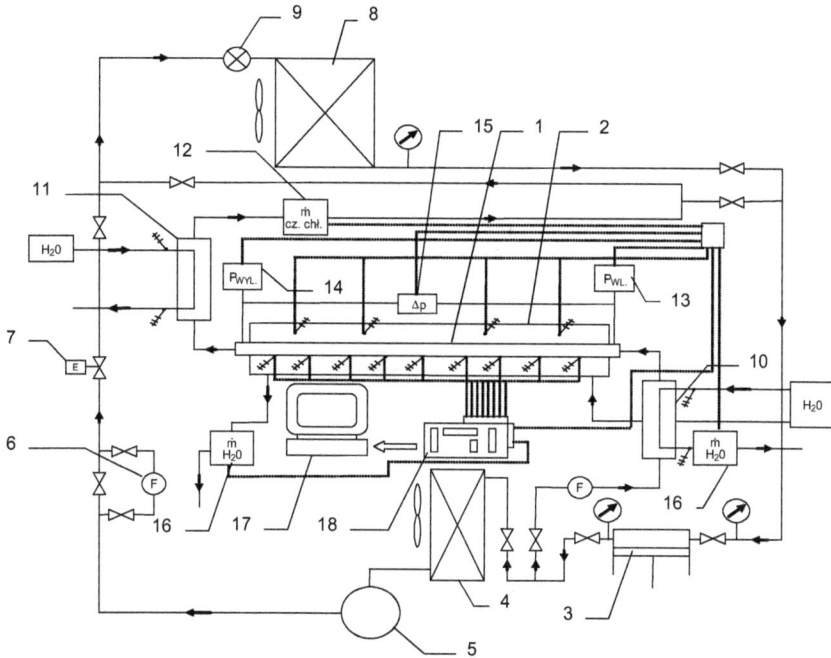

FIGURE 18.1
Schematic diagram of the test stand; 1—measuring section of pipe minichannel, 2—water duct, 3—cooling compressor unit, 4—air-cooled condenser, 5—liquid refrigerant tank, 6—filter–dryer of refrigerant, 7—electromagnetic valve, 8—finned air cooler, 9—expansion valve supplying, 10—heat exchanger for the pickup of the superheat of the refrigerant, 11—liquid refrigerant subcooler, 12—electronic refrigerant flow meter, 13—medium-pressure sensor on the inlet to the measuring section, 14—medium-pressure sensor on the outflow from the measuring section, 15—differential pressure sensor, 16—electronic water flow meter, 17—computer, and 18—data acquisition system.

The schematic diagram of the test stand (Figure 18.1) should be distinguished by two cooling systems with refrigerant, cooperating in parallel, that is installation of a single-stage refrigeration system supplied from a refrigerating compressor unit and cooling installation feeding the measuring section of pipe minichannel 1. The superheated steam of the refrigerant after compression in the piston compressor 3 was passed through the subsequent elements of the installation from position 3 to position 9 in Figure 18.1.

The condensing process of the refrigerant took place in the flow inside the pipe minichannel 1 (Figure 18.2). The superheated vapor of the refrigerant after discharge from the compressor 3 was directed by a control valve both to the basic system and to the supply system of the measuring section 1. Before the inflow of the refrigerant to the inlet section of the measuring section, the water-cooled heat exchanger 10 was installed (for receiving the superheat and for the preparation of the refrigerant condition at the inlet to

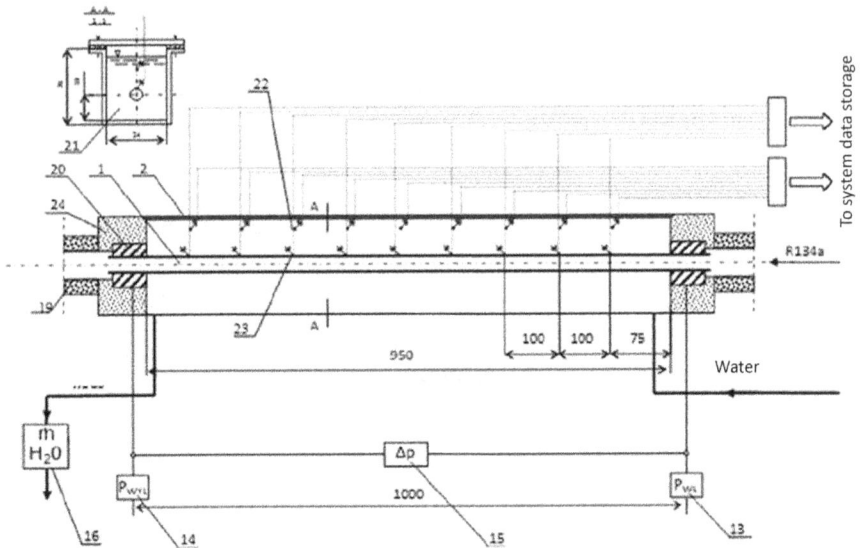

FIGURE 18.2
Schematic diagram of the measuring section of the pipe minichannels; 19—copper pipe (diameter 8/10 mm), 20—connector, 21—water-cooling channel, 22—thermocouple for measuring water temperature, 23—thermocouple for measuring minichannel surface temperature, and 24—insulation (other markings as in Figure 18.3).

the minichannel). The pressure of the refrigerant at the inflow and outflow of the measuring section was measured using piezoresistive sensors 13 and 14 with PMP 131-A1401A1W-type transmitters produced by Endress + Hauser; the measuring range of the sensors is 0–2.5 MPa. The pressure drop of the refrigerant over the length of a 1,000-mm pipe minichannel was additionally measured using the differential pressure sensor 15 with a Deltabar SPMP-type transducer. The liquid of refrigerant leaving the measurement section was subcooled in the heat exchanger 11. The flow rate of the refrigerant liquid was measured using a Coriolis flow meter. (The same type of flow meter was used for cooling water.) The liquid of refrigerant after leaving the measuring section was discharged to the installation feeding the air cooler 8 (Charun 2012).

Figure 18.2 shows a schematic diagram of the pipe minichannel section with control and measurement instrumentation. The basic element of the measurement system was a minichannel section with an internal diameter $d_w = 0.21$–3.30 mm and an overall length 1,000 mm. (The pressure drop of the refrigerant was measured throughout this length.) A section of the pipe minichannel was placed in the water channel 21 (Figure 18.2). It was a channel made of aluminum, with a rectangular section (dimensions 28 × 24 mm) and an internal length of 950 mm (active length of the measuring section). In nine cross sections along the effective length of the minichannel, the temperature of its external surface was measured using K-type thermocouples installed at a distance of 100 mm. In the same sections, thermoelectric sensors were placed to measure the temperature of the cooling water flowing in the water channel. All thermoelectric sensors, before installation on the test section, were sized against a standard glass thermometer with an elemental scale of 0.1°C, performing their individual characteristics.

Indirectly, the vapor quality x and heat flux density q were determined according to a specially developed methodology. To determine the heat flow Q discharged during condensation into cooling water (flowing in the water channel 2 in Figure 18.1), the concept presented in the authors' work was used (Shin and Kim 2004a,b), with air as a cooling factor. Comments included in the authors' works (Del Col et al. 2014, Kandlikar et al. 2013) were concluded, also those regarding the investigations of zeotropic mixtures.

Before performing the basic tests (on the test stand shown in Figure 18.1), testing investigations of the indirect method were carried out on a separate test station. The schematic diagram of this test section is shown in Figure 18.3. In Figure 18.3b is shown the test section made from minichannel with the same internal diameter and an active length of 950 mm. There is the same arrangement of thermocouple sensors to measure the temperature of

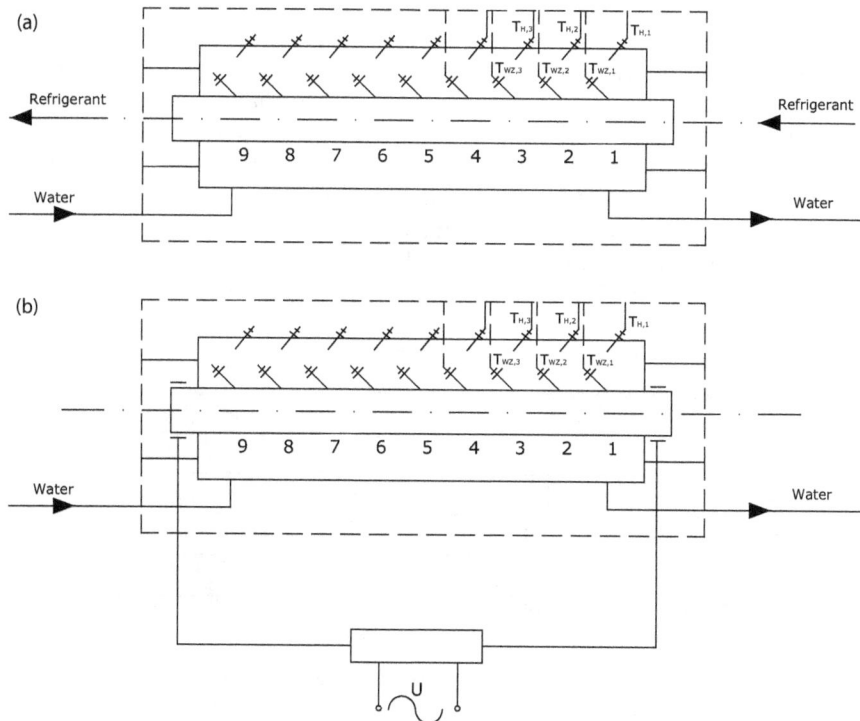

FIGURE 18.3
Schematic diagram of the comparative method to determine the heat flux discharged during condensation in the minichannel; (a) basic measurement section according to Figure 18.1 and (b) additional measuring section heated by electricity.

the refrigerant and the cooling water. Figure 18.3a shows the section with a part of the installation from Figures 18.1 and 18.2.

The refrigerant did not flow inside the section (b) of the minichannel ($m = 0$) during the testing measurements, while the cooling water was flowing outside, with the same flow rate as during the basic tests in section (a). The minichannel section (b) was included in the independent electric heating system, where it works as an electrical resistor. The electric current parameters such as the supplied electrical power Q_{el} were measured in each case along with the values of the outer wall surface temperature of the minichannel T_{wz} and the cooling water temperature T_H in each subsequent measurement cross section (in nine subsequent cross sections, for $i = 1, 2, ..., 9$). The supplied electric power Q_{el} was transmitted in the form of heat flux Q_i (according to Joule's law), causing an increase in the outer surface temperature of the minichannel in each of its cross sections. Knowing the values of $Q_{el} = Q_i$, the internal diameter d, and the length L_i of each part of the measurement section, the heat flux density q_i related to its internal surface was determined:

$$q_i = \frac{\dot{Q}_i}{\pi d L_i} \tag{18.1}$$

On the basis of the measured values of the wall temperature $T_{wz,i}$ and the water temperature $T_{H,i}$ in the ith cross section, thermal characteristics were made in the form of $q_i = f(T_{wz,i} - T_{H,i}) = f(\Delta T)$, where $\Delta T = T_{wz,i} - T_{H,i}$. From the experimental characteristics, the local heat flux density was determined on the inside surface of the minichannel in a given measurement cross section. Figure 18.4 shows an example characteristic $q_i = f(\Delta T)$ for one of the cross sections. Such auxiliary experimental characteristics were prepared for all cross sections of the tested pipe minichannels.

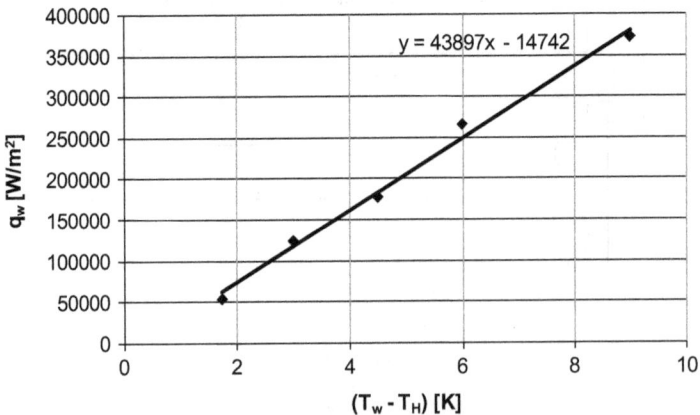

FIGURE 18.4
Example characteristic of the local heat flux density $q_w = f(\Delta t)$ for the cross section no. 8 of the test section of the pipe minichannel.

Knowing the value of q_i in a given cross section of minichannel, the local value of the heat transfer coefficient α_x was determined:

$$\alpha_x = \frac{q_i}{T_s - T_{w,i}} \tag{18.2}$$

where $T_{w,i}$ is the temperature of the internal surface of the minichannel in a given cross section and T_s is the saturation temperature.

The vapor quality x of the refrigerant at the inflow to the test section of the minichannel was determined by an indirect method from the energy balance equation of the exchanger 10 shown in Figure 18.1. The exchanger was cooled by water flowing countercurrent to the refrigerant in the amount of m_H. The temperature of the cooling water at the inflow $T_{H,en}$ and the outflow $T_{H,ex}$, the mass flow rate m_R, and the thermal parameters of the refrigerant on the inflow (superheated vapor) and outflow (wet saturated vapor with vapor quality x) were measured. Heat exchange between the refrigerant and cooling water in the exchanger 10 took place in a sensible and latent heat transfer, and the energy balance equation takes the following form:

$$\dot{m}_H \cdot c_{p,H} \cdot \Delta T_H = \dot{m}_R \cdot c_{p,R}\left(T_R - T_s\right) + \dot{m}_R \cdot r \cdot \Delta x \tag{18.3}$$

$$\Delta x = \frac{1}{\dot{m}_R \cdot r}\left[\dot{m}_H c_{p,H} \cdot \Delta T_H - \dot{m}_R c_{p,R}\left(T_R - T_s\right)\right], \tag{18.4}$$

where ΔT_H is the increase in cooling water temperature exchanger, T_R the temperature of the superheated refrigerant vapor at the inflow to the exchanger, r the heat of evaporation, $c_{p,H}$ and $c_{p,R}$ the specific heat of the water and the refrigerant, and Δx the change in vapor quality in the flow through the exchanger, from the value $x = 1$ (saturated dry steam) to the value x on the outflow from the exchanger (and on the inflow to the measuring section of the mini-channel), i.e.,

$$\Delta x = 1 - x. \tag{18.5}$$

By adjusting the cooling water flow rate, it was possible to obtain the value of $x = 1$ on the outflow from the exchanger 10. The local values of the vapor quality x in individual cross section of the minichannel were determined using the dependence:

$$x_i = \frac{Q_i}{\dot{m}_R \cdot r} \tag{18.6}$$

The presented experimental method allowed determining the local and average values of the heat transfer coefficient with an accuracy of ±10%.

On the basis of the measurements made on the test stand described above, according to the developed methodology, it was possible to implement research programs for thermal and flow characteristics of condensation. Under the thermal condensation characteristics in minichannels should be understood the dependence of the local heat transfer coefficient on vapor quality x in the form $\alpha_x = f(x)$ for G = const and the average heat transfer coefficient on the mass flux density: $\alpha_a = f(G)$, for x_a = const.

Flow condensation characteristics determine the dependence of the local pressure drop per unit of minichannel length $(\Delta p/L)_x$ or the average pressure drop $(\Delta p/L)_a$ from local vapor quality x and mass flux density G. Figure 18.5 presents the examples of thermal and flow characteristics of condensation in a minichannel.

18.2.4 Investigations of Two-Phase Flow Structures during Refrigerants Condensation in Minichannels

Knowledge of two-phase flow structures plays an important role in the selection of the adequate correlation for the calculation of the heat transfer coefficient and flow resistance. From the physical point of view, the following two-phase flow structures, characteristic of refrigerants condensation in minichannels, can be distinguished:

- *Bubble flow*: The structure of gas phase (intermittent gas phase) in the form of bubbles in a continuous liquid phase.
- *Plug flow*: Gas bubbles can reach a size comparable to the internal diameter of the channel and move mainly in the upper part of its cross section.
- *Slug flow*: With increasing flow rate, shearing stresses cause an increase in the range of waves forming gas bubbles along the flow channel,
- *Stratified flow*: The liquid and gas phases are separated by a smooth phase separation surface. This usually occurs at low speeds of both phases.
- *Wave flow*: Along with the increase in the gas phase velocity, there are disturbances at the phase separation boundary, which results in the formation of waves moving in the direction of flow.
- *Annular flow*: The gas phase has a high velocity in the gas core, and on the wall surface of the channel, a liquid condensate film is formed whose thickness is generally asymmetrical in relation to the cross section of the channel, with a larger thickness at the bottom of the channel.

Figures 18.6 and 18.7 show examples of two-phase condensation flow structures in vertical and horizontal minichannels. As it can be seen, two-phase flow structures in a vertical minichannel differ slightly from those observed in horizontal channels (Figure 18.7).

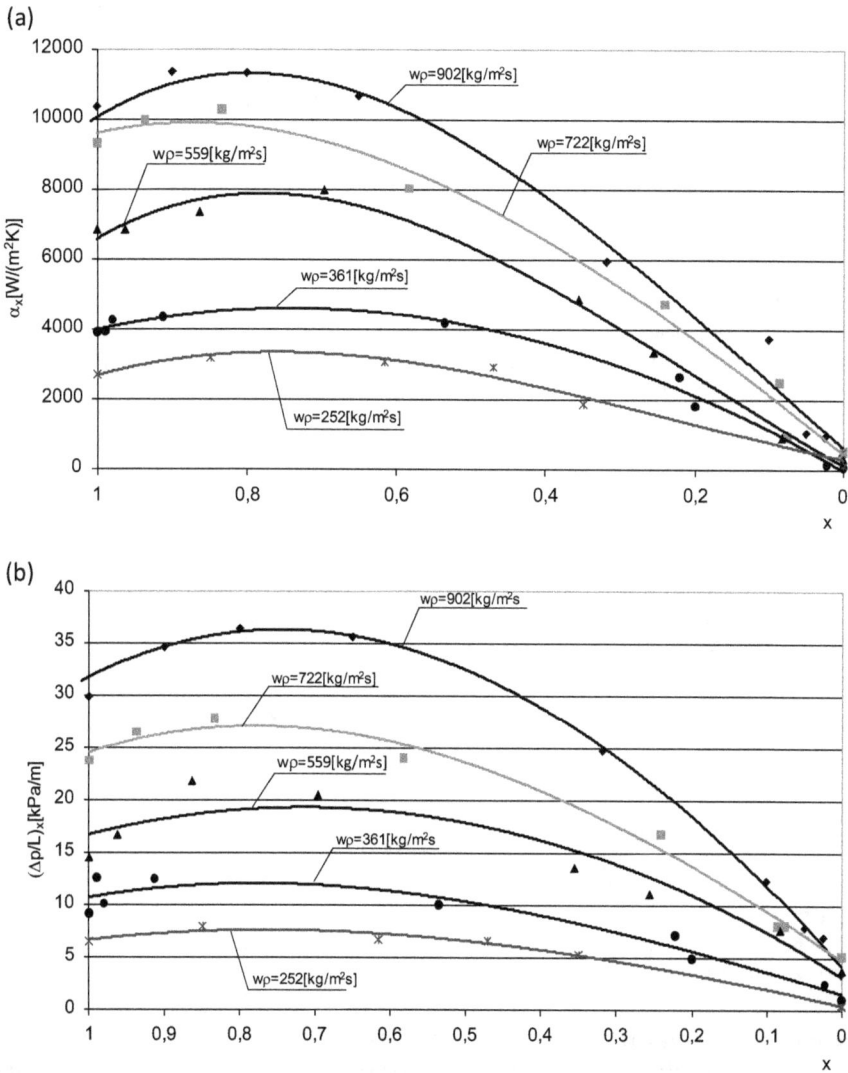

(a)

(b)

FIGURE 18.5
Sample experimental thermal (a) and flow (b) characteristics of R404A refrigerant condensation in a pipe minichannel tube with an internal diameter of 1.4 mm prepared on the basis of measurements results made on the test stand shown in Figure 18.1 [17].

On the test stand presented in Sikora (2017), experimental measurements were carried out in the scope of flow structures during the condensation of HFE-7100 and Novec 649 refrigerants in horizontal minichannels with an internal diameter $d = 0.5$–2.5 mm. Using the authors' own database of condensation experimental results, the so-called "calculation method" of flow structures identification was used. The idea of this method is to compare the

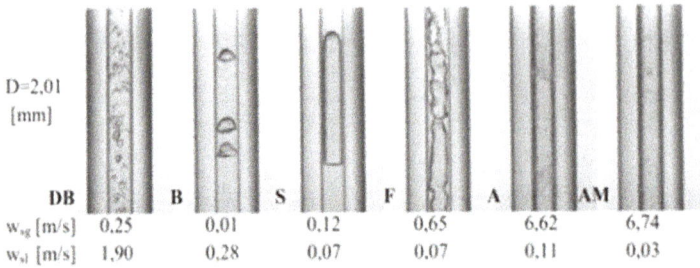

FIGURE 18.6

Examples of two-phase flow structures in a vertical minichannel (minichannel with an internal diameter of 2.01 mm) according to Chen and Tian (2006): DB—bubble dispersion flow, B—bubble flow, S—slug flow, F—frothy, A—annular, AM—annular–misty.

FIGURE 18.7

Classification of two-phase flow structures in a horizontal minichannel by Coleman and Garimella (2003).

own experimental results with the results of calculations according to the criteria of two-phase flow structure maps of other authors. It was necessary to determine the boundaries of individual flow structures, using a map of two-phase flow structures by Thome (2004–2008) described in Figure 18.8 and Coleman and Garimella (2003) (Figure 18.9) The parameters describing, in this case, the limits of the flow structure boundaries are the vapor quality x and the mass flow density G.

Figure 18.8 presents the results of a computational analysis of flow structures according to Thome's map (Thome 2004–2008) for both the studied refrigerants. Figure 18.9 shows the same kind of comparison, but according to the map of Coleman and Garimella (2003). It was found that in the range of parameters in which own experimental investigations of the described refrigerants condensation in horizontal minichannels were carried out,

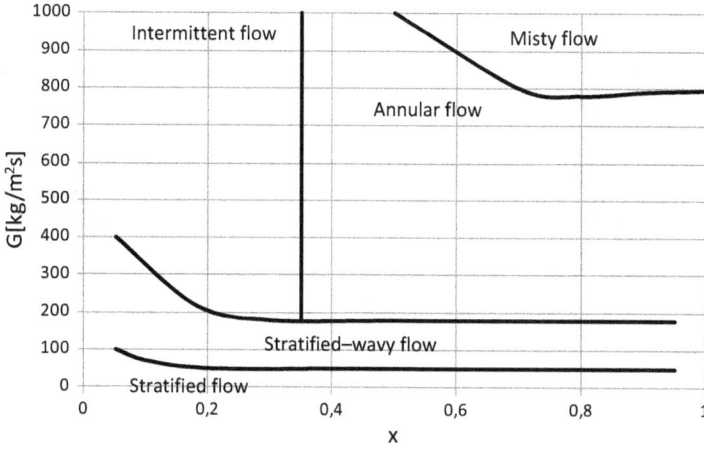

FIGURE 18.8
The flow structure map by Thome (2004–2008).

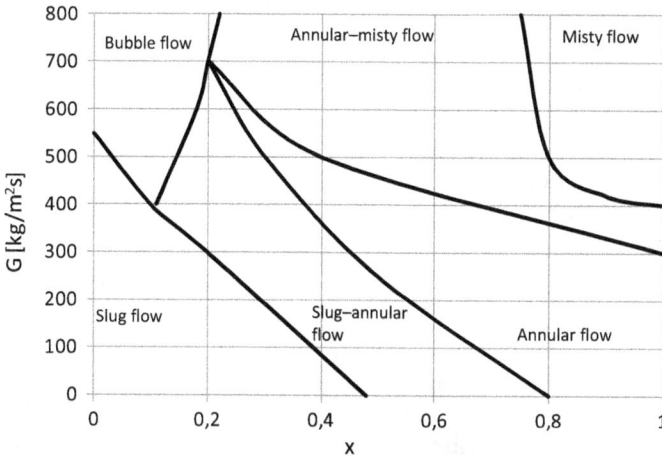

FIGURE 18.9
The flow structure map by Coleman and Garimella (2003).

annular, plug, and slug flow structures are often observed. Unfortunately, the used maps have clear limitations in terms of low mass flux density values. For this reason, most of the results for the Novec 649 condensation are outside the scale of those maps. Hence, there have been difficulties with the clear definition of the flow structure. Another difficulty was the occurrence of the so-called transitional structures, such as annular–wavy structures. Most maps of two-phase flow structures were created for the R134a refrigerant, and therefore, there may be discrepancies during the identification of the flow structures for the investigated low-pressure refrigerants.

(a)   (b)

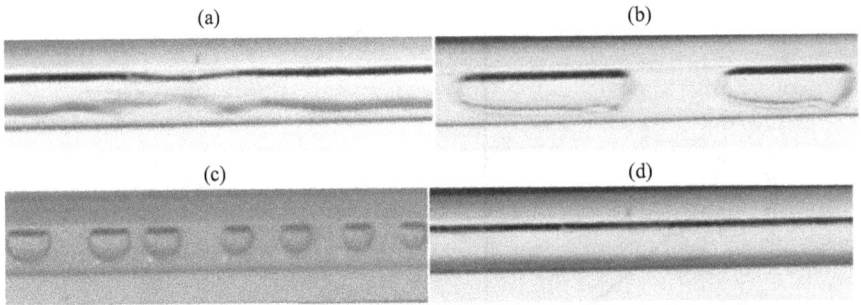

(c)   (d)

FIGURE 18.10
Exemplary flow structures of the HFE-7100 refrigerant during condensation in a minichannel with an internal diameter $d = 2.0$ mm: (a) wavy—$T_s = 80°C$, $G = 1,060$ kg/m²s, $x = 0.62$; (b) slug—$T_s = 82°C$, $G = 1,060$ kg/m²s, $x = 0.40$; (c) bubbly—$T_s = 61°C$, $G = 1,060$ kg/m²s, $x = 0.20$; (d) annular—$T_s = 75°C$, $G = 1,060$ kg/m²s, $x = 0.70$.

(a)   (b)

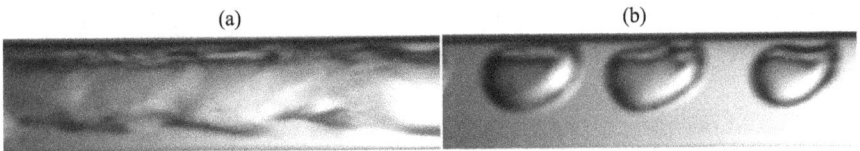

FIGURE 18.11
Exemplary flow structures of Novec 649 refrigerant during condensation in a minichannel with an internal diameter $d = 1.0$ mm: (a) wavy—$T_s = 55°C$, $G = 1,698$ kg/m²s, $x = 0.68$; (b) bubbly—$T_s = 44°C$, $G = 1,521$ kg/m²s, $x = 0.27$.

The experimental visualization of two-phase flow structures during the condensation of HFE-7100 refrigerant in minichannels made of glass was carried out on the experimental test stand. Figure 18.10 shows a picture of a two-phase structure recorded with a high-speed camera. Figure 18.11 presents the visualization results for the Novec 649 refrigerant. The mass flow density G in the experimental research was higher than 1,000 kg/m²s in most cases. A shift of bubble and slug flow occurrence was observed. The most frequent structures in this case were the annular–wavy and wavy structures.

18.3 Summary

This chapter presented some of the more important problems related to experimental investigations of refrigerants condensation in minichannels. Two of them were taken into account, i.e., measurements concerning the construction of thermal and flow characteristics of condensation in minichannels and investigations of two-phase flow structures during this process. A characteristic feature of the characteristics investigations is the use of indirect methods to determine the coefficient of heat transfer and flow resistance in the two-phase area. The determination of the heat flux density, the vapor quality, and the heat transfer coefficient requires the development of a heat collection methodology in the condensation process. The original methodology of the experimental solution to this problem was presented. This method is based on the use of the "heating method." The basics of this method are presented in the case of water as a coolant. (This method can also be used in the case of air.) The second important research problem of condensation in minichannels is the need to know the two-phase flow structures in this process. The values of the heat transfer coefficient and flow resistance depend on the kind of the flow structure. It should be emphasized that in the light of literature analysis, it is a development problem, especially for new refrigerants. The methodology and research results in this field are presented, for example for refrigerants HFE-7100 and Novec 649. It also turns out that the published two-phase structure maps for R134a refrigerant condensation are of little use. Works on the maps of condensation flow structures in minichannels are conducted in many global scientific centers, and generalized results are hopefully expected.

References

Baummer T., Cetegen E., Ohadi M., Dessiatoun S. 2008. Force fed evaporation and condensation utilizing advanced microstructured surfaces and microchannels. *Microelectronics Journal* vol. 39, no 7, 975–980.

Bohdal T., Charun H., Sikora M. 2011. Comparative investigations of the condensation of R134a and R404A refrigerants in pipe minichannels. *International Journal of Heat and Mass Transfer* vol. 54, 1963–1974.

Bohdal T., Charun H., Sikora M. 2012. Heat transfer during the condensation of refrigerants in pipe minichannels. *MFTP-2012-ID56: International Symposium Multiphase Flow and Transport Phenomena*, Agadir-Marocco.

Bohdal T., Charun H., Sikora M. 2015. Empirical study of heterogeneous refrigerant condensation in pipe minichannels. *International Journal of Refrigeration* vol. 59, 210–223.

Cavallini A., Doretti L., Matkovic M., Rossetto L. 2005. Update on condensation heat transfer and pressure drop inside minichannels. *Proceedings of ICMM, 3rd International Conference on Microchannel and Minichannel*, Toronto.

Charun H. 2012. Thermal and flow characteristics of the condensation of R404A refrigerant in pipe minichannels. *International Journal of Heat and Mass Transfer* vol. 55, 2692–2701.

Chen L., Tian Y.S. 2006. The effect of tube diameter on vertical two-phase flow regimes in small tubes. *International Journal of Heat and Mass Transfer* vol. 49, 4220–4230.

Coleman J., Garimella S. 2003. Two-phase flow regimes in round, square and rectangular tubes during condensation of refrigerant R134a. *International Journal of Refrigeration* vol. 26, no 1, 117–128.

Del Col D., Azzolin M., Bortolin S. 2014. Two-phase flow and heat transfer of a non-azeotropic mixture inside a single microchannel. *Proceedings of International Refrigeration and Air Conditioning Conference*, Purdue.

Ewim D., Meyer J. 2019. Condensation heat transfer coefficients in an inclined smooth tube at low mass fluxes. *International Journal of Heat and Mass Transfer* vol. 133, 686–701.

Ewim D., Meyer J, Abadi S., 2018. Heat transfer coefficients during the condensation of low mass fluxes in smooth horizontal tubes. *International Journal of Heat and Mass Transfer* vol. 123, 455–467.

Garcia-Cascales J.R., Vera-Garcia F., Gonzalez-Macia J., Corberan-Salvador J.M., Johnson M.W., Kohler G.T. 2010. Compact heat exchangers modeling: Condensation. *International Journal of Refrigeration* vol. 33, no 1, 135–147.

Kandlikar S.G. 2003. Microchannels and minichannels: History, terminilogy, classification and current research needs. *First International Conference on Microchannels and Minichannels*, New York.

Kandlikar S.G., Colin S., Peles Y., Garimella S., Pease R.F., Brandner J.J., Tuckermann D.B. 2013. Haet transfer in microchannels: 2012, status and research needs. *Journal of Heat Transfer* vol. 135, 061401.

Leducq D., Macchi-Tejeda H., Jabbour O., Serghini T. 2003. Experimental study and thermal modeling of a R404A small channel air condenser. *Proceedings of 21th II R International Congress of Refrigeration*, Washington, DC.

Macdonald M., Garimella S. 2016. Hydrocarbon mixture condensation inside horizontal smooth tubes. *International Journal of Heat and Mass Transfer* vol. 1000, 139–149.

Mehendale S.S., Jacobi A.M., Shah R.K. 2000. Fluid flow and heat transfer at micro- and mesoscales with application to heat exchanger design. *Applied Mechanics Reviews* vol. 53, no 7, 175–193.

Mikielewicz J. 1995. Modelowanie procesów cieplno – przepływowych. Wyd. Zakład Narodowy im. Ossilińskich. Wrocław–Warszawa – Kraków (In Polish).

Obhan C.B., Garimella S. 2001. A comparative analysis of studies on heat transfer and fluid flow in microchannels. *Microscale Thermophys* vol. 5, no 4, 293–311.

Ohadi M.M. 2007. A self: Contained system for the thermal meagement of next generation high heat flux electronics. *Proceedings of 7th Annual Busiess and Technology Summit*, Natick.

Shin J.S., Kim M.H. 2004a. An experimental study of condensation heat transfer inside a minichannel with a new measurement technique. *International Journal of Multiphase Flow* vol. 30, 311–325.

Shin J.S., Kim M.H. 2004b. An experimental study of condensation heat transferin sub-millimeter rectangular tubes. *Journal of Thermal Science* vol. 13, no 4, 350–357.

Sikora M. 2017. Flow structures during refrigerants condensation. *Journal of Mechanical and Energy Engineering* vol. 41, no. 1, 101–106.

Sun L., Mishima K. 2009. Evaluation analysis of prediction method for two-phase flow pressure drop in minichannels. *International Journal of Multiphase Flow* vol. 35, 47–54.

Thome J.R. 2004–2008. *Engineering Data Book III: Chapter 8 – Condensation Inside Tubes.* Wolverine Tube, Shawnee, OK.

Thome J.R. 2005. Condensation in plain horizontal tubes: Recent advances in modeling of heat transfer to pure fluids and mixtures. *Journal of the Brazilian Society of Mechanical Sciences and Engineering* vol. 27, 23–30.

Thome J.R., El Hajal J., Cavallini, A. 2003a. Condensation in horizontal tubes, part 1: Two-phase flow pat tern map. *International Journal of Heat and Mass Transfer* vol. 46, no 18, 3349–3363.

Thome J.R., El Hajal J., Cavallini A. 2003b. Condensation in horizontal tubes, part 2: New heat transfer model based on flow regimes. *International Journal of Heat and Mass Transfer* vol. 46, no 18, 3365–3387.

Wilson E. 1915. A basis for rational design of heat transfer apparatus. *Transactions of the ASME* vol. 37, 47–70.

Index

For Product Safety Concerns and Information please contact our EU
representative GPSR@taylorandfrancis.com
Taylor & Francis Verlag GmbH, Kaufingerstraße 24, 80331 München, Germany

www.ingramcontent.com/pod-product-compliance
Lightning Source LLC
Chambersburg PA
CBHW060424220326
41598CB00021BA/2281